Chemistry today

Euan S. Henderson

M
Macmillan Education

© Euan S. Henderson 1976

All rights reserved. No part of this publication
may be reproduced or transmitted, in any form or by
any means, without permission.

First published 1976
Reprinted 1977, 1978

Published by
Macmillan Education Ltd
Houndmills Basingstoke Hampshire RG21 2XS
and London
Associated companies in Delhi Dublin
Hong Kong Johannesburg Lagos Melbourne
New York Singapore Tokyo

Printed in Great Britain by
A. Wheaton and Co. Ltd
Exeter

To Irini, Christopher and Anna

Contents

	Preface	vi
	Acknowledgements	vii
1	Elements—the building blocks of chemistry	1
2	Atoms in chemistry	29
3	Chemical reactions	58
4	Inside the atom	82
5	Atoms joining together	110
6	Metals	137
7	Acids	195
8	Sulphuric acid—the most important chemical in industry	222
9	Carbon	246
10	Energy in chemistry	317
11	Silicon	340
12	Nitrogen and other elements important to life	355
13	Limestone	380
14	Salt	401
15	Chemical analysis	424
	Appendices	
A	Useful addresses	439
B	Background reading and wall charts	440
C	Films and film loops	442
	Index	444

Preface

This book provides a chemistry course to 16+. It assumes the background of a more general science course in the first two years of secondary school, or in a middle school, which has provided some of the basic language of the subject together with a study of air and water and some knowledge of separation techniques. It covers the requirements of all the CSE boards' mode 1 syllabuses and extends somewhat beyond them, so that it could equally provide the basis for a mode 3 or for a GCE course.

The approach is strongly rooted in practical work and a large number of class experiments are included. These are cast in an investigative framework and sufficient information is given so that most pupils will be able to proceed without additional instructions. Experiments are interspersed with questions which encourage the pupil to draw his or her own conclusions as the work proceeds. Teachers' notes appended to each chapter give guidance on setting up class experiments, include sources of materials, and suggest demonstrations.

At the same time it is recognised that not all areas of the subject lend themselves to first-hand experimental work. Many such topics are covered in the text. No book of this length could, however, provide adequately for all pupils' interests. To permit further exploration on a project basis, appendices list sources for a wealth of background material, mostly free, including books, pamphlets, wall charts, films and film loops, which can be assembled into a project resource bank.

A very large number of questions and problems of all types is provided throughout the text.

Acknowledgements

The author wishes to express his grateful thanks to the very many industrial, commercial and other organisations who have provided every assistance in supplying information and checking facts:

Albright and Wilson, Allied Breweries, Aluminium Federation, Blue Circle Group, Borax Consolidated, Brick Development Association, British Aqua-Chem, British Broadcasting Corporation, British Man-Made Fibres Federation, British Petroleum, British Pottery Promotion Service, British Steel Corporation, Building Research Station Advisory Service, Chemical Construction, Consolidated Gold Fields, W. T. Copeland, Copper Development Association, Council of Industrial Design, Courtaulds, Cray Valley Products, Department of Trade and Industry, Dunlop, Edwards Instruments, Esso, Fibreglass, Fisons, Flour Advisory Bureau, Formica, Fry's Metals, Gas Council, Glass Manufacturers Federation, Griffin and George, Guild of Sound and Vision, Hadrian Paints, H. J. Heinz, Her Majesty's Customs and Excise, Imperial Chemical Industries, James A. Jobling, Johnson Matthey, Lead Development Association, Lythe Minerals, McDougalls Foods, Magadi Soda, Malaysian Tin, Metropolitan Water Board, Mining Journal, Mobil, Monsanto Chemicals, Mullard, Nalfloc, National Coal Board, National Sulphuric Acid Association, North Thames Gas, R. F. D. Parkinson, Permutit, Proctor and Gamble, Pyrene, Rediweld, Reed Engineering and Development Services, Rentokil Advice Centre, Rowney, Royal Mint, Samuel Montagu, Sandeman, Schools Information Centre on the Chemical Industry, Scotch Whisky Association, Shell International, Spiring Enterprises, Tate and Lyle, Tin Industry Board, Tin Publications, Tin Research Institute, Unilever, United Kingdom Atomic Energy Authority, Vitreous Enamel Development Council, Wallpamur, Wedgwood, World Bureau of Metal Statistics, Zinc Development Association.

Thanks are also due to the following examinations boards for permission to reprint questions from past papers. Each question is individually credited in the body of the book.

Associated Lancashire Schools Examining Board (*Lancashire*)
East Anglian Examinations Board (*East Anglia*)
Metropolitan Regional Examinations Board (*London*)
Middlesex Regional Examining Board (*Middlesex*)
North-Western Secondary School Examinations Board (*North West*)
South-East Regional Examinations Board (*South East*)
Southern Regional Examinations Board (*South*)

Welsh Joint Education Committee (*Wales*)
West Midlands Examination Board (*West Midlands*)
West Yorkshire and Lindsey Regional Examining Board (*West Yorkshire*).

The author and publishers are grateful to the following for permission to reproduce photographs:

Associated Press 2.3; Bowaters (United Kingdom) Paper Co Ltd 9.11; British Aluminium Corporation and Short Brothers and Harland 6.46; The Trustees of the British Museum 1.3; British Petroleum Co Ltd 9.24; British Steel Corporation 6.13, 6.26, 6.27, 6.38, 6.39; British Tourist Authority 6.45; Central Electricity Generating Board 10.1; Courtauld Ltd 9.13; Copper Development Association 6.19, 6.25a, 6.25b, 6.29a, 6.29b, 6.30, 6.31, 6.48; Clarendon Laboratory, Oxford University 6.32; Crown Copyright (Building Research Laboratory) 9.10; Alan Curtis 3.1(right); De Beers 9.5; Dunlop Rubber Co Ltd 9.53; Esso Petroleum Co Ltd 10.7; French Government Tourist Office 14.2; ICI Ltd Agricultural Division 7.1; ICI Ltd 11.7, 12.10; ILEA 6.23; Lead Development Association 6.50; Mullard Ltd 11.1; National Coal Board 9.19; National Society for Clean Air 8.14; Permutit Ltd 13.8; Pirelli Ltd 11.8; Popperfoto 4.1, 13.7; Radcliffe Observatory per the Royal Astronomical Society 1.4; Radio Times Hulton Picture Library 2.11; Scotch Whisky Association 9.18; Shell 3.1(left), 12.2; Tate and Lyle Ltd 9.9; Tetley Walker Ltd 9.17; Texasgulf Inc 8.8; Unilever Ltd 9.55; United Kingdom Atomic Energy Authority 4.18, 4.20, 4.21, 4.22, 4.23, 4.24; Wilmot Breedon 5.12; Zinc Development Association 6.41, 6.52.

1 Elements—the building blocks of chemistry

Builders use about 65 different types of brick of various shapes and sizes. If you pulled a house to pieces you would probably not find all the different types. A simple house might have only two or three kinds. A very elaborate house might use twenty or thirty. The difference between one house and another depends on three things.

The type of bricks used;
how many of each type are used;
the patterns in which they are cemented together.

Chemistry is rather like this. There are millions of different substances, just as there are many different kinds of house. By taking substances to pieces (ANALYSING them), chemists have found that they are all made from only 105 fundamental types of substance. These 105 substances, which cannot be split chemically into anything simpler, are called THE CHEMICAL ELEMENTS. Figure 1.1 gives a list of their names.

Everything in the universe is made up of the 105 elements. Rubber, water, steel, wood, washing powder, the human body, and the rocks on the moon are all very different substances. But their differences depend only on three things.

Which elements they are made of;
how much of each element they contain;
the way in which the elements are joined together.

Questions

1 What do chemists mean by *elements*?

2 Use Figure 1.1 to make a list of the elements you have met before.

3 Choose 6 elements that you are familiar with. Write a sentence or two about each of them. You could say what they look like, what they are used for, and so on.

actinium	europium	molybdenum	scandium
aluminium	fermium	neodymium	selenium
americium	fluorine	neon	silicon
antimony	francium	neptunium	silver
argon	gadolinium	nickel	sodium
arsenic	gallium	niobium	strontium
astatine	germanium	nitrogen	sulphur
barium	gold	nobelium	tantalum
berkelium	hafnium	osmium	technetium
beryllium	hahnium	oxygen	tellurium
bismuth	helium	palladium	terbium
boron	holmium	phosphorus	thallium
bromine	hydrogen	platinum	thorium
cadmium	indium	plutonium	thulium
caesium	iodine	polonium	tin
calcium	iridium	potassium	titanium
californium	iron	praseodymium	tungsten
carbon	krypton	promethium	uranium
cerium	lanthanum	protactinium	vanadium
chlorine	lawrencium	radium	xenon
chromium	lead	radon	ytterbium
cobalt	lithium	rhenium	yttrium
copper	lutetium	rhodium	zinc
curium	magnesium	rubidium	zirconium
dysprosium	manganese	ruthenium	
einsteinium	mendelevium	rutherfordium	
erbium	mercury	samarium	

Figure 1.1

4 Examine some elements you have not seen before and describe what they look like.

5 The element polonium was discovered by Madame Curie and was named after her native country, Poland. Some years later another element, curium, was named in honour of Madame Curie herself. Which other elements are named after people and places?

6 The names of some elements were taken from Greek words. Use a dictionary or encyclopedia to find out where the names argon, bromine, chlorine, helium, hydrogen, neon, oxygen, and xenon came from.

7 Here are descriptions of some substances.
Substance A is a yellow solid. When heated it melts to a yellow liquid.

Substance B is a blue solid. When heated it gives off a brown gas and a black solid is left.

Substance C is a colourless solution. When an electric current is passed through it the gases hydrogen and chlorine are produced.

Substance D is a reddish-brown solid. It conducts electricity very well.

Substance E is a colourless gas. When a taper is brought close to it, it explodes.

(a) Which two of these substances are definitely not elements?
(b) Can you guess which elements the other three might be?

8 *To analyse* means to take a substance to pieces to find out what simpler substances it is made from. What might a chemist use analysis for?

9 From memory write down a list of all the elements you can think of. When you have finished check your list with Figure 1.1. Over thirty elements and you have done extremely well. Twenty is very good.

.1 Elements joining together

Experiment 1.1   Looking at iron

(a) Examine some powdered iron. What colour is this element?
(b) Put half a spatula-full of iron into half a test tube of water. Does it sink, float, or dissolve?
(c) Put a spatula-full of powdered iron on a piece of paper. Bring a magnet up close to it. What happens?
(d) Put half a spatula-full of powdered iron into half a test tube of dilute acid. Describe what you see happening.
(e) Put half a spatula-full of powdered iron into half a test tube of the liquid called carbon disulphide. Shake the tube well. Does any of the iron dissolve?

Experiment 1.2   Looking at sulphur

Repeat the five parts of experiment 1.1, but use powdered sulphur instead of iron.

Experiment 1.3   What are the properties of a mixture of iron and sulphur?

Put 2 spatulas-full of powdered iron and 2 spatulas-full of powdered sulphur into a dry beaker and mix them together well.

Repeat the five parts of experiment 1.1, using this mixture of iron and sulphur. Before you do each part of the experiment try to work out what will happen.

Were you correct? Did anything unexpected happen?

If you were given a mixture of iron and sulphur, how could you separate the two elements?

Experiment 1.4 Making a compound of iron and sulphur

Put a spatula-full of your mixture of iron and sulphur from experiment 1.3 on a piece of asbestos paper and heat it strongly.

Describe what happens when the mixture is heated.

The substance left on the asbestos paper is a compound called iron sulphide, formed by the two elements iron and sulphur joining together chemically.

Carry out the five parts of experiment 1.1 using powdered iron sulphide. Does the compound iron sulphide behave differently from the mixture of iron and sulphur? Have you found any way of separating iron and sulphur from iron sulphide?

Experiment 1.5 Making a compound and breaking it down again

(a) Fill a test tube with oxygen and cork it. Heat a small piece of carbon in a pair of tongs until it is red hot. Quickly drop the red-hot carbon into the tube of oxygen and replace the cork. Describe what you see happening.

(b) The carbon and oxygen have joined together chemically to produce a compound, a gas called carbon dioxide.

Hold a 5 cm piece of magnesium ribbon in a pair of tongs. Set it alight and quickly plunge it into the test tube. What do you see happening? Have you got any of the carbon back again?

Air: an important mixture of elements

Air is a MIXTURE of mainly two elements, nitrogen and oxygen, together with much smaller quantities of a few others.

Oxygen supports burning. A wooden splint, for instance, burns very brightly in pure oxygen. Nitrogen will not allow things to burn in it at all. Air behaves just as we might expect a mixture of oxygen and nitrogen to behave. A wooden splint burns much less brightly in air than in pure oxygen, because of the nitrogen.

The density of nitrogen is 1170 g m^{-3} (grams per cubic metre). The density of oxygen is 1330 g m^{-3}. The density of air is exactly what we would expect for a mixture of $\frac{4}{5}$ nitrogen and $\frac{1}{5}$ oxygen:

$$(\tfrac{4}{5} \times 1170) + (\tfrac{1}{5} \times 1330) = 1202 \text{ g m}^{-3}.$$

The properties of air are a mixture of the properties of nitrogen and oxygen.

A MIXTURE of elements has no special properties. The properties of the mixture are a blend of the properties of the elements in it.

Chemical compounds

When elements combine together chemically they make a new substance, called a COMPOUND. In a compound the elements are not just mixed. They are joined firmly in a definite pattern. A mixture of elements is like a pile of different kinds of bricks. A compound is like a house in which all the bricks are cemented together.

The properties of a compound are quite different from the properties of the elements of which it is made.

For example, the elements carbon and sulphur are brittle solids and hydrogen is a gas. If these three elements are combined together chemically in the right proportions they make rubber, a compound with entirely new properties.

Hydrogen cyanide (commonly called *prussic acid gas*) is a very poisonous compound. But it is composed of three elements, carbon, hydrogen, and nitrogen, all of which are quite harmless to human beings, either separately or mixed together.

Figure 1.2 shows some of the properties of two elements, iron and sulphur. It also shows the properties of a mixture of iron and sulphur, and the properties of the compound, called iron sulphide, which is made by combining iron and sulphur chemically.

The mixture of iron and sulphur has exactly the same properties as the separate elements.

The compound of iron and sulphur (iron sulphide) is entirely different from the individual elements.

Splitting up mixtures and compounds

Figure 1.2 tells us something else about mixtures and compounds also. The mixture of iron and sulphur can be separated in four ways.
By placing it in water;
by using a magnet;
by reacting with an acid;
by shaking with carbon disulphide.

It is usually not difficult to find a way of separating a mixture into its elements. Compounds are more difficult to split up. None of the methods of separating the mixture of iron and sulphur will separate the elements from the compound iron sulphide.

Some compounds are very hard to split up. Aluminium oxide, for

	Iron	Sulphur	Mixture of iron and sulphur	Compound of iron and sulphur (iron sulphide)
Colour	Black	Yellow	Grey	Black
In water	Sinks	Floats	Iron sinks, sulphur floats	Sinks
Effect of magnet	Attracted	Not attracted	Iron attracted, sulphur not attracted	Not attracted
With dilute acid	Reacts, giving hydrogen gas	Does not react	Iron reacts giving hydrogen gas, sulphur does not react	Reacts, giving foul-smelling gas called hydrogen sulphide
With carbon disulphide	Insoluble	Soluble	Sulphur dissolves, iron does not	Insoluble

Figure 1.2

example, is such a strong compound of aluminium and oxygen that chemists once thought it was an element, which they called *alumina*. For over a century alumina was included in lists of elements until a Danish scientist, Hans Oersted, succeeded in splitting it up (DECOMPOSING it) into aluminium and oxygen.

Like aluminium oxide, most compounds are very difficult to decompose into their elements. But some can be decomposed much more easily. Mercury oxide, for instance, splits up into its elements, mercury and oxygen, when it is heated quite gently.

Differences Between Mixtures and Compounds

1 The properties of a mixture are a blend of the properties of its elements. The properties of a compound are entirely different from the properties of its elements.

2 A mixture is usually quite easy to separate into its elements. A compound is usually difficult to split up into its elements.

Questions

1 Which of the following statements about mixtures of elements is *untrue*?
A   The colour of a mixture is a blend of the colours of its elements.

TABLEAU DES SUBSTANCES SIMPLES.

	Noms nouveaux.	Noms anciens correspondans.
Substances simples qui appartiennent aux trois règnes & qu'on peut regarder comme les élémens des corps.	Lumière..........	Lumière.
	Calorique.........	Chaleur. Principe de la chaleur. Fluide igné. Feu. Matière du feu & de la chaleur.
	Oxygène.........	Air déphlogistiqué. Air empiréal. Air vital. Base de l'air vital.
	Azote............	Gaz phlogistiqué. Mofete. Base de la mofete.
	Hydrogène.......	Gaz inflammable. Base du gaz inflammable.
Substances simples non métalliques oxidables & acidifiables.	Soufre...........	Soufre.
	Phosphore.......	Phosphore.
	Carbone.........	Charbon pur.
	Radical muriatique.	Inconnu.
	Radical fluorique..	Inconnu.
	Radical boracique,.	Inconnu.
Substances simples métalliques oxidables & acidifiables.	Antimoine........	Antimoine.
	Argent...........	Argent.
	Arsenic..........	Arsenic.
	Bismuth..........	Bismuth.
	Cobolt...........	Cobolt.
	Cuivre...........	Cuivre.
	Etain............	Etain.
	Fer..............	Fer.
	Manganèse.......	Manganèse.
	Mercure.........	Mercure.
	Molybdène.......	Molybdène.
	Nickel...........	Nickel.
	Or...............	Or.
	Platine...........	Platine.
	Plomb...........	Plomb.
	Tungstène........	Tungstène.
	Zinc.............	Zinc.
Substances simples salifiables terreuses.	Chaux...........	Terre calcaire, chaux.
	Magnésie........	Magnésie, base du sel d'Epsom.
	Baryte...........	Barote, terre pésante.
	Alumine.........	Argile, terre de l'alun, base de l'alun.
	Silice............	Terre siliceuse, terre vitrifiable.

Figure 1.3 A list of elements known in 1789, according to the French chemist, Antoine Lavoisier. He was wrong about quite a few!

Elements—the building blocks of chemistry

B   The properties of a mixture can be quite different from the properties of its elements.
C   A mixture of two gaseous elements is a gas.
D   The density of a mixture is the average of the densities of its elements.
E   The elements can easily be separated from a mixture.

2 Which of the following best describes a chemical compound?
A   A substance formed by mixing two elements.
B   A substance formed by mixing two or more elements.
C   A substance with special properties.
D   A substance formed by combining two elements.
E   A substance formed by combining two or more elements.

3 Look at the following lists.
A   Iron, steel, copper.
B   Salt, rubber, water.
C   Mercury, bromine, water.
D   Air, brass, bronze.
E   Mercury, iron, oxygen.
   (a) Which list consists of elements?
   (b) Which list consists of mixtures of elements?
   (c) Which list consists of compounds?

4 Hydrogen and helium are both colourless gases. The density of hydrogen is 80 g m^{-3} and the density of helium is 170 g m^{-3}. Hydrogen burns, but helium does not. Both gases are insoluble in water. Answer the following questions about a mixture of 50 per cent hydrogen and 50 per cent helium.
   (a) What will be the colour of the mixture?
   (b) Will the mixture burn?
   (c) Will the mixture be soluble in water?
   (d) What will be the density of the mixture?

5 Choose one of the words *combine*, *decompose*, *dissolve*, *melt*, *mix* to fill each of the gaps in the following sentences.
   (a) Always _____ some salt in water used for boiling potatoes.
   (b) You can _____ mercury oxide by heating it gently.
   (c) Solder is easy to _____ .
   (d) To make a cake you _____ the ingredients together.
   (e) When iron and sulphur are heated together they _____ with one another.

6 Make a list of some of the things in your kitchen at home, such as sugar, baked beans, frying pan. Say whether you think each one is made of an element, a mixture, or a compound.

1.2 Common and rare elements

In the universe

Astronomers have calculated that the universe is made up of 90 per cent hydrogen, 9 per cent helium, and only 1 per cent of all the other elements put together (Figure 1.4). The two commonest elements in the universe are the two gaseous elements with the lowest densities. When the earth was first formed it was so hot that it was a spinning ball of vapour. Most of the lightest elements, like hydrogen and helium, were spun off into space, so the earth now has a far larger proportion of the denser elements than has the universe as a whole.

In the earth

We do not know very much about the centre of the earth. But as a result of drilling for oil, mining for minerals, and measuring the speed at which earthquake shock waves travel, geologists have been able to find out a good deal about the composition of the earth down to a depth of about 30 kilometres. Ninety of the 105 elements are found in this outer CRUST (the other 15 are man-made elements which are not found in nature). A few of these 90 elements, such as gold and sulphur, are found free and uncombined, but most of them combined chemically with each other to form compounds while the earth was cooling down.

The elements are by no means equally common in the earth's crust. Ninety-nine per cent of the mass of the crust is made up of only ten elements. The other 80 elements together make up the remaining 1 per cent. Figure 1.6 shows the proportions of some of the more common elements.

The outer crust which geologists have been able to study makes up only about one-hundredth of the total mass of the earth. Little is known about what makes up the other $\frac{99}{100}$. Beneath the crust lies the MANTLE, which is very hard, dense rock. It is believed that the central CORE is composed mainly of molten iron and nickel, but scientists have not been able to make sure of this yet.

In the sea

All the elements which are found in the earth's crust are also found in the sea, but the proportions are different. Water itself is a compound of hydrogen and oxygen, so these are the main elements in the sea. The other elements which are common in the sea are those whose compounds are soluble in water. They have been dissolved from the rocks and washed into the sea by rain and rivers. Figure 1.5 shows the average proportions of some of the commoner elements in the sea.

The total amount of dissolved substances varies. Seas in hot climates and inland seas (such as the Dead Sea) lose more water by evaporation and have higher concentrations.

The greater concentration of some elements in the sea has proved very useful to man. Chlorine, for instance, forms only 0.2 per cent of the earth's crust, but in the form of one of its compounds, *salt* (sodium chloride), it makes up over 2 per cent of the sea. So it is much easier to extract chlorine from the sea than from the earth.

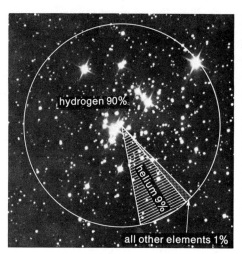

Figure 1.4 The composition of the universe

Figure 1.5 The composition of the sea

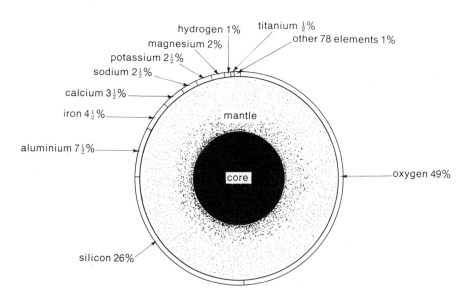

Figure 1.6 The composition of the earth's crust

10 Chemistry today

In the air

Air is a mixture of gases. Most of the air is made up of two elements, nitrogen and oxygen, but it also contains small amounts of five other gaseous elements: argon, neon, helium, krypton, and xenon. There is also some carbon dioxide (a compound of carbon and oxygen) in air, together with varying amounts of water vapour and very often other impurities. Figure 1.7 shows the composition by volume of pure, dry air.

Nitrogen	78%
Oxygen	21%
Argon	0.9%
Carbon (mainly as carbon dioxide gas)	0.01%–0.1%
Neon	0.0015%
Helium	0.0005%
Krypton	0.000005%
Xenon	0.0000006%

Figure 1.7

Questions

1 (a) What are the two commonest elements in the earth's crust?
  (b) What are the two commonest elements in the universe?
  (c) Why are your answers to (a) and (b) not the same?

2 Why is it not surprising that hydrogen and oxygen are the two commonest elements in the sea and chlorine and sodium are the two next commonest?

3 Of the ten most common elements in the earth's crust, which two do you think are most important to man? (Refer to Figure 1.6 if necessary.)

4 Make a list of six elements which are *not* among the ten commonest in the earth's crust but which are very important to man. Try to find out how many parts per million of these elements there are in the crust.

5 The mass of the earth has been calculated to be
$$6\,000\,000\,000\,000\,000\,000\,000\ (6 \times 10^{21})\ \text{tonnes.}$$
  (a) What is the mass of the earth's crust? (Assume that the density of the earth is the same right through.)
  (b) The concentration of gold in the earth's crust is 1 part in 800 000 000 (8×10^8). How many tonnes of gold are there in the earth's crust?

Elements—the building blocks of chemistry

1.3 The discovery of the elements

The first elements to be discovered were the ones that are found free and uncombined in the earth's crust. Silver and gold were among the earliest. They have been used for decorative purposes and for coinage since before history was recorded, when primitive men were still making their tools and weapons from stone. The first really useful metal to be discovered was *bronze*, a mixture of the elements copper and tin. These metals are usually found in the earth as compounds which are quite easy to break down. Bronze was used for making tools until about 1 200 BC, when iron, which is much stronger, was discovered.

Apart from gold, silver, copper, tin, and iron, four other elements were known to Egyptian chemists before the birth of Christ: carbon, lead, mercury, and sulphur.

From the time of the Pharaohs until the middle of the seventeenth century, chemistry made very little progress and only four more elements were discovered: zinc, antimony, arsenic, and bismuth. At this time chemistry was closely tied up with magic. The *alchemists*, as they were then called, spent their time trying to turn common metals into gold and looking for the secret of eternal life.

The thirteen elements that were known by 1650 had been discovered mainly by accident, and chemists still did not understand what an element was. In 1661 an English chemist, Robert Boyle, defined an element as *a substance which cannot be broken down into anything simpler*. Chemists then knew what they were looking for and set about trying to break substances down in a scientific way.

They examined pieces of rock, heated them, and treated them with various chemicals to see what would happen. If they managed to make a new substance from the rock, they tried to break it down further. When they obtained a substance which no one could break down into simpler substances it was accepted as an element. Of course some mistakes were made. Up to the end of the eighteenth century *alumina* and *lime* were thought to be elements, and it was many years before they were found to be compounds of aluminium with oxygen and calcium with oxygen. But by 1805 twenty-seven new elements had been discovered, including the gases oxygen, nitrogen, hydrogen, and chlorine and the metals nickel, chromium, tungsten, and uranium.

Towards the end of the eighteenth century, electricity was discovered and a young English chemist, Humphry Davy, was the first to use an electric current to try to break down substances to find new elements. In two years he discovered potassium, sodium, barium, boron, calcium, and magnesium. Other chemists followed Davy's lead, and by 1850 fifty-nine elements were known, including all the common ones.

Up to this time the problem for chemists had been to find ways of splitting up compounds to discover what elements they were made of. During the second half of the nineteenth century the problem became the discovery of the very rare elements, those which form less than one part per million of the earth's crust. By 1900 most of these rarer elements had been discovered, including the noble gases of the atmosphere (helium, neon, argon, krypton, and xenon) and some radioactive elements such as radium.

This century a few more elements have been added to the list. Four more very rare elements have been found in the earth: europium, hafnium, lutetium, and rhenium. In addition, several elements which are not found in nature have been produced in atomic reactors and cyclotrons. These are all radioactive elements such as plutonium, americium, curium, and most recently the 105th element, hahnium, discovered in March 1970.

Questions

1 Silver and gold are two of the rarer elements. Why do you think they were among the first to be discovered?

2 Certain types of rock contain a compound of copper with sulphur which is easily decomposed by heating with *charcoal* (partly burnt wood). How might stone age men have come to discover copper? Put your answer in the form of an imaginary story if you like.

3 Aluminium is the third most common element in the earth's crust. Why do you think it was not discovered until 1825?

4 Why do you think it was that helium, neon, argon, krypton, and xenon were all discovered at about the same time, between 1894 and 1898?

5 Why have gold, silver, and copper been the favourite substances for making coins and jewellery for thousands of years?

6 Choose any six elements and try to find out who discovered them and when.

7 Which element's discovery do you consider was the most important to man? Explain your choice.

4 Classifying the elements

When a great many new things are discovered there is a lot of new information to be remembered. Most people find that one way of helping themselves to remember things is to classify them into groups. Before the middle of the eighteenth century very few elements were known. A system for classifying them was not really necessary. But as chemistry became more scientific and new elements were discovered almost every year, chemists needed to find a way of dividing them into groups.

Metals and non-metals

At first the elements were divided into two groups: those which are METALS, and those which are not, called NON-METALS. Some of the differences between metals and non-metals are shown in Figure 1.8.

	Metals	Non-metals
Appearance	Bright and shiny	Dull
Conduction of heat	Good	Bad
Conduction of electricity	Good	Bad
Workability	Can be bent without breaking; can be hammered into shape (malleable), and drawn into wires (ductile)	Brittle, not malleable or ductile
Strength	Strong	Weak
Melting points	High	Low
Boiling points	High	Low
Densities	High	Low
When struck	Give a ringing sound	No ringing sound

Figure 1.8

With many elements it is very easy to decide whether they should be classified as metals or as non-metals. Silver, for instance, has all the characteristic properties of a metal. Oxygen has all the properties of a non-metal. Some elements, however, are more difficult to classify exactly. Boron has most of the properties of a non-metal, yet melts at over 2000°C. Mercury has many metallic properties, yet melts below room temperature. But taking all the properties listed in Figure 1.8 together, the classification is straightforward enough. It is usually easy to decide which group a particular element belongs to, even if one or two properties do not seem to fit in.

The disadvantage of the metal/non-metal classification is that it does not tell us much about the chemical behaviour of elements, though there are some chemical differences between metals and non-metals, shown in Figure 1.9.

Once again, not all elements fit the classification perfectly. For example, silicon is a non-metal, but silicon oxide (*sand*) is a solid which does not dissolve in water. Several metals, such as silver and gold, do not react with acids.

	Metals	Non-metals
With dilute acid	Many (but not all) react, producing hydrogen gas	Never react
Properties of compounds with oxygen (oxides)	Either insoluble or dissolve in water to form an alkaline solution	Dissolve in water to form an acidic solution
Properties of compounds with chlorine (chlorides)	Unreactive solids	Very reactive gases or liquids
Properties of compounds with hydrogen (hydrides)	Do not form stable compounds with hydrogen	Form stable compounds with hydrogen

Figure 1.9

In spite of this, the classification of elements into metals and non-metals is useful. Metals resemble one another in so many ways that in this book we shall study them all together in chapter 6. But a better, more detailed, classification of elements is necessary as well.

Experiment 1.6   Classifying elements

Examine the five elements carbon, copper, iron, lead, and sulphur.
   (a) Which have a shiny appearance? Which are dull?
   (b) Use a bulb and battery to find out which of the elements conduct electricity and which do not.
   (c) Hold one end of a piece of each element in a bunsen flame. Be prepared to let go when the other end becomes too hot to hold! Which conduct heat well and which do not? **Do not try this experiment with sulphur**.
   (d) Try to estimate which have a high density and which have a low density.
   (e) Try bending a piece of each element. Which are brittle, and which bend without breaking?
   (f) Hit a piece of each element with the base of a bunsen burner. Which are strong and which are weak?
   Use Figure 1.8 and the results of your experiments to decide which of these five elements are metals and which are non-metals.

Experiment 1.7   Some more ways to classify metals and non-metals

   (a) Put a little of each of the following elements separately in half a test tube of dilute acid: carbon, copper, magnesium, silicon, sulphur,

Elements—the building blocks of chemistry

and zinc. Which ones react with the acid and which do not? What gas is produced by the ones which do react?

(b) Look at samples of carbon chloride, copper chloride, etc. Describe briefly what each one looks like (e.g. colourless liquid, or white crystalline solid).

(c) Test a solution of carbon oxide with indicator paper. Is it an acid, an alkali, or neither? Repeat with a solution of the oxide of each of the other elements in turn.

Use Figure 1.9 and the results of your experiments to decide which of these elements are metals and which are non-metals.

Experiment 1.8 Investigating the properties of lithium

Your teacher will have demonstrated some of the properties of sodium and potassium. These elements are too dangerous for you to do the experiments yourself. Carry out the following experiments with lithium and see in what ways it is similar to sodium and potassium and in what ways it is different.

(a) Put a small piece of lithium on a piece of asbestos paper and heat it with a small bunsen flame. Describe what happens.

(b) Put a small piece of lithium in half a test tube of water. Describe what happens.

Hold a lighted splint at the mouth of the tube. What gas is being produced?

When the reaction has finished drop a piece of indicator paper into the tube. What does this tell you?

Experiment 1.9 Investigating the properties of iodine

Your teacher will have demonstrated some of the properties of chlorine and bromine, which are very poisonous and difficult to handle. Carry out the following experiments with iodine to compare it with chlorine and bromine.

(a) Shake a very small crystal of iodine with half a test tube of water. Does any dissolve?

Warm the tube gently over a small bunsen flame (**do not boil**). Does this help the iodine to dissolve?

Drop a piece of indicator paper into the tube. What happens?

(b) Shake a small crystal of iodine with half a test tube of sodium hydroxide solution. Describe what happens.

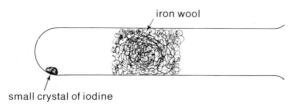

Figure 1.10

(c) Put a small crystal of iodine in a hard-glass test tube and push a tuft of iron wool halfway down the tube as shown in Figure 1.10.

Hold the tube in tongs and heat the iron strongly until it is red hot. Then quickly warm the iodine crystal to vaporise it. Does iodine react with hot iron? What is the compound left in the middle of the tube at the end of the experiment? How does iodine compare with chlorine and bromine in its reactivity towards iron?

Experiment 1.10 Comparing the combining power of chlorine, bromine, and iodine

(a) Sodium iodide is a compound of sodium and iodine. Dissolve two or three crystals of sodium iodide in half a test tube of water. Now add a few drops of a solution of bromine and shake the tube. The blackish-brown substance produced is iodine. We can describe what has happened by means of a WORD EQUATION:

bromine + sodium iodide → iodine + sodium bromide.

Which combines more strongly with sodium: bromine or iodine?

(b) Sodium bromide is a compound of sodium and bromine. Dissolve two or three crystals in half a tube of water. Add a few drops of a solution of chlorine and shake the tube.

What is the yellowish-brown substance produced? Write a word equation to describe what has happened. Which combines more strongly with sodium: chlorine or bromine?

(c) What would you expect to happen if you added a solution of chlorine to a solution of sodium iodide? Try it out and see if your prediction is correct.

(d) What would you expect to happen if you added a solution of bromine to a solution of sodium chloride? Try it out.

Experiment 1.11 Investigating the properties of magnesium, calcium, and barium

WARNING: Barium and its compounds are extremely poisonous. Wash your hands after the experiment.

(a) Drop small pieces of each element separately into half a test tube of water. For magnesium use a 5 cm piece of ribbon and clean it thoroughly with sandpaper first (*why*?).

Describe what you see in each case. Try to find out what gas is produced. Test each tube with indicator paper afterwards.

(b) Drop small pieces of magnesium and calcium separately into half a test tube of dilute acid. **Do not try this experiment with barium.**

Describe what happens. What gas is produced?

(c) Look at samples of magnesium oxide, calcium oxide, and barium oxide. What colour are they? Find out if they are soluble in water. Test the solutions with indicator paper.

(d) Look at samples of magnesium sulphate, calcium sulphate, and barium sulphate. What colour are they? Find out if they are soluble in water.

Should these three elements be grouped as a family? In what ways do they differ from each other?

Experiment 1.12   Looking at some more elements

(a) Put small pieces of each of the following elements separately in half a test tube of water: copper, iron, nickel, silver, and zinc. Are they more or less dense than water? Do they react with water?

(b) Hold small pieces of each of these five metals in tongs in a small bunsen flame. Do any of them burn? Describe any changes that you see.

(c) Have a look at some compounds of cobalt, copper, chromium, manganese, iron, nickel, silver, and zinc. What do you notice about most of these compounds?

Make a list of the differences between these metals and the chemical family lithium, sodium, and potassium.

Chemical families

At the beginning of the nineteenth century, chemists found that some elements resembled one another very closely and they grouped similar elements into FAMILIES.

Lithium, sodium, and potassium belong to the family known as the ALKALI METALS, because they react with water to form an alkaline solution. They resemble one another in many ways.

1 A freshly-cut surface is silvery in colour, but tarnishes quickly.
2 Very soft (can be cut with a knife).
3 Float on water (all have densities less than 1 g cm^{-3}).
4 React vigorously with water, producing hydrogen gas and an alkaline solution.
5 Burn vigorously in oxygen, forming the white metal oxide.
6 Burn vigorously in chlorine, forming the white metal chloride.

The only difference between these three elements is the speed with which they react with substances like water, oxygen, and chlorine. Potassium reacts more quickly than sodium and sodium reacts more quickly than lithium. The alkali metal family also includes three other elements, which you have probably not seen: rubidium and caesium, which are very rare, and francium, which is radioactive.

Another family is the HALOGENS, of which you have probably met chlorine, bromine, and iodine. Two less familiar elements also belong to this family: fluorine, a highly poisonous gas, and astatine, which is radioactive. All the members of the halogen family have similar properties, some of which are shown in Figure 1.11.

	Chlorine	Bromine	Iodine
Colour	Yellow-green	Dark red-brown	Very dark purple, almost black
State at room temperature	Gas	Liquid	Solid
Solubility in water	Very soluble	Fairly soluble	Slightly soluble
Effect of solution on indicator paper	Bleaches	Bleaches	Bleaches
Solubility in solution of sodium hydroxide	Very soluble	Very soluble	Very soluble
Reaction with iron	Reacts violently to form iron chloride	Reacts quickly to form iron bromide	Reacts slowly to form iron iodide

Figure 1.11

As with the alkali metal family, the halogens differ from one another mainly in the speed with which they react.

The NOBLE GASES, helium, argon, neon, krypton and xenon, which are present in small quantities in the air, form another family, together with radon, a radioactive gas. The members of this family resemble one another in being almost totally unable to take part in any chemical reactions.

The periodic classification

In 1869 a Russian chemist, Dmitri Mendeléev, worked out a table in which he arranged families of elements in vertical columns. This became known as the PERIODIC CLASSIFICATION (Figure 1.12). Mendeléev's classification soon proved to be very useful to chemists, and even today is considered the best way of grouping the elements.

EACH OF THE EIGHT MAIN VERTICAL COLUMNS OF ELEMENTS IS CALLED A GROUP. These groups are numbered I, II, III, IV, V, VI, VII, and O in Figure 1.12. Group I contains the alkali metals, Group VII the halogens, Group O the noble gases, and so on. All the elements in a group have very similar chemical properties. So if you know about the properties of one of the elements in a group you can form a fairly accurate picture of what another element in the same group must be like, even if you have never seen it. Fluorine, for instance, is one of the less familiar members of Group VII. From Figure 1.11 we can deduce that it will be a coloured gas, very

Elements—the building blocks of chemistry

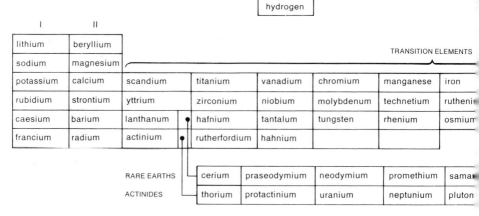

Figure 1.12 The periodic classification

soluble in water and in sodium hydroxide solution, whose solution in water bleaches, and which reacts violently with iron.

The thirty elements forming a block in the middle of the classification are called the TRANSITION ELEMENTS. They are a special, large group, all with similar properties.

1 They are hard, strong metals.
2 They sink in water (densities are greater than 3 g cm^{-3}).
3 They do not react easily with oxygen (they form oxides only when heated strongly in oxygen).
4 They do not react easily with water (some do not react with water at all).
5 Their compounds with other elements are almost always brightly coloured.

There are two other special groups in the periodic classification: the RARE EARTHS, which are all found in the same kind of rock in the earth's crust, and the ACTINIDES, which are all radioactive. Hydrogen has such special properties that it forms a group of its own.

THE METALS AND THE NON-METALS ARE IN SEPARATE AREAS OF THE PERIODIC CLASSIFICATION, SHOWN IN FIGURE 1.13.

The non-metals all lie in the upper right-hand corner. The metals are on the left and in the centre. The elements which are not easy to classify because they have some metallic and some non-metallic properties, are known as *metalloids.* They are found on the border between the metal and non-metal areas.

THE PERIODIC CLASSIFICATION SHOWS HOW FAST THE INDIVIDUAL MEMBERS OF A GROUP REACT. On the left-hand side the speed of reaction INCREASES

				III	IV	V	VI	VII	0
									helium
				boron	carbon	nitrogen	oxygen	fluorine	neon
				aluminium	silicon	phosphorus	sulphur	chlorine	argon
alt	nickel	copper	zinc	gallium	germanium	arsenic	selenium	bromine	krypton
dium	palladium	silver	cadmium	indium	tin	antimony	tellurium	iodine	xenon
um	platinum	gold	mercury	thallium	lead	bismuth	polonium	astatine	radon

pium	gadolinium	terbium	dysprosium	holmium	erbium	thulium	ytterbium	lutetium
ricium	curium	berkelium	californium	einsteinium	fermium	mendelevium	nobelium	lawrencium

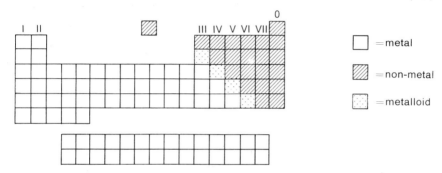

Figure 1.13

towards the bottom of a group. On the right-hand side the speed of reaction DECREASES towards the bottom of a group. For the groups in the middle there is very little difference between the speed of reaction of the various members of the group.

Using the periodic classification

In 1916 Thomas Midgley, a mechanical engineer working for an American company, was given the job of finding a way to prevent *knocking* (sometimes called *pinking*) in petrol engines. The mixture of petrol vapour and air in a car cylinder should explode when the sparking plug fires, at the instant when the piston is right at the top of the cylinder. Knocking is caused by the gas mixture exploding while the piston is still moving up the cylinder.

Midgley tried adding various substances to the petrol. By pure chance

Elements—the building blocks of chemistry   21

the element iodine was among the first substances he tried. It was noticeably effective in reducing knocking. Various compounds of iodine turned out to be just as effective. But iodine and its compounds are rather expensive, so Midgley tried other members of the halogen family. He found that chlorine and bromine were not nearly as good as iodine. Next he tried Group VI of the periodic classification. He found that selenium compounds reduced knocking dramatically. Since anti-knock properties had improved in going down Group VII, Midgley wondered if the same would be true for Group VI. Would tellurium be even better than selenium? It was, but unfortunately it could not be used commercially because tellurium compounds are extremely poisonous.

So Midgley looked again at the periodic classification. He had found that anti-knock properties increased down a group and also that Group VI compounds were better than Group VII compounds (Figure 1.14).

IV	V	VI	VII	
carbon	nitrogen	oxygen	fluorine	
silicon	phosphorus	sulphur	chlorine	increasing
germanium	arsenic	selenium	bromine	anti-knock
tin	antimony	tellurium	iodine	effect
lead	bismuth	polonium	astatine	

increasing anti-knock effect

Figure 1.14

If these trends continued, he thought, the best anti-knock substance would be a compound of lead. This prediction from the periodic classification turned out to be correct. A compound of lead is now added to most petrol to reduce knocking. Had Midgley not used the periodic classification intelligently he might have had to try millions of substances before he stumbled on the best one. As it was, with the aid of Mendeléev's work of fifty years before, the solution to the problem was found very quickly. Midgley's work is just one example of the usefulness of the periodic classification.

Questions

1 An element has the following properties: silver colour, melting point below 100°C, good conductor of electricity, floats on water, and when heated it burns in air and the product when added to water turns litmus blue.
   (a) Which of these properties are those of a metal?
   (b) Name one element which has properties similar to the above.

(c) After reacting this element with oxygen what is the name of the product?

(d) What sort of substance is formed when this product is added to water? *(West Yorkshire)*

2 Classify each of the following elements as metals or non-metals, giving reasons for your decision in each case: carbon, gold, hydrogen, mercury, oxygen, tin.

3 Figure 1.15 gives some information about six elements, called A, B, C, D, E, and F. Use this information to decide which are metals and which are non-metals.

	Density g cm^{-3}	Melting point °C	Boiling point °C	Electrical conductivity
A	8.9	1083	2582	Good
B	0.0013	-219	-183	Bad
C	1.8	44	281	Bad
D	0.5	181	1331	Good
E	13.6	-39	357	Good
F	2.2	over 3000	over 3000	Good

Figure 1.15

4 The elements carbon and silicon belong to the same group of the periodic classification (Group IV).

(a) What does the term *group* mean in this context?

(b) Name or give the number of another group of elements and give the names of three of its members.

(c) What can you say about the properties of compounds of elements in the same group? *(London)*

5 Figure 1.16 shows an outline of the periodic classification with the elements identified by numbers.

(a) Which numbers represent the alkali metals?

(b) Which numbers represent the halogens?

(c) Give the numbers of three elements which do not usually form any compounds.

(d) Which of the following sets of elements is *not* a Group?

A 4, 12, 20, 38, 56, 88.   C 13, 14, 15, 16, 17, 18.
B 5, 13, 31, 49, 81.   D 2, 10, 18, 36, 54, 86.

Elements—the building blocks of chemistry

					1												2
3	4									5	6	7	8	9	10		
11	12											13	14	15	16	17	18
19	20	21		22	23	24	25	26	27	28	29	30	31	32	33	34	35	36
37	38	39		40	41	42	43	44	45	46	47	48	49	50	51	52	53	54
55	56	57–71	72	73	74	75	76	77	78	79	80	81	82	83	84	85	86
87	88	89–103	104	105													

Figure 1.16

(e) Which of the sets of elements in (d) are all gases at room temperature?

(f) Which of the sets of elements in (d) contains at least one element which is solid, one which is liquid, and one which is a gas at room temperature?

(g) Which of the following elements is most reactive: 4, 12, 20, 38, 56, or 88?

(h) Which of the following elements is least reactive: 8, 16, 34, 52, or 84?

(j) Give the numbers of five elements whose compounds you would expect to be coloured.

(k) Which of the following elements are metals and which are non-metals: 7, 17, 27, 37, 47?

6 A friend who has missed some of your lessons about the periodic classification argues that since chlorine is a greenish-yellow gas and bromine is a reddish-brown liquid they cannot possibly belong to the same family of elements. Describe one short experiment you could show him to persuade him that he is wrong.

7 A silver-coloured solid element reacts with nitric acid to form a green solution. Which of the following could it be?
A   An alkali metal.
B   A halogen.
C   A noble gas.
D   A member of Group II of the periodic classification.
E   A transition element.

8 Group II of the periodic classification, reading from top to bottom, consists of the elements beryllium, magnesium, calcium, strontium,

barium, and radium. Which of the following statements about strontium is *not* true?
A It reacts more slowly with dilute acid than does magnesium.
B Its compounds are white.
C It is a solid at room temperature.
D It reacts more slowly with water than does barium.
E It burns when heated in air.

9 Use Figure 1.11 to write as full a description as you can of what you would expect the element astatine to be like.

10 Three bottles A, B and C are known to contain a solution of chlorine, a solution of potassium bromide, and a solution of potassium iodide, but the labels have become mixed up, so no one knows which is which. Mixing the solutions produces the following results:

A + B → an orange solution;
B + C → a colourless solution;
A + C → a dark brown solution.

Which solution is in bottle A, which in B, and which in C?

11 In the periodic classification sodium is placed below lithium in Group I and fluorine is placed above chlorine in Group VII. Which of the following pairs of elements will react together most violently?
A sodium and fluorine
B sodium and chlorine
C lithium and sodium
D lithium and fluorine
E lithium and chlorine

12 Figure 1.17 gives some information about 10 elements, called J, K, L, M, N, O, P, Q, R and S.
  (a) Give the letters of TWO sets of *three* elements which belong to the same group (family) in the periodic classification of elements.
  (b) Give reasons for your answer to part (a).
  (c) How might R be stored?
  (d) Which one might be chlorine?
  (e) Which one could be iron?
  (f) Suggest a name for element M.
  (g) How might N be stored?
  (h) Which element could be used for pipes which carry drinking water?
  (i) Element K is used in making cars and ships. What could be done to increase its lasting qualities?
  (j) Explain your answer to (i).
  (k) Which one might be bromine?

Elements—the building blocks of chemistry   25

Element	Boiling point °C	Melting point °C	Physical appearance	Effect of water	Effect of air
J	2310	1083	Orange-metallic	None	Very slowly tarnishes
K	2450	1525	Metallic	Rusts	Rusts
L	1400	186	Silvery when freshly cut	Reacts and effervesces slowly	Very rapidly tarnishes
M	357	−38	Silvery	None	None
N	730	62	Silvery when freshly cut	Reacts vigorously, effervescing and burning	Very rapidly tarnishes
O	−33	−102	Greenish-yellow	Dissolves	None
P	784	97	Silvery when freshly cut	Reacts vigorously, effervescing	Very rapidly tarnishes
Q	59	−7.3	Reddish-brown	Dissolves	None
R	290	43	Yellowish-white	None	Immediately catches fire
S	184	114	Almost black	Dissolves very slightly	None

Figure 1.17

    (l) Which one might be sodium?
  (m) Which one is a gas at room temperature? (20°C)
  (n) Which two are liquids at room temperature?   (*London*)

13 Find out as much as you can about the uses of either the noble gases or the halogens and their compounds. You could either write about them or build up a chart, using drawings and pictures cut out of magazines, to illustrate them.

14 Why are people now worried about the effects of Thomas Midgley's discovery? What might be done about it?

For the teacher

A reference collection of as many elements as possible, in a variety of forms (e.g. zinc powder, granulated zinc, zinc foil), is useful.

Experiments 1.1–1.4 Iron reduced by hydrogen is preferable: commercial iron filings may be greasy, may give hydrogen sulphide with dilute acid, and often do not react readily with sulphur. Powdered roll sulphur should be used: flowers of sulphur are largely insoluble in carbon disulphide. 4M hydrochloric acid or 2M sulphuric acid. Powdered iron(II) sulphide needs to be provided: pupils will not make enough.

By way of introduction or amplification the differences between a number of elements and compounds can usefully be examined in a series of short demonstrations e.g. hydrogen/oxygen/water, hydrogen/chlorine/hydrogen chloride, mercury/oxygen/mercury(II) oxide, sodium/chlorine/sodium chloride, magnesium/oxygen/magnesium oxide.

Experiment 1.5 Wood charcoal. Hard glass 150 × 25 mm tubes. An oxygen cylinder is an advantage (obtainable from BOC), otherwise tubes of oxygen can be prepared in advance from hydrogen peroxide/manganese(IV) oxide.

Complementary experiments are (a) making zinc sulphate from zinc/dilute sulphuric acid, then electrolysing a solution to recover the zinc, and (b) making copper(II) nitrate from copper/concentrated nitric acid, then heating to produce copper(II) oxide and reducing with town gas or hydrogen.

Experiment 1.6 Carbon—charcoal sticks or graphite rods. Sulphur—lumps of roll sulphur. Metal foils.

Experiment 1.7 4M hydrochloric acid or 2M sulphuric acid and the powdered elements. Litmus is less likely to lead to misleading results than universal indicator. Carbon dioxide and sulphur dioxide can conveniently be provided as ready-made solutions. Cylinders of carbon dioxide can be obtained from the Distillers Company, and sulphur dioxide demonstration bottles from BDH.

Experiment 1.8 Before pupils carry out tests on lithium, similar tests can be demonstrated with sodium and potassium. Their reaction with water should be carried out in a large trough and a safety screen is advisable. It is dangerous to attempt to collect the hydrogen: with potassium it will burn in any case, and with sodium it can be made to burn by dropping the sodium on a piece of filter paper floating in the trough.

For a further comparison the three metals burning in chlorine can be demonstrated. Gas jars are most conveniently filled from a chlorine demonstration bottle (obtainable from BDH), otherwise from potassium manganate(VII)/concentrated hydrochloric acid. Reaction of the deflagrating spoon with chlorine tends to colour the otherwise white smoke: placing asbestos paper between the metal and the spoon helps.

Experiment 1.9 Before pupils investigate iodine, similar tests can be demonstrated with chlorine and bromine. For bromine (c) can be demonstrated as for iodine. For chlorine the apparatus shown in Figure 1.18 is suitable (a fume cupboard is desirable).

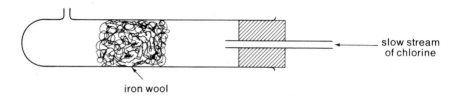

Figure 1.18

Experiment 1.10 Chlorine—saturated solution, freshly prepared. Bromine—saturated solution diluted ×5.

Experiment 1.11 M hydrochloric acid or 0.5M sulphuric acid.

2 Atoms in chemistry

An element is made up of tiny individual particles called *atoms*, in much the same way as a house is made up of individual bricks or a packet of salt is made up of separate small crystals.

An atom is the smallest individual particle of an element.

It is possible to split atoms into even smaller particles. Uranium atoms, for example, are disintegrated in the atomic bomb and, in a more controlled manner, in atomic power stations. But after an atom of uranium has been split it is no longer uranium. Atoms of two different elements, barium and krypton, are left. The smallest particle of uranium is an atom of uranium.

2.1 Evidence for the existence of atoms

We often hear the word *atom* nowadays. Most people have a fairly good idea of what an atom is, yet no one has ever seen one. They are too small to be seen, even with the most powerful microscope. If 'seeing is believing' there is no *proof* that atoms exist. Yet almost everyone, certainly every scientist, is convinced that they do exist.

Brownian motion

One of the most striking pieces of evidence for the existence of atoms was first observed by a Scottish botanist, Robert Brown, in 1827. Looking through a microscope at pollen grains floating in water, he noticed that the particles of pollen were in constant erratic motion. He knew that something must be moving the grains, as they were not living organisms and could not move by themselves, but he did not know what.

We can explain BROWNIAN MOTION quite easily in terms of the movement of atoms. The water surrounding the pollen grains is made up of small groups of atoms, called water MOLECULES, moving erratically in all directions. These collide with the pollen grains. It is collisions between the very small water molecules and the very much larger pollen grains which cause the grains to move.

Experiment 2.1 Are the particles of a gas in motion?

(a) Remove the cork from a test tube of carbon dioxide and quickly put your thumb over the end. Invert this tube of carbon dioxide over an 'empty' test tube, with the mouths of the two tubes held close together as in Figure 2.1(A). Hold the tubes like this for 3 minutes, then cork them both quickly.

Opening the tubes for as short a time as possible, pour $\frac{1}{2}$ cm depth of lime water into each tube and shake.

What was in the 'empty' tube at the beginning of the experiment?

What happens to the lime water in the lower tube? What does this tell you about the contents of the lower tube at the end of the experiment?

Is there any carbon dioxide left in the upper tube?

Carbon dioxide is more dense than air. Are the results of this experiment what you would expect?

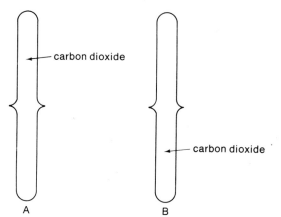

Figure 2.1

(b) Repeat experiment (a), but this time have the tube of carbon dioxide *below* the empty tube as in Figure 2.1(B).

Is there any carbon dioxide left in the lower tube at the end of the experiment?

Does any carbon dioxide find its way into the upper tube?

What does this experiment tell us about the particles in carbon dioxide gas?

(c) Repeat experiments (a) and (b), using hydrogen instead of carbon dioxide. The lime water test was used for carbon dioxide in the first two experiments. Decide how you are going to test for hydrogen in this experiment.

Is hydrogen more or less dense than air? Would you expect it to move upwards or downwards?

What does this experiment show about the way hydrogen gas moves?

Diffusion

If a bottle of perfume is opened for a moment in the middle of a room, a little of the vapour escapes. Within a few minutes the smell of the perfume is noticeable all over the room. The small amount of perfume vapour released from the bottle has spread throughout the air. It is impossible to explain how the perfume spread in this way except by using the idea of atoms.

The perfume vapour is made up of millions of small groups of atoms (perfume molecules), which are moving about in a random way. The result of this random motion is that they soon become spread more or less evenly through all the air in the room.

Diffusion of a gas (or vapour) is the spreading out of the particles so that they distribute themselves through the whole space available.

Brownian motion and the diffusion of gases are just two of the many pieces of evidence which lead us to believe that atoms exist. They also provide us with another important piece of information: *atoms are in constant motion.*

Questions

1 What is the difference between an atom and a molecule?

2 Describe any one experiment that suggests that substances consist of small particles in random motion.

3 Explain the following observations.
  (a) If a drop of liquid bromine is placed in the bottom of a gas jar and the lid is put on, the whole jar becomes uniformly filled with bromine vapour within a few minutes.
  (b) If a small crystal of potassium permanganate is put in the bottom of a test tube of water and the tube is allowed to stand for several hours, all of the water eventually becomes coloured purple by the potassium permanganate.
  (c) In a shaft of sunlight through the window of a room, particles of dust in the air can be seen dancing about.
  (d) A block of gold and a block of silver are left clamped tightly together for several years. When they are separated chemical analysis detects the presence of gold a few millimetres below the surface of the silver and silver below the surface of the gold.

4 In question 3, why is (a) complete in a few minutes, and (b) complete in a few hours, while (d) shows only a small effect after several years?

5 What other evidence can you think of for the existence of atoms?

2.2 The history of the atomic theory

As far as we know, some of the ancient Greek philosophers were the first to suggest that substances are made up of tiny particles. The most famous of these was Democritus, who was born in about 460 BC. Unfortunately, he had no evidence to back up his theory and, as a result, it was forgotten for many centuries.

In the seventeenth century a number of scientists, including Sir Isaac Newton, the discoverer of the law of gravity, began to realise that Democritus must have been right. But it was not until the beginning of the nineteenth century that an ATOMIC THEORY was put forward in a clear way.

The originator of modern atomic ideas was an English chemist, John Dalton, who published his atomic theory in 1803. He made the following suggestions.

---

1 All substances are made up of small particles (to which he gave the name *atoms*, from a Greek word meaning *impossible to cut*).
2 Atoms cannot be divided.
3 Atoms can neither be created nor destroyed.
4 Atoms of one particular element all have the same mass.
5 The atoms of different elements have different masses.
6 Reaction between elements to form compounds involves atoms joining together. Compounds are made up of small groups of atoms called molecules. The mass of a molecule (which Dalton called a *compound atom*) is the sum of the masses of its atoms.

It is now known that statements 2, 3, 4, and 5 are not exactly correct.
2 In special circumstances atoms *can* be divided. In fact, the atoms of radioactive elements like radium and uranium split of their own accord.
3 New atoms *can* be created. This is happening all the time in the sun.
4 Atoms of an element do not always have exactly the same mass.
5 It is possible for atoms of different elements to have the same mass.

Dalton's other two statements are the most important, and have contributed most to the development of chemistry as a science.

All substances are made up of small particles, called atoms.

Reaction between elements to form compounds involves atoms joining together.

It is difficult to realise nowadays, when the idea of atoms is so generally accepted, and seems so obvious, what a remarkable scientific advance Dalton made. His atomic theory is as important in the history of science as Newton's discovery of the law of gravity, Faraday's invention of the electric motor, and Rutherford's splitting of the atom.

2.3 How big are atoms?

Atoms are roughly spherical in shape. Their diameter ranges from $\frac{1}{40\,000\,000}$

of a centimetre (2.5×10^{-8} cm) for the smallest atom (hydrogen) to about $\frac{1}{20\,000\,000}$ of a centimetre (5×10^{-8} cm) for large atoms such as those of potassium.

The centimetre is obviously not a very convenient unit of length for measuring atoms, just as it would not be very convenient to measure the size of this page in kilometres ($\frac{1}{4000}$ km × $\frac{1}{7000}$ km). We choose measurements of length to suit the job we are going to use them for: centimetres or millimetres for the size of a piece of paper, metres for the height of a building, kilometres for the distance between two towns, and so on.

A convenient unit of length for measuring atoms is the NANOMETRE (abbreviated *nm*).

1 nanometre = $\frac{1}{1\,000\,000\,000}$ metre (10^{-9} m) = $\frac{1}{10\,000\,000}$ centimetre (10^{-7} cm)
1 centimetre = $10\,000\,000$ nanometres (10^7 nm)
1 metre = $1\,000\,000\,000$ nanometres (10^9 nm).

It is difficult for anyone to appreciate just how small atoms are. Figure 2.2 shows the comparative size of some other small objects.

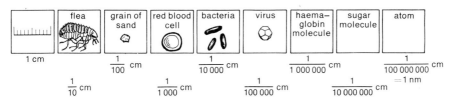

Figure 2.2

Experiment 2.2 Getting an idea of the size of particles

Dissolve a small crystal of *potassium permanganate* in a test tube of water (about 10 cm³).

Now pour away about $\frac{9}{10}$ of your solution, fill the tube with water, and mix the solution. You have diluted the original solution 10 times (to 100 cm³).

Again pour away $\frac{9}{10}$, fill up with water, and mix. You have now diluted the original solution 100 times (to 1000 cm³).

Repeat this process over and over again, until you can only just see the colour of the potassium permanganate in the water. (When the solution gets very dilute holding the tube against a piece of white paper may help you to see the colour.)

How many times have you now diluted the original solution?

How many cubic centimetres of solution would you now have if you had not thrown any away?

You can still see the colour due to the potassium permanganate in every one of these cubic centimetres, so there must be at least one potassium permanganate particle in every cubic centimetre. How many particles *at least* must there have been in the original potassium permanganate crystal?

Experiment 2.3 Estimating the size of an oil particle

The plan is to let a drop of oil spread out over the surface of water and to measure the thickness of the oil layer. If we assume that the oil spreads out as far as it can, the layer will be one oil particle thick. The thickness of the layer is the same as the diameter of the oil particles.

You are provided with a solution of an oil. 1 cm³ of this solution contains 1/10000 cm³ of the oil. First measure how many drops of this solution make 1 cm³. Count how many drops from a teat pipette are needed to fill a measuring cylinder to the 1 cm³ mark. Make a note of the result.

Now put water into a flat tray to a depth of ½ cm. Cover the surface of the water with a fine dusting of talcum powder from a 'pepper pot'. Use your teat pipette to allow *one drop* of the oil solution to fall on to the surface of the water in the middle of the tray. Hold the nozzle of the pipette as close as possible to the surface of the water without actually touching it.

As quickly as you can, measure the diameter of the oil patch in centimetres. Make a note of the result.

Empty the tray and prepare a new powdered water layer. Repeat the experiment twice more, so that you have three results.

Calculate the average diameter of your oil patch from the results of your three experiments. The patches are an irregular shape but since this is only a very rough experiment assume they are square. Calculate the average area of your oil patch in square centimetres.

Now calculate the volume of one drop of the oil solution. Divide the volume of oil in 1 cubic centimetre of the solution (1/10000 cm³) by the number of drops in 1 cm³. This gives the volume of oil in your oil patch.

The thickness of the oil patch is its volume divided by its area. Calculate the thickness of your oil patch. Your result is the approximate diameter of one oil particle.

How do we know about the size of atoms?

Oil leakages from tankers at sea cause serious pollution problems because the oil spreads out over the surface of the water to cover a vast area (Figure 2.3).

You can see the same effect of oil spreading over water in the streets on a wet day with motor oil on puddles. According to the atomic theory we can think of oil as a collection of millions of individual particles, rather like a bag of very tiny marbles. If marbles are poured on to the ground they spread out to make a layer only one marble thick. In the same way, if

Figure 2.3 Oil (from the wreckage of the giant tanker Torrey Canyon) being burnt at sea in an attempt to prevent the oil from reaching nearby beaches

oil is poured on to water the individual particles spread out into a layer one particle thick.

When one drop of oil, with a volume of $\frac{1}{25}$th of a cubic centimetre, is put on to the surface of some water it spreads out into a layer with an area of 400 000 square centimetres (a circle with a diameter of over 7 metres). We can think of this oil layer as a very thin cylinder of oil lying on the surface of the water (Figure 2.4).

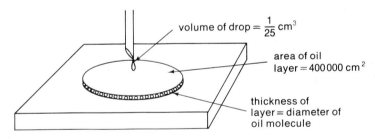

Figure 2.4

The volume of the cylinder is the volume of the oil ($\frac{1}{25}$ cm³). This is equal to the area of the top of the cylinder (400 000 cm²) multiplied by its depth:

$$\text{volume} = \text{area} \times \text{depth}$$
$$\tfrac{1}{25} \text{ cm}^3 = 400\,000 \text{ cm}^2 \times \text{depth in cm.}$$

We can easily calculate the depth of the cylinder:

$$\text{depth} = \frac{1}{25} \div 400\,000 = \frac{1}{25} \times \frac{1}{400\,000} = \frac{1}{10\,000\,000} \text{ cm} = 1 \text{ nanometre.}$$

Since the layer of oil is one particle thick, we can say that the diameter of a single oil particle is about 1 nanometre. Oil is a compound, not an element, so an oil particle is a molecule (a small group of atoms). *Atoms must be even less than 1 nanometre in diameter.*

Using other techniques, which are more accurate than the oil drop method, the sizes of atoms can now be measured with considerable accuracy. The diameters of atoms of some of the commoner elements are given in Figure 2.5. Most atoms have a diameter between $\frac{1}{4}$ nm and $\frac{1}{2}$ nm.

Element	Diameter of atom (nanometres)
Bromine	0.39
Chlorine	0.36
Copper	0.26
Gold	0.29
Hydrogen	0.24
Iodine	0.43
Iron	0.25
Lead	0.35
Lithium	0.30
Mercury	0.30
Oxygen	0.28
Potassium	0.45
Silver	0.29
Sodium	0.37
Sulphur	0.37

Figure 2.5

Questions

1 What is a nanometre?

2 If the diameter of a hydrogen atom is $\frac{1}{4}$ nm, how many hydrogen atoms would be needed to make a row 1 mm long (1 mm = $\frac{1}{10}$ cm)?

How many hydrogen atoms would be needed to cover a pinhead with a single layer (area of pinhead = 1 mm²)?

3 1 cm³ of an evil-smelling gas called hydrogen sulphide was released in the centre of a room measuring $10 \times 4 \times 2\frac{1}{2}$ metres. Within two minutes the smell was noticeable all over the room.
    (a) What are the dimensions of the room in centimetres?
    (b) What is the volume of the room in cubic centimetres?

The volume of your nasal cavity is about 10 cm³. This is the volume of air taken in by a single sniff.

(c) How many 'sniffs-full' of air are there in the room?

In order to be able to smell the hydrogen sulphide there must be at least one molecule of it in each sniff-full.

(d) How many molecules, *at least*, must there have been in the original 1 cm³ of gas?

(e) How did the hydrogen sulphide spread through all the air in the room?

(f) Make an estimate of the speed at which the hydrogen sulphide molecules move through the air in the room.

4 In a tanker accident 10 000 tonnes of oil leak into the sea. If the oil spreads into a layer one particle thick how many square kilometres of sea will be polluted? (Assume 1 tonne = 1 cubic metre; diameter of oil particle = 1 nm = 10^{-9} m.)

5 In Figure 2.5 what do you notice about the diameters of atoms of the following elements: bromine, chlorine, iodine, lithium, potassium, and sodium?

2.4 How heavy are atoms?

Atoms are very small. They are also very light. We cannot see them, so we certainly cannot put one on a balance to weigh it. During the nineteenth century chemists devised all sorts of ingenious schemes for estimating the masses of atoms, but we know now that the results they obtained were often wildly wrong.

The mass spectrometer

The first correct measurements of the masses of atoms were made by Sir Joseph Thomson in 1912 by means of a piece of apparatus called a MASS SPECTROMETER (Figure 2.6). The same kind of apparatus, with some improvements, is still used today.

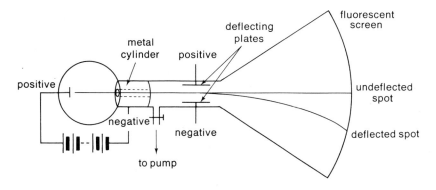

Figure 2.6

Atoms in chemistry

All the air is removed from the apparatus by means of a vacuum pump. A little of the vapour of the element is put in the round glass bulb and subjected to a very high voltage. This gives the atoms a positive electrical charge. They are repelled from the positive metal plate and attracted towards the negative metal cylinder. Some of them pass through the hole in the cylinder at high speed and produce a bright spot on the fluorescent screen, which is rather like a television screen.

A voltage is then applied between the two metal deflecting plates. The stream of positively charged atoms is repelled from the positive plate and attracted towards the negative plate, and the spot on the screen is deflected. A stream of heavy atoms will be deflected less than a stream of light atoms. By measuring the amount of deflection of the spot on the screen the mass of the atoms can be calculated. Nowadays mass spectrometers are so accurate that the masses of atoms can be measured to an accuracy of greater than one part in a million.

The masses of atoms

Even the heaviest natural atom, an atom of uranium, has a mass of less than $\frac{1}{2\,000\,000\,000\,000\,000\,000\,000}$ of a gram (5×10^{-22} g). Just as the centimetre is not a convenient length measurement for atoms, the gram is obviously not a convenient unit for measuring their mass. The unit we use is called the ATOMIC MASS UNIT. This unit is chosen so that *the lightest atom, the hydrogen atom, has a mass of* 1 *atomic mass unit. We say that the atomic mass of hydrogen is* 1. A uranium atom weighs 238 atomic mass units; the atomic mass of uranium is 238. Figure 2.7 gives a list of the atomic masses of all the elements.

$$1 \text{ atomic mass unit} = \frac{1}{602\,000\,000\,000\,000\,000\,000\,000} \text{ gram } (1.66 \times 10^{-24} \text{ g}).$$

1 gram = 602 000 000 000 000 000 000 000 (6.02×10^{23}) atomic mass units.

2.5 The composition of compounds

Chemists are interested in how atoms join together: how compounds are formed from elements.

Copper sulphide is a compound of two elements, copper and sulphur. It is quite easy to find experimentally the proportions by weight in which copper and sulphur combine. Analysis of 96 g of copper sulphide shows that it is made up of 64 g of copper and 32 g of sulphur. This is a useful piece of information. It tells us, for instance, exactly the correct proportions in which to mix copper and sulphur before heating them together to make copper sulphide.

A manufacturer needs to know the proportions *by weight* in which elements combine. But chemists want to know rather more than this. In order to understand how atoms join together they want to know the proportions in which atoms combine *by number*. Is copper sulphide, for

Element and mass in atomic mass units

actinium	227	Ac	hafnium	178	Hf	praseodymium	141	Pr
aluminium	**27**	**Al**	hahnium	258	Ha	promethium	145	Pm
americium	243	Am	helium	4	He	protactinium	231	Pa
antimony	122	Sb	holmium	165	Ho	radium	226	Ra
(*stibium*)			**hydrogen**	**1**	**H**	radon	222	Rn
argon	40	Ar	indium	115	In	rhenium	186	Re
arsenic	75	As	**iodine**	**127**	**I**	rhodium	103	Rh
astatine	210	At	iridium	192	Ir	rubidium	85	Rb
barium	137	Ba	**iron** (*ferrum*)	**56**	**Fe**	ruthenium	101	Ru
berkelium	249	Bk	krypton	84	Kr	rutherfordium	257	Rf
beryllium	9	Be	lanthanum	139	La	samarium	150	Sm
bismuth	209	Bi	lawrencium	257	Lw	scandium	45	Sc
boron	11	B	**lead**	**207**	**Pb**	selenium	79	Se
bromine	**80**	**Br**	(*plumbum*)			silicon	28	Si
cadmium	112	Cd	**lithium**	**7**	**Li**	silver	108	Ag
caesium	133	Cs	lutetium	175	Lu	(*argentum*)		
calcium	**40**	**Ca**	**magnesium**	**24**	**Mg**	sodium	23	Na
californium	251	Cf	manganese	55	Mn	(*natrium*)		
carbon	**12**	**C**	**mendelevium**	**256**	**Mv**	strontium	88	Sr
cerium	140	Ce	**mercury**	**201**	**Hg**	**sulphur**	**32**	**S**
chlorine	**35**	**Cl**	(*hydragyrum*)			tantalum	181	Ta
chromium	52	Cr	molybdenum	96	Mo	technetium	99	Tc
cobalt	59	Co	neodymium	144	Nd	tellurium	128	Te
copper	**64**	**Cu**	neon	20	Ne	terbium	159	Tb
(*cuprum*)			neptunium	237	Np	thallium	204	Tl
curium	247	Cm	nickel	59	Ni	thorium	232	Th
dysprosium	163	Dy	niobium	93	Nb	thulium	169	Tm
einsteinium	254	Ei	**nitrogen**	**14**	**N**	**tin** (*stannum*)	**119**	**Sn**
erbium	167	Er	nobelium	254	No	titanium	48	Ti
europium	152	Eu	osmium	190	Os	tungsten	184	W
fermium	253	Fm	**oxygen**	**16**	**O**	(*wolfram*)		
fluorine	19	F	palladium	106	Pd	**uranium**	**238**	**U**
francium	223	Fr	**phosphorus**	**31**	**P**	vanadium	51	V
gadolinium	157	Gd	platinum	195	Pt	xenon	131	Xe
gallium	70	Ga	plutonium	242	Pu	ytterbium	173	Yb
germanium	73	Ge	polonium	210	Po	yttrium	89	Y
gold	**197**	**Au**	**potassium**	**39**	**K**	zinc	65	Zn
(*aurum*)			(*kalium*)			zirconium	91	Zr

Figure 2.7 The symbol on the right of each column is that used to identify a mole of each of the elements. All atomic masses are given to the nearest whole number, based on the carbon-12 scale. For artificial elements the mass of the longest-lived isotope is given.

instance, formed by copper and sulphur atoms combining in the ratio of one copper atom to one sulphur atom; two copper atoms to one sulphur atom; one copper atom to three sulphur atoms; or what?

Counting atoms by weighing

A two-penny piece weighs exactly twice as much as a penny piece. Banks make use of this to count copper coins. If a cashier weighs out 1000 grams of penny pieces he knows that he has exactly 280 coins. If he weighs out 2000 grams of two-penny pieces he knows that he has exactly the same number of coins: 280 two-penny pieces. Similarly, 500 grams of two-penny pieces and 250 grams of penny pieces contain the same number of coins. Weighing out coins like this is much quicker than counting them. In any bank you will see a special pair of scales used for this purpose.

Chemists work in the same way. A copper atom is twice as heavy as a sulphur atom. (Atomic mass of sulphur = 32; atomic mass of copper = 64.) So 32 grams of sulphur and 64 grams of copper (twice the mass) contain the same number of atoms. The actual number of atoms in 32 grams of sulphur or 64 grams of copper is very large: 602 000 000 000 000 000 000 000 (six-hundred-and-two thousand million million million atoms). This is also the number of atoms in 1 gram of hydrogen (atomic mass = 1), 12 grams of carbon (atomic mass = 12), 238 grams of uranium (atomic mass = 238), and so on.

We often give special names to important numbers, such as dozen (for 12), score (for 20), and gross (for 144). The number 602 000 000 000 000 000 000 000 (6.02×10^{23}) is very important in chemistry, so it is given a special name: a MOLE. Just as we may refer to a dozen eggs or a gross of screws, in chemistry we often refer to a mole of atoms.

The atomic mass in grams of any element contains the same number of atoms: a mole of atoms. For example, 64 grams of copper contain a mole of copper atoms, 32 grams of sulphur contain a mole of sulphur atoms, 1 gram of hydrogen contains a mole of hydrogen atoms.

The composition of copper sulphide

By making use of the mole idea we can quite easily work out, from the results of a chemical analysis, the proportions by number in which atoms combine.

Analysis of copper sulphide shows that 64 grams of copper combine with 32 grams of sulphur.

Atomic mass of copper = 64. Atomic mass of sulphur = 32.
64 g of copper is 1 mole.   32 g of sulphur is 1 mole.
1 mole of copper atoms combines with 1 mole of sulphur atoms.
6.02×10^{23} copper atoms combine with 6.02×10^{23} sulphur atoms.
1 atom of copper combines with 1 atom of sulphur

Another example: water

Water is a compound of two elements, hydrogen and oxygen. Analysis of water shows that 1 g of hydrogen combines with 8 g of oxygen.

Atomic mass of hydrogen = 1   Atomic mass of oxygen = 16
1 g of hydrogen is 1 mole.   8 g of oxygen is $\frac{1}{2}$ mole.
1 mole of hydrogen atoms combines with $\frac{1}{2}$ mole of oxygen atoms.
2 atoms of hydrogen combine with 1 atom of oxygen.

Chemist's shorthand

Shorthand is a method of writing used by secretaries to save time. It makes use of symbols to represent words. The symbols are quicker to write than the words themselves. Chemists have their own special shorthand to save time and space in writing down chemical information. For example:

1 mole of sulphur atoms is represented by the symbol *S*;
1 mole of copper atoms is represented by the symbol *Cu*.

The symbols for one mole of atoms of each of the elements are shown in Figure 2.7.

Sometimes the initial letter of the element's name is used (e.g. C, S, P, N). Sometimes, when the names of several elements begin with the same letter, the initial letter is used together with another letter (e.g. C, Ca, Cl, Cs, Cm). And with some of the earliest elements to be discovered, before about 1800, when chemists often wrote in Latin, the symbol is taken from the Latin name for the element (e.g. K from *kalium*, Cu from *cuprum*, Ag from *argentum*).

Just as the secretary who did not know the exact meaning of her own shorthand would not be much use, chemists must know exactly what their shorthand symbols mean. Cu, for instance, means one mole of copper atoms: *the symbol tells us the name of the element and the quantity of it.* If the quantity is not one mole the shorthand must show this. For example, three moles of hydrogen atoms is written 3H; half a mole of sodium atoms is written $\frac{1}{2}$Na (or 0.5Na).

Shorthand for compounds

Chemical shorthand is also used to represent information about compounds. In copper sulphide copper and sulphur are combined in the proportion of 1 mole of copper atoms to 1 mole of sulphur atoms. In shorthand this can be written CuS, which means 1 mole of copper atoms and 1 mole of sulphur atoms combined together.

CuS is called the FORMULA of copper sulphide. *It tells us what elements are present in the compound and the proportions in which the atoms of*

Atoms in chemistry   41

these elements have combined. In writing a formula it is usual to write first the element which is further to the left in the periodic classification. For a compound between a metal and non-metal, for instance, the metal is written first.

The proportions in which hydrogen and oxygen combine to form water are 2 moles of hydrogen atoms to 1 mole of oxygen atoms. The formula for water is written H_2O. We do not need to write H_2O_1 because O means 1 mole of oxygen atoms and the subscript 1 is unnecessary.

Working out formulas: some more examples

1 Analysis of lead oxide shows that 23.9 g of lead oxide contain 20.7 g of lead. What is the formula of lead oxide?
The mass of oxygen in 23.9 g of lead oxide is $(23.9 - 20.7) = 3.2$ g.

>20.7 g of lead combine with 3.2 g of oxygen.

(It makes things easier if the decimals are removed: the proportions are still the same. In this case multiply by 10.)

>207 g of lead combine with 32 g of oxygen.

To convert grams of an element to moles divide by the atomic mass.

>Atomic mass of lead = 207   Atomic mass of oxygen = 16
>207 g of lead is 1 mole   32 g of oxygen is 2 moles
>   1 mole of lead combines with 2 moles of oxygen.
>   *The formula of lead oxide is PbO_2.*

2 An experiment shows that 2.93 g of aluminium chloride can be made from 0.60 g of aluminium. What is the formula of aluminium chloride?
The mass of chlorine in 2.93 g of aluminium chloride is $(2.93 - 0.60) = 2.33$ g.

>0.60 g of aluminium combine with 2.33 g of chlorine,

or

>60 g of aluminium combine with 233 g of chlorine.

Atomic mass of aluminium = 27   Atomic mass of chlorine = 35
60 g of aluminium is $\frac{60}{27} = 2.22$ moles.   233 g of chlorine is $\frac{233}{35}$
$\qquad\qquad\qquad\qquad\qquad\qquad\qquad\qquad\qquad\qquad = 6.67$ moles.
>2.22 moles of aluminium combine with 6.67 moles of chlorine,
>or   1 mole of aluminium combines with 3 moles of chlorine.
>   *The formula of aluminium chloride is $AlCl_3$.*

Experiment 2.4   Finding the formula of magnesium oxide

Weigh a crucible with its lid. Write down the weight (result 1).

Clean 15 cm of magnesium ribbon with sandpaper (why?). Wind it into a tight coil and put it in the crucible. Weigh the crucible, lid, and magnesium ribbon. Write down the weight (result 2).

Support the crucible on a pipeclay triangle on a tripod. Heat it with a fierce bunsen flame until the magnesium begins to burn (you can see the glow through the side of the crucible). When it is burning lift the crucible lid a little with tongs every now and then to let in air (why?). Avoid allowing any more white smoke to escape than you can help (why?).

When the magnesium seems to be completely burnt remove the crucible lid and continue heating strongly for a minute or so (why?).

Replace the lid and allow the crucible to cool. Weigh the crucible, lid, and magnesium oxide. Write down the weight (result 3).

From your three results you can calculate the masses of magnesium and oxygen combining together to form magnesium oxide:

mass of magnesium = result 2 − result 1;
mass of oxygen = result 3 − result 2.

Now use the method shown in the examples on page 42 to work out the formula of magnesium oxide.

Experiment 2.5 Finding the formula of copper oxide

Weigh a reduction tube (a test tube with a small hole in the end). Write the weight down (result 1).

Carefully put 3 spatulas-full of copper oxide in a neat heap in the middle of the reduction tube. Weigh the tube and copper oxide. Write down the weight (result 2).

Clamp the reduction tube and connect it to the gas supply as shown in Figure 2.8.

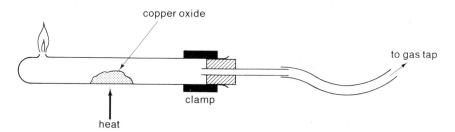

Figure 2.8

Pass a slow stream of gas through the tube, and light it at the small hole. You should have a flame not more than 3 cm high. Now heat the copper oxide with a small bunsen flame. When all the copper oxide has been reduced to copper stop heating it. Leave the apparatus to cool with gas continuing to pass through (why?). When the tube is quite cool switch off the gas. Weigh the reduction tube with the copper. Write down the weight (result 3).

From your three results you can calculate the masses of copper and oxygen combining together to form copper oxide:

$$\text{mass of copper} = \text{result 3} - \text{result 1};$$
$$\text{mass of oxygen} = \text{result 2} - \text{result 3}.$$

Now use the method shown in the examples on page 42 to work out the formula of copper oxide.

Note In order to find the formula of any chemical compound it is first necessary to carry out an experiment such as either of the ones above. As you will have discovered it is quite a time-consuming business. No one would have time to find experimentally the formula of every chemical they had to use. From now on therefore we will use the formulas that other people have worked out. But whenever you see a formula in a book, remember that someone has done an experiment to work it out.

Questions (Refer to Figure 2.7 for atomic masses where necessary.)

1 What is an atomic mass unit?

2 How many atoms are there in 1 g of hydrogen?
A   1?
B   about 1 000?
C   about 1 000 000?
D   about 1 000 000 000 000?
E   more than 1 000 000 000 000 000 000?

3 Which one of the following best describes a mole?
A   It is a dozen atoms.
B   It is the number of atoms in hydrogen.
C   It is the number of atoms in a unit of hydrogen.
D   It is the number of atoms in a gram of hydrogen.
E   It is the number of atoms in a kilogram of hydrogen.

4 If 1 mole of oxygen contains 6.02×10^{23} atoms, how many atoms will there be in (a) 1 mole of technetium, (b) 2 moles of scandium, (c) $\frac{1}{10}$ mole of tin?

5 *Copy and fill in the blanks*. Uranium is an _____. The symbol for 1 mole of uranium is ___. This symbol represents _____ grams of uranium which contain _____ atoms.

6 Write out in full the meaning of the following symbols: C, Ca, Cu, 2Na, 10H, $\frac{1}{2}$Br, $\frac{1}{10}$Cl, 0.5Al, 0.02N, 0.61Zn.

7 Write out the full meaning of the formulas (a) PbS, (b) $MgCl_2$, (c) Fe_2O_3.

8 How many moles of atoms are there in:
    (a) 139 g of lanthanum,   (b) 402 g of mercury,
    (c) 300 g of arsenic,   (d) 730 g of germanium,
    (e) 240 g of titanium,   (f) 10 g of helium,
    (g) 110 g of neon,   (h) 35 g of gallium,
    (i) 5.9 g of cobalt,   (j) 32 g of tellurium,
    (k) 17 g of rubidium,   (l) 1.9 g of osmium,
    (m) 3 g of polonium,   (n) 0.5 g of argon,
    (o) 4.4 g of strontium?

9 Calculate the formulas of the following compounds:
    (a) nitrogen oxide, containing 28 g of nitrogen combined with 80 g of oxygen;
    (b) copper fluoride, containing 32 g of copper combined with 19 g of fluorine;
    (c) silver oxide, containing 27 g of silver combined with 2 g of oxygen;
    (d) sodium phosphide, containing 6.9 g of sodium combined with 3.1 g of phosphorus;
    (e) lead chloride, containing 2.07 g of lead combined with 1.4 g of chlorine;
    (f) aluminium oxide, containing 13.5 g of aluminium combined with 12 g of oxygen;
    (g) boron chloride, containing 33 g of boron combined with 315 g of chlorine;
    (h) lithium chloride, containing 0.1 g of lithium combined with 0.5 g of chlorine;
    (i) magnesium nitride, containing 3.6 g of magnesium combined with 1.4 g of nitrogen;
    (j) manganese oxide, containing 4.95 g of manganese combined with 1.92 g of oxygen.

10 3.9 g of potassium burns in oxygen to form 4.7 g of potassium oxide. What is the formula of potassium oxide?

11 0.378 g of iron reacts with bromine to produce 2 g of iron bromide. What is the formula of iron bromide?

12 A compound of carbon and hydrogen is found to contain exactly 75 per cent of carbon by weight. What is its formula?

13 Chalk contains 40 per cent calcium, 12 per cent carbon, and 48 per cent oxygen by weight. What is its formula?

14 Some clean, bright copper turnings were heated in air. The pieces turned black and increased in weight.
   (a) What was the black substance which formed on the surface of the copper?
   (b) Account for the increase in weight.
   Fresh copper was placed in a tube and connected to a gas syringe which contained 50 cm^3 of air. When the copper was heated it again turned black and volume changes were seen in the syringe. At first the volume increased, then on cooling it decreased to 40 cm^3.
   (c) Explain why the volume increased at first then decreased.
   (d) Name the principal substance in the gas left in the syringe.
   (e) What would happen if some fresh copper turnings were heated in this gas?
   (f) Explain your answer to (e).
   Some of the substance formed when copper was heated in air was placed in a tube, and hydrogen was passed over the heated material. The black substance changed to a pinkish-coloured material which conducted electricity.
   (g) What is the pinkish-coloured material?
   (h) What else is produced in this reaction?
   (i) Explain how this change has come about.
   The black substance was placed in a porcelain boat and weighed. The boat was placed in a tube, heated, and hydrogen was passed over it. The same change occurred as above. After cooling with the hydrogen still passing over it, the boat was reweighed.

Results   *Before heating*
   Weight of boat + contents = 23.379 grams
   Weight of empty boat   = 15.429 grams
   Weight of black material =   7.950 grams
                    After heating
   Weight of boat + contents = 21.779 grams

   (j) What weight of pinkish-coloured material was produced?
   (k) If the atomic mass of the element formed is 63.5, how many moles have been produced?
   (l) The loss in weight is due to the removal of an element called X. What weight of X was lost?
   (m) If the atomic mass of X is 16, how many moles of X were lost?
   (n) What is the ratio of the number of atoms of each element in the black substance?
   (o) Suggest a reason why the hydrogen was made to flow over the pinkish-coloured material as it cooled.
   The element sulphur (symbol S) is in the same group of the periodic table as element X.

(p) What would you expect to happen if copper was heated in sulphur vapour?
(q) What would be the name of the product? (*London*)

6 The states of matter: solid, liquid, and gas

Almost any substance can exist as a solid, a liquid, or a gas. Solid, liquid, and gas are known as the THREE STATES OF MATTER. We are familiar with some substances in all three states. We have all seen solid water (ice), liquid water, and gaseous water (steam). With most substances, however, we are familiar with only one state, usually because the other states only exist at extreme temperatures. We think of copper, for instance, as a solid. But if it is heated to 1083°C it melts to a liquid, and if liquid copper is heated to 2562°C it boils, giving copper gas. We think of oxygen as a gas, but if it is cooled below -183°C it becomes a liquid, and below -219°C it is a solid.

Experiment 2.6 Changes of state

Half-fill a beaker with small pieces of ice. Stand it on a gauze and tripod and heat it with a *small* bunsen flame.

Stir *carefully* with a thermometer and make a note of the temperature every 1 minute. Continue until the water has been boiling for 5 minutes.

Use your results to draw a graph with temperature (in °C) on the vertical axis and time (in minutes) on the horizontal axis. Use your graph to answer the following questions.

(a) What is happening during the two periods when the graph is horizontal?
(b) What is the melting point of ice?
(c) What is the boiling point of water?
(d) During the time when the temperature is rising from the melting point of ice to the boiling point of water the heat of the bunsen flame is being used to raise the temperature of the water. What is the heat of the bunsen flame being used for while the ice is melting and while the water is boiling?
(e) How long did it take for all the ice to melt?
(f) Which takes more heat: melting a certain mass of ice or raising the temperature of the same mass of water from freezing point to boiling point?
(g) Which takes more heat: melting a certain mass of ice or boiling the same mass of water?

Solids

In a block of ice the individual groups of atoms (water molecules) are held together in a regular pattern (Figure 2.9). It is this regular pattern of

Atoms in chemistry

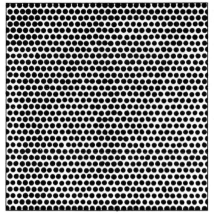
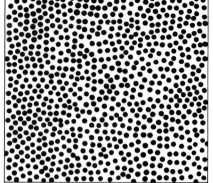

Figure 2.9 In a solid the atoms or molecules are packed close together in a regular pattern

Figure 2.10 In a liquid the atoms or molecules are packed close together but have no regular pattern and are free to move

molecules which gives snow crystals their beautiful regular shapes (Figure 2.11). The molecules cannot move from place to place, although they are vibrating 'on the spot'. If ice is cooled, the molecules vibrate more and more slowly. In theory they should stop moving completely at the absolute zero of temperature ($-273°C$), but in practice this temperature has never been reached. When ice is heated, the energy provided makes the molecules vibrate faster, but as long as the temperature does not go above 0°C they still remain in the same positions relative to one another and the regular pattern of the particles is not upset.

Solids into liquids: melting

If ice is heated sufficiently, the vibration of the molecules becomes so vigorous that the forces holding them together in a regular pattern are overcome. The pattern breaks down. For ice this happens at 0°C: the solid melts into a liquid. Different solids melt at different temperatures. The MELTING POINT depends on the strength of the forces holding the particles together.

Heating a solid to its melting point is rather like a house being shaken by an earthquake. If the earthquake tremor is a slight one the structure of the house vibrates slightly, but no permanent damage is done. If the tremor is strong enough to overcome the binding power of the mortar holding the bricks together the whole house collapses. The regular pattern of the bricks is destroyed and a jumbled-up heap of bricks remains. The particles (atoms or molecules) in a solid are, like the bricks in a house, held close together in a regular pattern. A liquid is like the pile of bricks after an earthquake: the particles are not much further apart than in the solid but they have no regular pattern (Figure 2.10).

But there is one difference. The bricks in the pile are stationary. The

Figure 2.11 Snow crystals

molecules in liquid water are moving about in a random way, as we know from observing Brownian motion.

Experiment 2.7 Finding melting points

Seal the end of a melting point tube (a thin glass tube about 8 cm long and 2 mm in diameter) by holding one end in a bunsen flame.

Fill the tube to a depth of about $\frac{1}{2}$ cm with finely-powdered naphthalene. This is most easily done by pushing the open end of the tube into the naphthalene, then rubbing the tube against the edge of a wire gauze to shake the solid down to the bottom.

Attach the melting point tube to a thermometer with a rubber band and set up the apparatus shown in Figure 2.12.

Heat the water *slowly* with a small bunsen flame and stir continuously. Watch the solid in the melting point tube carefully and note the temperature at which it melts.

If you have time you can find the melting point of another solid as well.

What particular advantage does this method of finding melting points have?

Why is the melting point tube attached to the bulb of the thermometer?

How could this method for finding melting points be adapted for use with solids which melt above 100°C?

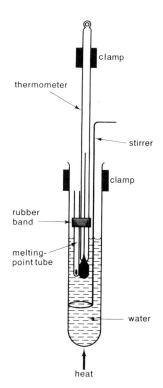

Figure 2.12

Atoms in chemistry

Liquids into gases: boiling

As liquid water is heated up, the molecules move faster and faster. Two things prevent the molecules escaping from the body of the liquid: the atmosphere pushing down on the surface, and the forces between the molecules. Although the forces which held the solid together are no longer strong enough to maintain a regular pattern, they still have the effect of keeping the liquid together.

As the temperature increases and the molecules move faster and faster, the point is eventually reached when they are moving rapidly enough to overcome both of the restraining influences, and the liquid boils. For water, under normal conditions, this happens at 100°C. This is called the BOILING POINT of the liquid; the temperature at which it changes into a gas. Different liquids have different boiling points depending on the strength of the forces between the particles.

The boiling point of a liquid also depends on the pressure exerted by the atmosphere. Water only boils at 100°C under normal atmospheric pressure. If the pressure is less than normal the particles can escape more easily from the liquid and the boiling point is lower. If the pressure is greater than normal the particles find it more difficult to escape and the boiling point is higher. Figure 2.13 shows the boiling point of water at various pressures.

Pressure (atmospheres)	Boiling point of water (°C)
$\frac{1}{4}$	62
$\frac{1}{2}$	84
$\frac{3}{4}$	94
1	100
2	125
3	139
4	150
5	156

Figure 2.13

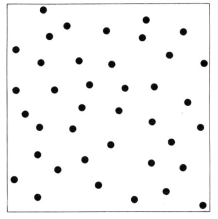

Figure 2.14 In a gas the atoms or molecules are far apart and move freely

In a gas the individual particles are moving very rapidly. They are so far away from one another that there are no forces acting between them to hold them together, and each particle moves completely independently of the others. A gas therefore DIFFUSES (spreads out) to fill the whole space available to it. Figure 2.14 represents the particles in a gas.

Experiment 2.8 An unusual change of state

(a) Put one small crystal of iodine in a test tube and warm it gently over a small bunsen flame. Describe what you see happening.
Does the iodine melt? Explain what has happened.
(b) Repeat (a), using a spatula-full of ammonium chloride in place of iodine.

Solids into gases: sublimation

A few solids behave in an exceptional way. When they are heated, they do not melt, but change directly into a gas. This is called SUBLIMATION. Examples of substances that sublime, and which therefore cannot normally exist in the liquid state, are iodine and carbon dioxide.
Figure 2.15 is a summary of all the changes of state.

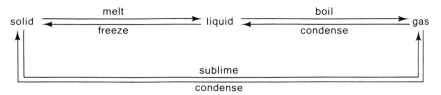

Figure 2.15

The volumes of solids, liquids, and gases

Figure 2.16 shows the volumes of 1 mole of particles of various substances at 25°C. The volume of 1 mole of a solid is usually quite small, because the particles are packed close together. The actual volume occupied by 1 mole of any solid depends on two things: how big the particles are, and how closely they are packed together.

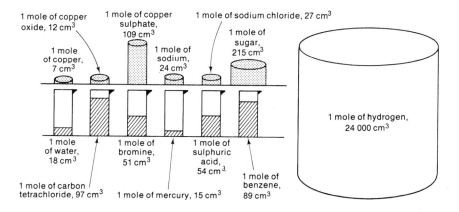

Figure 2.16

Atoms in chemistry 51

The volume of 1 mole of a liquid is generally a bit bigger than the volume of 1 mole of a solid. When a solid melts its volume usually increases by about 10 per cent. The particles are still the same size, but in melting the regular structure has broken down so the particles are slightly further apart.

By comparison with solids and liquids, gases have very large volumes, because the particles are over a hundred times further apart. The volume of a gas depends mainly on the distance between the particles. The volume of the particles themselves contributes very little to the total volume of a gas.

The distance between the particles in a gas depends on the temperature and pressure. At high pressures the particles are pushed closer together and the volume of 1 mole is smaller. At high temperatures the particles move faster and further apart so the volume of 1 mole is larger. But at any one temperature and pressure the volume of 1 mole of *any* gas is roughly the same. Figure 2.17 shows the volume occupied by some common gases at normal atmospheric pressure and at a temperature of 25°C.

Gas	Volume (cm^3) of 1 mole at normal atmospheric pressure and 25°C
argon (Ar)	24 000
neon (Ne)	24 000
helium (He)	24 100
chlorine (Cl_2)	23 600
hydrogen (H_2)	24 000
nitrogen (N_2)	24 000
oxygen (O_2)	24 000
ammonia (NH_3)	24 110
carbon dioxide (CO_2)	24 300
methane (CH_4)	24 430
sulphur dioxide (SO_2)	23 900

Figure 2.17

Some gaseous elements, the noble gases, are composed of separate atoms. In others, such as hydrogen, oxygen, nitrogen, and chlorine, there are no separate atoms. These gases are made up of molecules, each molecule consisting of two atoms joined together (Figure 2.18).

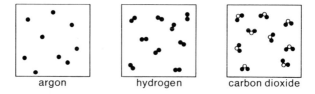

argon   hydrogen   carbon dioxide

Figure 2.18

A mole of hydrogen molecules, for instance, is written in chemical shorthand as H_2, a mole of chlorine as Cl_2, and so on. Gaseous compounds are also made up of molecules, each consisting of a small group of atoms. Carbon dioxide molecules, for example, are each made up of one carbon atom and two oxygen atoms, so a mole of carbon dioxide is represented in shorthand as CO_2.

But whether it is made up of atoms or molecules, *1 mole of any gas has a volume of about 24 000 cubic centimetres (cm^3) at room temperature (25°C) and normal atmospheric pressure.*

The structure of solids, liquids, and gases

Solids	Liquids	Gases
Particles close together	Particles almost as close together as in solids	Particles a long way apart
Particles in regular pattern	No regular pattern	No regular pattern
Particles not moving from place to place but vibrating on the spot	Particles moving about	Particles moving about very rapidly
Have a definite shape	Take the shape of the container	No shape at all

Questions

1 *Copy and fill in the blanks.* Carbon and silicon are _____ in the same group of the _____ _____. Many of their properties are similar but their oxides are quite different in some ways. Silicon dioxide is a solid at room temperature. It _____ at 1610°C and _____ at 2230°C. Carbon dioxide is a _____ at room temperature which _____ to a solid when it is cooled to −78°C. Solid carbon dioxide does not melt. At −78°C it _____ to a gas, without first changing into a _____.

2 Which of the following best describes what happens when a liquid freezes?
A The particles stop moving.
B The particles form a regular arrangement.

C The particles stop moving and form a regular arrangement.
D The particles continue to vibrate.
E The particles continue to vibrate and form a regular arrangement.

3 Which of the following statements about the boiling point of a liquid is true?
A It never varies.
B It rises if the atmospheric pressure rises.
C It falls if the atmospheric pressure rises.
D It falls if the atmospheric pressure falls.
E It depends on the temperature of the room.

4 Which *two* of the following statements do you consider best describe the state of the 'particles' which make up the solid structure of a metal element?
   (i) They move quickly from place to place, covering a large distance.
   (ii) They are at great distances apart from each other.
   (iii) They are close to each other.
   (iv) They are all the same type.
   (v) They are always decomposing. (*West Yorkshire*)

5 A certain substance sublimes when heated. What would you expect to see if it were heated:
   (a) in a long dry tube;
   (b) in an open evaporating basin? (*Wales*)

6 The graph in Figure 2.19 shows how the temperature of some benzene changes as it is cooled.
   (a) What happens at point B?
   (b) What happens at point C?

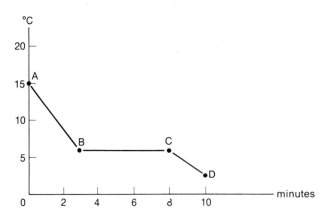

Figure 2.19

54 Chemistry today

(c) What is happening between A and B?
(d) What is happening between C and D?
(e) What is the melting point of benzene?
(f) How long does it take for the benzene to freeze?

7 Draw a graph of the change in the boiling point of water with pressure, using the data in Figure 2.13 (page 50). Plot boiling point (in °C) on the vertical axis and pressure (in atmospheres) on the horizontal axis. Use your graph to answer the following questions.
(a) What is the boiling point of water when the pressure is 1.5 atmospheres?
(b) At what pressure does water boil at 105°C?
(c) Estimate the pressure at which water would boil at 50°C.

8 Use Figure 2.16 (page 51) to find the volumes of 1 mole of copper and 1 mole of sodium. Why are they so different, since the diameters of copper and sodium atoms are similar?

9 Which of the following substances (A, B, C, D and E) are solids, which are liquids, and which are gases at room temperature (about 20°C)?

	Melting point (°C)	Boiling point (°C)
A	6	80
B	80	218
C	−93	4
D	−93	−6
E	−112	91

10 (1 mole of any gas, at room temperature and pressure, has a volume of 24 000 cm^3.)
(a) What is the formula of water if 48 000 cm^3 of hydrogen combine exactly with 24 000 cm^3 of oxygen?
(b) What is the formula of hydrogen chloride if 25 cm^3 of hydrogen combine exactly with 25 cm^3 of chlorine?

For the teacher

Brownian motion (a) In liquids: dilute 1 drop of Johnson's photographic opaque or Aquadag (colloidal graphite) in a test tube of water and view 1 drop of the suspension at × 100 magnification. (b) In gases: a smoke cell with integral illumination that fits on a microscope stage is available from most suppliers. Both demonstrations are even more effective on a microprojector.

Experiment 2.1 Hydrogen and carbon dioxide cylinders are an advantage (obtainable from BOC and Distillers Company respectively), otherwise tubes of the gases can be prepared in advance from granulated zinc/4M hydrochloric acid/few drops copper (II) sulphate solution and marble/2M hydrochloric acid.

The diffusion of bromine can be demonstrated by placing 1 drop in the bottom of a gas jar, covering, and watching against a white background. Another useful demonstration is shown in Figure 2.20. A white ring of ammonium chloride smoke shows the boundary where the two gases meet and an estimate of the speed of diffusion can be obtained.

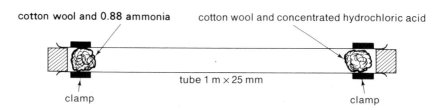

Figure 2.20

Experiment 2.2 Another useful experiment: put 1 drop of ether on a watch glass in the middle of the room and ask pupils to put up their hands when they can smell it. Calculation follows question 3 on page 36.

Experiment 2.3 The oil solution is 0.01 per cent palmitic acid, stearic acid, or oleic acid in ether, petroleum ether, or industrial methylated spirit. Trays should be at least 25 cm square: many school housecraft departments have enamelled meat trays which are quite suitable. 'Pepper pots' can be made from plastic-capped specimen tubes with holes pierced in the lid, and filled with talc.

Counting atoms by weighing $\frac{1}{2}$p, 1p, and 2p pieces (or 5p and 10p pieces) provide a useful demonstration. Weigh out, say, 100 g of $\frac{1}{2}$p pieces, 200 g of 1p pieces, and 400 g of 2p pieces and ask pupils to count the coins.

A set of identical clear bottles containing 1 mole of the commoner elements and labelled with the symbol for 1 mole is useful: 'all of these bottles contain the same number of atoms'. To reinforce the mole concept it may be worthwhile spending part of a lesson letting pupils weigh out 1 mole of iron, $\frac{1}{10}$ mole of sulphur, etc.

Experiments 2.4 and 2.5 An automatic balance weighing to two places of decimals (preferably a top-pan) is essential for quantitative experiments such as these, otherwise they become weighing exercises and pupils lose sight of the main objective. If such a balance is not available demonstration is probably preferable or, if the class is small, the teacher could do all the weighings. For experiment 2.5 AR copper(II) oxide gives much better results. Reduction tubes can be made by heating a soda-glass tube where the hole is required and closing the end with a cork or with the thumb when the glass is red hot: the build-up of pressure inside blows a neat hole.

As an introduction to these experiments mercury(II) chloride or zinc iodide (which are too expensive and/or poisonous for class practical) can be demonstrated.

Mercury(II) chloride. Weigh a 100 cm^3 beaker. Re-weigh with 5 g mercury(II) chloride. Heat on water bath with 25 cm^3 water and 25 cm^3 50 per cent phosphinic acid. Stir and, when mercury has collected in globules, wash by decantation with distilled water and acetone. Heat on water bath to dry and re-weigh.

Zinc iodide. Weigh a centrifuge tube and re-weigh with 0.5 g of powdered zinc. Add 1 g of iodine and re-weigh. Add 2 cm^3 industrial methylated spirit dropwise. When reaction is complete (colour of iodine disappears) centrifuge, decant, wash twice with industrial methylated spirit and dry over bunsen or in oven. Re-weigh tube with excess zinc.

Experiment 2.6 100 cm^3 beaker, crushed ice, stirring thermometer preferable.

Experiment 2.7 150 × 25 mm test tube as water bath. Suitable solids, other than naphthalene, are phenol and 4-methylnitrobenzene.

Boiling point and pressure Fit a round-bottomed flask with a rubber bung carrying a thermometer with bulb just protruding through the bung and a short piece of glass tubing, to which is attached a length of rubber tubing that can be closed by a clip. Third-fill the flask with water and boil it for about a minute to ensure that all the air has been displaced by steam. Remove the heat and seal the flask by closing the clip. Invert the flask and cool the bottom with a cloth soaked in cold water. The water will boil vigorously at a temperature well below 100°C.

Experiment 2.8 150 mm tubes are preferable to ensure condensation at the upper end.

Volumes of gases A box with a volume of approximately 24 000 cm^3 (29 × 29 × 29 cm) is useful to illustrate the volume occupied by a mole of gas.

3 Chemical reactions

Chemical reactions take place as a result of collisions between atoms and molecules. If a piece of iron is allowed to come into contact with water and oxygen it rusts. Molecules of water and oxygen collide with atoms of iron and as a result of the collisions iron oxide (*rust*) is formed. If the iron is covered with a coat of paint, water and oxygen molecules cannot come into contact with iron atoms so no chemical reaction is possible: the iron is protected from rusting.

3.1 The rate of chemical reactions

The rusting of iron is a very slow reaction. Some other examples of slow reactions are baking a cake (which may take several hours), human digestion (which takes about four hours), and the decay of animal and vegetable material (which may take years to complete).

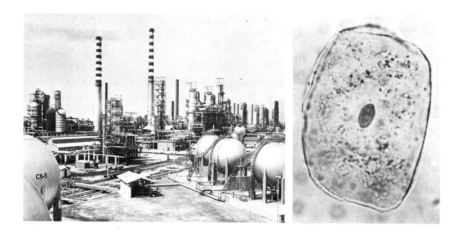

Figure 3.1 *Chemical reactions may occur on a huge scale, as in an oil refinery (left), or on a very small scale, as in a single cell of the human body, magnified 150 times in this photograph (right)*

Other reactions are very fast. When a stick of dynamite is detonated the chemical decomposition of the explosive is completed in less than a thousandth of a second. A match head burns in less than a second.

Chemists often find it convenient to make reactions go slower or faster than they would do if left to themselves. Ammonia gas, which is needed for the manufacture of fertilisers, is made by reacting hydrogen with nitrogen. These elements react together extremely slowly under normal conditions, and fertiliser production on a large scale would be commercially impossible without speeding up the reaction. Some explosives react so fast that the explosion is too violent for purposes such as quarrying: they would shatter the rock to powder. The explosion has to be 'slowed down' so that the rock is broken up into pieces of a convenient size.

In order to change the rate of a reaction, chemists need to find ways of increasing or decreasing the frequency with which the atoms or molecules collide.

Experiment 3.1   Investigating the effect of temperature on the reaction between zinc and an acid

Half-fill two test tubes with dilute acid. Heat one tube until the acid is nearly boiling, and stand the two tubes in a rack. Drop a small piece of zinc into each tube.

Is there any difference between the rates of reaction in the two tubes? If so, which one reacts faster? What gas is being produced?

Experiment 3.2   Investigating the reaction between sodium thiosulphate and hydrochloric acid

Half-fill a test tube with sodium thiosulphate solution. Using a teat pipette, add 2 drops of dilute hydrochloric acid, then shake the tube.

Describe what you see happening. Is this a slow reaction or a fast reaction?

The whitish solid which is thrown out of (precipitated from) the solution is sulphur. The reaction which takes place is:

sodium thiosulphate + hydrochloric acid →
    sodium chloride + sulphur + sulphur dioxide + water.

The sodium chloride produced remains in solution. The sulphur dioxide gas also dissolves in the water: you may be able to smell it.

Experiment 3.3   Investigating the effect of temperature on the reaction between sodium thiosulphate and hydrochloric acid

Measure 50 cm^3 of sodium thiosulphate solution into a conical flask. Find the temperature of the solution and write it down.

Measure out 5 cm^3 of dilute hydrochloric acid in a small measuring cylinder. Have a stopwatch ready. Add the acid to the sodium thiosulphate solution, start the stopwatch *at once*, and quickly swirl the contents of the flask to mix them. Write down the time it takes for the first appearance of cloudiness in the solution due to precipitated sulphur.

Empty the flask and *wash it out well with water.* Measure another 50 cm^3 of sodium thiosulphate solution into the flask, stand it on a gauze on a tripod and heat it to a temperature between 25°C and 30°C. Remove the flask from the heat, measure the temperature accurately, and write it down. Add 5 cm^3 of acid and again find the time taken for the sulphur to appear and write it down.

Repeat the experiment three times more, at temperatures between 35°C and 40°C, between 45°C and 50°C, and between 55°C and 60°C. Do not forget to record all your results.

Draw a graph of the temperature at which the reaction was carried out against the time taken for the first precipitate to appear. Put temperature in °C on the vertical axis and time in seconds on the horizontal axis.

Describe what your graph tells you about the effect of temperature on this reaction. Does doubling the temperature double the rate of the reaction? If not, why not?

The effect of temperature on the rate of reactions

If hydrogen and oxygen gases are mixed at room temperature nothing happens, but if a flame is applied to the gas mixture the reaction takes place with explosive speed. A cake cooks faster in a hot oven than in a warm oven. Food decays faster in hot countries than in colder countries. In fact, *all* chemical reactions take place faster at higher temperatures than at lower temperatures.

The atomic theory provides an explanation for this. Hydrogen and oxygen gases are made up of tiny particles moving around at random. At room temperature particles of the two gases are colliding, but the collisions are not vigorous enough to cause reaction. If the gases are heated by a flame the particles move faster. Collisions take place more frequently and more vigorously and are therefore more likely to lead to reaction. *An increase in temperature speeds up* ALL *chemical reactions; the extra heat energy makes the particles move faster and collide more frequently.* In the same way, a decrease in temperature slows down any chemical reaction.

If the temperature is not high enough some reactions will not start at all. A piece of coal cannot be lit with a match. The temperature of the burning match is not high enough to start the reaction between coal and oxygen. In lighting a coal fire paper and wood are used. A match is hot enough to start the combustion of the paper, the burning paper is hot enough to set fire to the wood, and the burning wood is hot enough to ignite the coal. Different substances have different IGNITION TEMPERATURES.

The effect of pressure on the rate of reactions involving gases

The reaction between hydrogen and nitrogen to make ammonia is very slow under normal conditions. Its speed is considerably increased by compressing the gases. In the manufacture of ammonia, hydrogen and nitrogen gases are reacted together under a pressure 250 times greater than normal atmospheric pressure. *All reactions involving gases are speeded up by increasing the pressure*, but the speed of reactions involving solids and liquids is hardly affected at all by altering the pressure.

The atomic theory provides an explanation. When a gas is compressed the particles are pushed closer together. The more crowded the gas particles, the more frequently collisions will take place and the faster will be the reaction. In solids and liquids the particles are already very close together. Increasing the pressure cannot affect the crowding of the particles very much so it has very little effect.

Experiment 3.4   Investigating the reaction of acid with zinc foil and powdered zinc

Half-fill two test tubes with dilute acid and stand them in a rack. Add 1 spatula-full of powdered zinc to one tube, and at the same time drop a piece of zinc foil into the other tube.

Describe what you see happening in the two tubes. Is there any difference between the rates of reaction in the two tubes? If so, which one reacts faster?

The effect of surface area on the rate of reactions involving solids

A large solid block of iron takes a very much longer time to rust away completely than the same weight of iron in the form of thin nails. A pile of sticks can be burnt more easily than a single log of wood of the same weight.

When a log of wood burns the oxygen molecules involved in the reaction can only collide with the outermost layer of wood particles. The inner layers are only exposed for reaction when the outer layers have burned away. The speed with which a solid reacts depends upon the surface area which is exposed for reaction. The larger the surface area the more particles are available for collision and the faster is the reaction.

A 2 cm cube of wood has six faces each of area 4 square centimetres (Figure 3.2).

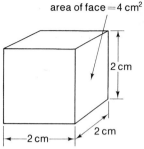

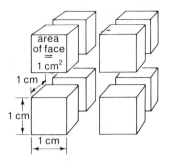

Figure 3.2   *Figure 3.3*

The total surface area of the block is $6 \times 4 \text{ cm}^2 = 24 \text{ cm}^2$. The 2 cm cube can be cut into eight 1 cm cubes (Figure 3.3). Each smaller cube has a surface area of 6 cm² and the total surface area of all the small cubes is $8 \times 6 \text{ cm}^2 = 48 \text{ cm}^2$. The smaller blocks have twice as large a surface area as the single large block, and might be expected to burn twice as fast.

The smaller the size of the pieces of a solid the faster they will react.

Experiment 3.5 Investigating the effect of acid concentration on the reaction between zinc and an acid

Half-fill a test tube with dilute acid and stand it in a rack. Use a measuring cylinder to dilute 5 cm³ of this acid with 15 cm³ of water, making sure it is mixed well. Half-fill another test tube with this very dilute acid and stand it in the rack.

Drop a piece of zinc into each of the two tubes.

Is there any difference between the rates of reaction in the two tubes? If so, which one reacts faster?

Experiment 3.6 Investigating the effect of concentration on the reaction between sodium thiosulphate and hydrochloric acid

Measure 50 cm³ of sodium thiosulphate solution into a conical flask. Have ready 5 cm³ of dilute hydrochloric acid in a small measuring cylinder and a stopwatch.

Add the acid to the sodium thiosulphate solution, start the stopwatch *at once*, and quickly swirl the contents of the flask to mix them.

Write down the time it takes for the precipitate of sulphur to begin to appear.

Empty the flask and *wash it out well with water*. Measure into the flask 40 cm³ of sodium thiosulphate solution and 10 cm³ of water. Again add 5 cm³ of acid and measure the time for the precipitate to appear.

Repeat the experiment 3 more times using 30 cm³ of sodium thiosulphate solution and 20 cm³ of water, 20 cm³ of sodium thiosulphate solution and 30 cm³ of water, and 10 cm³ of sodium thiosulphate solution and 40 cm³ of water.

Draw a graph of the number of cm³ of sodium thiosulphate used (on the vertical axis) against the time in seconds for the precipitate to appear (on the horizontal axis).

What does the shape of your graph tell you about the effect of changing the concentration of sodium thiosulphate on the rate of this reaction?

The effect of concentration on the rate of reactions involving solutions

If a housewife wants to remove stains from clothing she may use bleach. If the stains are particularly bad she will probably put more bleach into the wash, knowing that the more she puts in the faster the stains will be removed. We frequently use dilute acids in the laboratory without taking any special precautions in handling them. When we use concentrated acids we have to be very careful indeed because they are so much more reactive.

A concentrated solution has more dissolved particles per cubic centimetre than a dilute solution. The particles are closer together in the more concentrated solution and collisions take place more frequently. *A concentrated solution always reacts faster than a dilute solution.*

Catalysts

A sugar lump will not burn if a match is put to it; it only melts and chars. But if the sugar lump is dipped in cigarette ash first, it burns easily. The cigarette ash acts as a CATALYST to this particular reaction, speeding up the reaction without being changed itself. Some catalysts have been in use for hundreds of years, for instance in brewing. Here yeast is used as a catalyst to speed up the decomposition of starch into alcohol. Yet even today the way in which catalysts work is not fully understood.

Catalysis is altering the speed of a reaction by adding a substance which is not itself used up in the reaction. The word *catalyst* comes from a Greek word meaning *loosener.*

Chemical industry is very dependent on catalysts to increase the speed of reactions so that they become profitable to carry out on a large scale. In the manufacture of ammonia to make fertilisers, iron is used as the catalyst to increase the speed of the reaction between nitrogen and hydrogen:

$$\text{nitrogen} + \text{hydrogen} \xrightarrow{\text{iron}} \text{ammonia.}$$

We shall be looking at a number of the important processes of chemical industry in this book and catalysts are used very frequently.

Chemical reactions

Catalysts are also very important to life. The human body is chemically very complex, with thousands of different chemical reactions going on continually. The rates of most of these reactions are controlled by catalysts called ENZYMES. Enzymes in our saliva, stomach, and intestines control the digestion of our food. Others in our muscles control the burning up of sugar with the oxygen we breathe to produce energy. An important group of enzymes called hormones, produced under the direction of our master gland, the pituitary, which lies at the base of the brain, control the size to which each part of our bodies grows when we are young.

Though catalysts are most often useful to speed up reactions, some catalysts slow reactions down. Nitroglycerine has a tendency to explode unexpectedly of its own accord. To make explosives containing nitroglycerine safer a small amount of another chemical, urea, is added, which acts as a slowing-down catalyst, or INHIBITOR. Rubber perishes as a result of slow reaction with the oxygen in the air. An inhibitor is added during manufacture to slow down this reaction.

Experiment 3.7 Investigating the effect of copper sulphate on the reaction between zinc and an acid

Half-fill two test tubes with dilute acid and stand them in a rack. Use a teat pipette to add 2 drops of copper sulphate solution to one of the tubes. Now drop a piece of zinc into each of the tubes.

Is there any difference between the rates of reaction in the two tubes? If so, which one reacts faster? What does the copper sulphate do in this reaction?

If you have time, you can repeat this experiment using another copper compound in place of the copper sulphate. Try copper chloride solution or copper nitrate solution. Is the effect the same?

Experiment 3.8 Investigating the effect of manganese dioxide on the decomposition of hydrogen peroxide

(a) Half-fill a test tube with hydrogen peroxide solution. Add a spatula-full of manganese dioxide.

What do you observe? What is the function of the manganese dioxide? What gas is being given off?

(b) Set up the apparatus shown in Figure 3.4.

Measure 48 cm^3 of water and 2 cm^3 of hydrogen peroxide into the flask. Push the syringe plunger fully in. Have a stopwatch ready.

Put 1 spatula-full of manganese dioxide into the flask then quickly replace the bung and start the stopwatch. Measure the volume of gas produced in 15 seconds.

To avoid the syringe plunger jamming, hold the barrel of the syringe in one hand, and with the other hand gently rotate the plunger, but without pulling or pushing, while gas is being evolved.

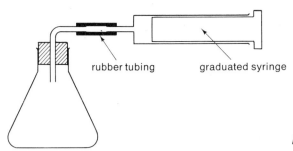

Figure 3.4

Empty and rinse out the flask. Measure out another 48 cm³ of water and 2 cm³ of hydrogen peroxide and repeat the experiment with 2 spatulas-full of manganese dioxide. If you have time, you can also try it with three, four and five spatulas-full.

What is the effect of increasing the quantity of catalyst on the rate of the reaction? Does doubling the quantity of catalyst double the rate?

(c) Repeat experiment (b) using a single lump instead of powdered manganese dioxide.

Does powdering the catalyst make it more or less effective? Why?

(d) Repeat experiment (b) using different concentrations of hydrogen peroxide solution. Try 46 cm³ of water with 4 cm³ of hydrogen peroxide and 42 cm³ of water with 8 cm³ of hydrogen peroxide, keeping the quantity of manganese dioxide the same each time.

What effect does varying the concentration of hydrogen peroxide have on the rate of the reaction? Why?

(e) Repeat experiment (b), but warm the mixture of 48 cm³ of water with 2 cm³ of hydrogen peroxide to about 40°C before adding the manganese dioxide.

What effect does raising the temperature have on the rate of the reaction? Why?

(f) Repeat experiment (b), using 1 spatula-full of other substances instead of manganese dioxide. You could try manganese sulphate, cobalt chloride, cobalt sulphate, copper oxide, copper sulphate, powdered copper, iron filings, dust, congealed blood, fresh potato, or other substances of your own choice.

Make a list of the substances you try in order of catalytic activity. Can you draw any general conclusions about the type of substances which catalyse this reaction?

Experiment 3.9 Investigating the catalytic effect of an enzyme

(a) Put a little starch solution in a test tube and add a few drops of iodine solution. What do you see?

This is a useful test for starch.

(b) Saliva contains an enzyme called amylase which begins the process of digesting starch.

First obtain a suitable sample of saliva. Rinse out your mouth. Then swill warm water around your mouth for a minute or so and spit it out into a beaker. This is a dilute solution of saliva.

Measure 10 cm^3 of starch solution into a boiling tube and stand it in a rack. Put 1 cm^3 of iodine solution into each of five test tubes and stand them beside the boiling tube.

Add 2 cm^3 of dilute saliva to the starch solution. Every minute for 5 minutes remove 2 drops of this mixture with a teat pipette and add them to one of the tubes of iodine solution.

Does the colour in the 5 tubes indicate that the starch is being broken down? Is it completely broken down at the end of 5 minutes? If so, how long did it take to break the starch down completely?

Repeat the experiment, but warm the starch solution to between 35°C and 40°C before adding the saliva.

How long does it take this time for the starch to be completely broken down? Is this what you would expect from your knowledge of reaction rates?

Repeat the experiment once more, this time heating the starch solution almost to boiling (above 90°C) before adding the saliva.

Is the result what you would expect? If not, can you explain why not?

Questions

1 Make a list of 5 reactions you have come across which are very fast, and 5 which are very slow. You could choose reactions you have seen in the laboratory or reactions which take place in nature (but do not include the examples on pages 58–9).

2 If a piece of coal is heated in a crucible it burns.
  (a) Which of the following would increase the rate of burning?
    (i) Heating the crucible less strongly.
    (ii) Heating the crucible more strongly.
    (iii) Crushing the coal to powder.
    (iv) Using a larger crucible.
    (v) Using a smaller crucible.
    (vi) Doing the experiment on top of Mount Everest.
    (vii) Doing the experiment at the bottom of a deep mineshaft.
    (viii) Putting a lid on the crucible.
    (ix) Blowing hot air on to the coal.
    (x) Blowing pure oxygen on to the coal.
  (b) In each of the cases where you think the coal would burn faster, explain why.

3 Potassium permanganate crystals react with hydrochloric acid to produce chlorine gas.
  (a) How would you measure the rate of this reaction? (If you wish,

you can answer this question by making a sketch of the apparatus, but say what quantities of substances you would use, and what measurements you would make.)

(b) Describe carefully how you would measure the effect of the following on the rate of this reaction:
   (i) the concentration of the acid;
   (ii) the temperature;
   (iii) a substance X, believed to act as a catalyst.

4 Explain the following.
   (a) Fresh food is best stored in a refrigerator.
   (b) Explosions sometimes occur in coal mines because of coal dust in the atmosphere.
   (c) Boiled potatoes take 20 minutes to cook in a pan, but only 2 minutes in a pressure cooker.
   (d) A concentrated solution of hydrogen peroxide is very unstable, and gives off oxygen gas. If a trace of acid is added it can be stored for a much longer time without decomposition.
   (e) If a red-hot iron nail is plunged into a jar of oxygen nothing happens; but if red-hot iron wool is plunged into oxygen it sparks brilliantly.
   (f) A mixture of hydrogen and chlorine is stable in the dark but if it is exposed to bright sunlight there is a violent explosion.
   (g) Bacteria are killed by high concentrations of oxygen, so cuts should be washed with hydrogen peroxide solution.
   (h) Concentrated solutions of hydrogen peroxide decompose faster than dilute ones.
   (i) The thyroid gland (which is in the throat) produces a hormone called thyroxin which helps the body to burn up fat. People who are abnormally thin sometimes have part of their thyroid gland removed surgically.

5 Which of the following is the best definition of a catalyst?
A   A substance which makes a reaction give better results.
B   A substance which increases the rate of a reaction.
C   A substance which decreases the rate of a reaction.
D   A substance which alters the rate of a reaction.
E   A substance which starts off a reaction.

6 Two gases X and Y are being reacted together in an industrial plant by passing a mixture of them through a steel vessel filled with lumps of catalyst. In what ways could the reaction be speeded up?

7 Two experiments are described below in which the speed of the reaction between calcite crystals and hydrochloric acid was measured.

Experiment 1
10 grams of large calcite crystals were placed in a flask and 200 cm³ of dilute hydrochloric acid were added. The flask and its contents were weighed at regular intervals of time.

Experiment 2
The same weight of small calcite crystals was used with the same volume of hydrochloric acid of the same strength. Again the flask and its contents were weighed at regular intervals of time. The results of these experiments are shown in the graph (Figure 3.5).

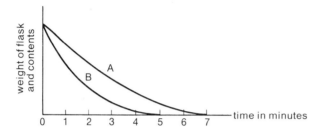

Figure 3.5

By reference to the graphs only say which experiment proceeds more quickly:
(a) during the first three-minute period—A or B?
(b) during the second three-minute period—A or B?
(c) Give scientific reasons for your answers to parts (a) and (b).
(d) By reference to the graphs and experiments say which graph refers to *Experiment 1:* A or B?
(e) Explain your answer to part (d).
(f) How does the speed of the reaction change with time?
(g) Suggest a way in which the reaction could be made to proceed more quickly than that recorded.
(h) During the reaction, the weight of the flask and its contents decreased. Suggest a reason for this.
(i) Briefly suggest another way in which the speed of the reaction could be measured.
(j) Using the same scales as in Figure 3.5 draw graphs to represent the changing weight of the flask and its contents under the following conditions.
  (i) Small pieces of calcite and acid of one-quarter the concentration at 20°C. Label this graph I.
  (ii) Small pieces of calcite as in (i) but using acid $1\frac{1}{2}$ times as concentrated as in the original experiments at 20°C. Label this graph II.
  (iii) Small pieces of calcite as in (i) and acid as in (ii) at 80°C. Label this graph III.
  (London)

.2 More chemical shorthand: describing reactions

Copper and sulphur react together to form copper sulphide. We can write:

copper + sulphur → copper sulphide.

This is a WORD EQUATION. It tells us what substances are reacting together and what new substances are formed. But it does *not* tell us anything about the proportions in which copper and sulphur react.

The formula of copper sulphide is CuS. This tells us that copper and sulphur combine in the proportion of one mole of copper to one mole of sulphur. We can write:

$$Cu + S \rightarrow CuS.$$

This is a CHEMICAL EQUATION. It is a shorthand way of writing *one mole of copper reacts with one mole of sulphur to form one mole of copper sulphide.*

A chemical equation has two advantages over a word equation: it is quicker to write, and it tells us the proportions in which the substances react.

A more complicated equation

The simple equation representing the formation of copper sulphide can be written from the results of one experiment to find the formula of copper sulphide. For more complicated reactions several experiments may be necessary to work out the equation.

For example, to work out the chemical equation for the reaction:

magnesium + hydrochloric acid →
                              magnesium chloride + hydrogen

experiments have to be carried out to find (a) the amount of hydrogen produced when a certain amount of magnesium reacts, (b) the amount of hydrochloric acid which reacts with a certain amount of magnesium, and (c) the amount of magnesium chloride produced from a certain amount of magnesium.

(a) By experiment it is found that when 24 g of magnesium (1 mole) reacts with sufficient hydrochloric acid to dissolve it all, the volume of hydrogen produced is 24 000 cm^3.

24 000 cm^3 of any gas is a mole of molecules. A molecule of hydrogen is made up of two atoms (page 52).

1 mole of magnesium (Mg) gives 1 mole of hydrogen (H_2).

(b) Hydrochloric acid is a compound of hydrogen and chlorine. By reacting together hydrogen and chlorine the formula of hydrochloric acid can be found: it is HCl.

The mass of 1 mole of hydrochloric acid is therefore $(1 + 35) = 36$ g. If 36 g of hydrochloric acid is put in a beaker and 50 g of magnesium is added it is found that 38 g of magnesium remain when the reaction stops.

So 12 g of magnesium ($\frac{1}{2}$ mole) has reacted with 36 g of hydrochloric acid (1 mole).

1 mole of magnesium (Mg) reacts with 2 moles of hydrochloric acid (2HCl).

(c) If the solution from experiment (b) is evaporated down, 47 g of magnesium chloride remain. 12 g of magnesium produce 47 g of magnesium chloride.

12 g of magnesium combine with $(47 - 12) = 35$ g of chlorine.
$\frac{1}{2}$ mole of magnesium combines with 1 mole of chlorine.
The formula of magnesium chloride is $MgCl_2$.
The mass of 1 mole of magnesium chloride is $(24 + 35 + 35) = 94$ g.
12g of magnesium ($\frac{1}{2}$ mole) produce 47g of magnesium chloride ($\frac{1}{2}$ mole).
1 mole of magnesium (Mg) gives 1 mole of magnesium chloride (MgCl$_2$)

All the information is now available to write the equation:

$$Mg + 2HCl \rightarrow MgCl_2 + H_2.$$

A chemical equation can only be written when the formulas of all the substances and the quantities of them involved in the reaction have been found by experiment. A lot of experimental work is required, so when a chemist wants to use an equation for a reaction he does not do all the experimental work himself unless he has to. If he can, he saves time by using the results of other people's work. But we must remember whenever we use an equation that someone else has worked it out *by experiment*.

Using chemical equations

With the help of a list of atomic masses (Figure 2.7, page 39) an equation enables us to work out the masses of substances reacting together.

Example 1 How much copper and how much sulphur would be needed to make 24 g of copper sulphide?
1 mole of copper sulphide (formula CuS) is $(64 + 32) = 96$ g. (This is called the FORMULA MASS of copper sulphide.)
   24 g of copper sulphide is $\frac{24}{96} = \frac{1}{4}$ mole. (*To convert grams of a compound to moles, divide by the formula mass.*)

The equation is	Cu	+	S	→	CuS
which tells us	1 mole	+	1 mole	→	1 mole;
so	$\frac{1}{4}$ mole	+	$\frac{1}{4}$ mole	→	$\frac{1}{4}$ mole.
This is	$\frac{1}{4} \times 64 = $ **16 g.**		$\frac{1}{4} \times 32 = $ **8 g.**		

To make 24 g of copper sulphide we need 16 g of copper and 8 g of sulphur.

Example 2 How much hydrochloric acid will react with 1 g of magnesium, and how much hydrogen will be produced?
1 g of magnesium is $\frac{1}{24}$ mole.

The equation is   Mg   +   2HCl   → $MgCl_2 + H_2$;
which tells us   1 mole +   2 moles   →   1 mole;
so $\frac{1}{24}$ mole + $\frac{2}{24} = \frac{1}{12}$ mole → $\frac{1}{24}$ mole.
This is $\frac{1}{12} \times (1 + 35)$ g $\frac{1}{24} \times 24\,000$ cm^3
                                     $= \frac{1}{12} \times 36 = 3$ g. $= 1\,000$ cm^3.

3 g of hydrochloric acid will react with 1 g of magnesium and 1 000 cm³ of hydrogen are produced.

An important example: making iron

Iron is made by heating *iron ore* (iron oxide) with *coke* (carbon):

   iron oxide   +   carbon   →   iron   +   carbon monoxide.

For each kilogram (1 000 g) of iron ore going into the furnace, the manufacturer wants to know how much coke is needed, and how much iron he can expect to get out.

The formula of iron oxide (found by experiment) is Fe_2O_3.
The formula mass is $(2 \times 56) + (3 \times 16) = 160$ g.
1 000 g of iron oxide is $\frac{1000}{160} = 6\frac{1}{4}$ moles.

The equation is Fe_2O_3 +   3C   →   2Fe   +   3CO;
which tells us   1 mole   +   3 moles   →   2 moles;
so $6\frac{1}{4}$ moles + $18\frac{3}{4}$ moles → $12\frac{1}{2}$ moles.
This is $18\frac{3}{4} \times 12 = 225$ g. $12\frac{1}{2} \times 56 = 700$ g.

225 g of carbon is required to react with 1 000 g of iron oxide and 700 g of iron is produced.

In practice the calculation will be a little more complicated than this because iron ore is not pure iron oxide. Nevertheless, this kind of calculation, using a chemical equation, is used in every part of chemical industry, both for large-scale and small-scale processes.

Other information which can be included in equations

We can give more information in an equation by writing in the states of the substances involved in the reaction: solids are represented (*s*); liquids are represented (*l*); and gases are represented (*g*).

In the manufacture of iron, for instance, the reaction between iron oxide and carbon is carried out at such a high temperature that the iron is produced in a molten state. To show this, the equation can be written:

$$Fe_2O_3(s) + 3C(s) \rightarrow 2Fe(l) + 3CO(g).$$

One other state symbol is also used: (*aq*), short for *aqueous*, meaning *in solution in water*. The reaction between magnesium and hydrochloric acid would be written:

$$Mg(s) + 2HCl(aq) \rightarrow MgCl_2(aq) + H_2(g).$$

Chemical reactions

The symbol (aq) tells us that the hydrochloric acid is dilute (a solution in water), and that the magnesium chloride is left in solution in water at the end of the reaction.

Information which is provided by equations

A full chemical equation tells us the following.

1. What substances react together and what new substances are produced.
2. The formulas of all the substances involved in the reaction.
3. The proportions (by moles) of all the substances involved.
4. The states of all the substances involved.

Information which is not provided by equations

Though equations are very useful as a shorthand way of writing down a lot of information about a chemical reaction, it is as well to remember that there are some things which they do *not* tell us.

1. They do not tell us under what conditions the reaction will take place (for instance, is heat or a catalyst needed?).
2. They do not tell us how fast the reaction is.
3. It must be remembered that equations are just a shorthand way of writing down *experimental* observations. An equation which has not been worked out by experiment is quite useless. For example:

$$Cu(s) + H_2O(l) \rightarrow CuO(s) + H_2(g)$$

looks like a perfectly reasonable equation. But in fact copper never reacts with water. Just because it is possible to invent a reasonable equation, it does not follow that the reaction actually takes place.

Questions (Refer to Figure 2.7, page 39 for atomic masses when necessary.)

1 Calculate the formula mass for the compounds represented by the formulas:

(a) CuO
(b) $NaCl$
(c) H_2O
(d) SO_3
(e) CCl_4
(f) Al_2O_3
(g) C_3H_8
(h) $CaCO_3$
(i) $CuSO_4$
(j) CH_2Cl_2
(k) C_2H_6O
(l) Na_2CO_3
(m) $Na_2S_2O_3$
(n) $C_{12}H_{22}O_{11}$
(o) $NaHCO_3$
(p) $Mg(OH)_2$
(q) $Zn(NO_3)_2$
(r) $(NH_4)_2CO_3$.

2 Iron displaces copper from copper sulphate solution as in the equation:

$$Fe + CuSO_4 \rightarrow Cu + FeSO_4.$$

What weight of pure copper would be displaced from excess copper sulphate solution by 7 g of pure iron? (*West Midlands*)

3 Zinc reacts with sulphuric acid according to the equation:

$$Zn + H_2SO_4 \rightarrow ZnSO_4 + H_2.$$

How much zinc would be needed to make $\tfrac{1}{5}$ g of hydrogen?

4 In one method for manufacturing sodium hydroxide, sodium carbonate is heated with calcium hydroxide:

$$Na_2CO_3 + Ca(OH)_2 \rightarrow 2NaOH + CaCO_3.$$

(Formula masses: sodium carbonate Na_2CO_3 = 106; sodium hydroxide NaOH = 40.) What weight of sodium hydroxide should be obtained from 212 tonnes of sodium carbonate? (*East Anglia*)

5 (a) Magnesium oxide may be prepared by burning magnesium in oxygen according to the equation:

$$2Mg + O_2 \rightarrow 2MgO.$$

(i) Calculate the formula mass of magnesium oxide.
(ii) Show all your working and calculate what weight of oxide could be produced from 16 g of magnesium.
(b) If magnesium oxide is reacted with warm dilute sulphuric acid a solution of magnesium sulphate results:

$$MgO + H_2SO_4 \rightarrow MgSO_4 + H_2O.$$

(i) Calculate the formula mass of magnesium sulphate.
(ii) Show all your working and calculate what weight of oxide would be needed to produce 40 g of magnesium sulphate.
(*Lancashire*)

6 Silver nitrate reacts with calcium chloride according to the equation:

$$2AgNO_3(aq) + CaCl_2(aq) \rightarrow 2AgCl(s) + Ca(NO_3)_2(aq).$$

(a) What weight of silver nitrate would react exactly with 11 g of calcium chloride?
(b) How many grams of the solid product would be formed in the same reaction?

7 An oxide of lead, commonly known as *red lead*, is reduced to the metal by hydrogen. The equation for this reaction is:

$$Pb_3O_4(s) + 4H_2(g) \rightarrow 3Pb(l) + 4H_2O(g).$$

(a) How much lead can be obtained by the reduction of 6.85 g of *red lead*?

(b) What volume of hydrogen (measured at room temperature and pressure) reacts with 6.85 g of *red lead*? (Take the volume of 1 mole of a gas at room temperature and pressure to be 24000 cm^3.)

(c) What do the state symbols in the above equation tell you about the conditions under which this reaction takes place?

8 The equation for the combination of oxygen and hydrogen is:

$$2H_2 + O_2 \rightarrow 2H_2O.$$

(a) What volume of oxygen would be required to completely burn 60 cm^3 of hydrogen?

(b) What volume of steam would be formed?

(Assume that all measurements are taken at 1 atm pressure and 110°C.)

(East Anglia)

9 When copper oxide is heated in a combustion tube in a stream of carbon monoxide gas, it reacts according to the equation:

$$CuO(s) + CO(g) \rightarrow Cu(s) + CO_2(g).$$

(a) Does the weight of the combustion tube and its contents increase, decrease, or stay the same during the reaction?

(b) What weight of carbon monoxide will react with 4 g of copper?

(c) What volume of carbon monoxide, measured at normal temperature and pressure, will react with 4 g of copper? (The volume of 1 mole of gas under these conditions is 24000 cm^3.)

10 $2NaOH + CuSO_4 \rightarrow Cu(OH)_2 + Na_2SO_4$

From the above equation calculate the weight of copper hydroxide precipitated when 80 g of sodium hydroxide are dissolved in distilled water and treated with excess copper sulphate solution.

(Middlesex)

11 $2NaHCO_3 \rightarrow Na_2CO_3 + H_2O + CO_2$

The above equation represents the effect of heat on sodium bicarbonate. Calculate the weight of sodium carbonate formed when 8.4 g of sodium bicarbonate are completely decomposed.   *(Middlesex)*

3.3 Reversible reactions

Experiment 3.10 Investigating the action of heat on zinc oxide

Put one spatula-full of zinc oxide in a small test tube and heat it. What do you observe?

Allow the tube to cool. What happens now?

Can this process be repeated over and over again?

Experiment 3.11 Investigating the action of acid and water on bismuth chloride

Put one spatula-full of bismuth chloride in a test tube. Add concentrated hydrochloric acid from a teat pipette until the bismuth chloride has *just* dissolved.

(Concentrated hydrochloric acid is very corrosive: take great care, and report to your teacher at once if any gets on your skin or clothes.)

Add water drop by drop from a teat pipette to the solution of bismuth chloride. What happens?

Once again, add concentrated hydrochloric acid, drop by drop, shaking the tube after each drop. What happens now?

See how many more times you can reverse the reaction by adding water, then concentrated hydrochloric acid, and so on.

Experiment 3.12 Investigating the reaction of copper sulphate with ammonia solution and acid

Put 1 cm depth of copper sulphate solution into a test tube. Add ammonia solution drop by drop from a teat pipette, shaking after each drop is added. A precipitate will be formed. What colour is it? Continue adding drops of ammonia solution until the precipitate dissolves. What colour is the solution now?

Now add dilute acid to the solution, drop by drop from a teat pipette, again shaking after each drop. Describe what you see happening.

See if you can repeat the whole process again.

One way or both ways?

Many chemical reactions will go only in one direction. If you bake a cake it cannot be unbaked. Once a match has burnt you can't get the match back. If magnesium is heated with copper oxide, copper and magnesium oxide are formed:

magnesium + copper oxide → magnesium oxide + copper.

There is no way of reversing this reaction. Magnesium oxide cannot be made to react with copper to give the magnesium and copper oxide back again.

In some cases, however, it is possible to reverse a reaction. For instance, blue copper sulphate crystals lose water when they are heated, leaving a white powder:

blue crystals → white powder + water.

When water is added to the white powder it turns blue again:

white powder + water → blue crystals.

If iron is heated in steam, iron oxide and hydrogen are produced (Figure 3.6).

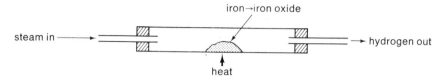

Figure 3.6

If the iron oxide is then heated in hydrogen, the iron and water are recovered (Figure 3.7).

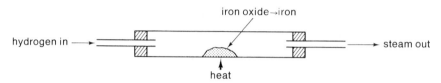

Figure 3.7

Reactions which can be made to go either way are called REVERSIBLE REACTIONS.

Equilibrium

If a reaction is reversible it is usually possible, by adjusting the conditions of the experiment, to reach a balance point where it neither goes completely one way nor the other. If iron and steam are heated in a closed container they react to form iron oxide and hydrogen. The hydrogen cannot escape so it starts reacting with the iron oxide to give iron and steam back again.

At first, there will only be a little iron oxide and hydrogen in the container, so the reaction:

$$\text{iron oxide} + \text{hydrogen} \rightarrow \text{iron} + \text{steam}$$

is slower than the reaction:

$$\text{iron} + \text{steam} \rightarrow \text{iron oxide} + \text{hydrogen}.$$

(Remember that the rate of a reaction depends on the concentration of the substances reacting together.) As time goes on, the quantity of iron oxide and hydrogen increases, and the quantity of iron and steam decreases. Eventually the reaction is going at the same rate in both directions. *Both reactions are continuing, but the composition of the mixture does not change any further.* We say that the reaction has reached EQUILIBRIUM. We can write:

$$\text{iron} + \text{steam} \rightleftharpoons \text{iron oxide} + \text{hydrogen}.$$

The sign ⇌ means that the reaction is proceeding in both directions at the same speed: it has reached equilibrium.

If we start off with iron oxide and hydrogen in the closed container, the composition of the final mixture will be the same. The equation means exactly the same if it is written the other way round:

iron oxide + hydrogen ⇌ iron + steam.

Disturbing equilibrium

An equilibrium does not usually balance at the point where the reaction is exactly half-way. Equilibrium may be reached when the reaction is 75 per cent one way and 25 per cent the other, or even when it is 99.999 per cent one way and only 0.001 per cent the other. The balance point of an equilibrium depends on the conditions: the temperature, the pressure, and the concentration of the reacting substances at the outset. If any of these conditions are altered the point of balance is shifted.

This is often very important in industrial processes. The reaction between nitrogen and hydrogen to make ammonia for fertilisers, for example, is reversible and reaches the equilibrium:

nitrogen + hydrogen ⇌ ammonia.

Under normal conditions of temperature and pressure the percentage of ammonia when equilibrium is reached is extremely small, and quite useless for large-scale manufacture. But by choosing the correct temperature, pressure, and proportions of nitrogen and hydrogen, the balance point can be adjusted so that there is at least 10 per cent of ammonia in the equilibrium mixture, making manufacture quite feasible.

Experiment 3.13   Investigating an equilibrium

(a) One-quarter-fill a test tube with carbon tetrachloride. Add a few drops of potassium iodide solution. Do the two liquids mix? Which is at the top?

(b) Choose two very small crystals of iodine of about the same size and put one into each of two test tubes. One-quarter-fill one tube with carbon tetrachloride and the other with potassium iodide solution. What colour is a solution of iodine in potassium iodide solution?

(c) Add an equal volume of potassium iodide solution to the tube containing a solution of iodine in carbon tetrachloride. Shake the tube *very gently*. What do you observe? Explain what has happened.

Now shake the tube more vigorously. Describe and explain any further change you see.

Shake the tube again. Is there any further change? Why?

(d) Add an equal volume of carbon tetrachloride to the tube containing a solution of iodine in potassium iodide solution. Shake *gently*, then harder, then harder again.

Describe and explain what you see after each shaking.

Compare the tube from the end of experiment (d) with the tube from the end of experiment (c). Do they look the same? Why?

(e) What do you think would happen if you removed the top layer (the solution of iodine in potassium iodide solution) from one of the tubes, added fresh potassium iodide solution, and shook again? Try doing this. Use a teat pipette to remove the top layer.

Experiment 3.14 Disturbing an equilibrium

Iron chloride solution is pale yellow in colour and potassium thiocyanate solution is colourless. When the two solutions are mixed a new complex substance is formed which is blood-red in colour. The reaction reaches the equilibrium:

iron chloride + potassium thiocyanate ⇌ red complex.

The position of the balance point of this equilibrium can be judged by the intensity of the red colour. The darker the colour the further the reaction has moved to the right; the paler the colour the further it has moved to the left.

Half-fill four 100 cm^3 beakers with water. To each beaker add 1 cm^3 of iron chloride solution and 1 cm^3 of potassium thiocyanate solution (use a measuring cylinder). Stir all the beakers; the density of the red colour should be the same in all of them.

Keep beaker 1 for comparison purposes.

To beaker 2 add 20 cm^3 of iron chloride solution.

To beaker 3 add 20 cm^3 of potassium thiocyanate solution.

To beaker 4 add 10 spatulas-full of ammonium chloride and dissolve by stirring.

What do you notice about the intensity of the red colour in the four beakers? In which beakers has the equilibrium balance-point been shifted to the right and in which has it been shifted to the left?

Questions

1 Make a list of some reactions which can only be made to go in one direction.

2 Make a list of some reactions which are easily reversible.

3 Nitrogen and hydrogen are heated together in a closed vessel. A reaction takes place which may be represented:

nitrogen + hydrogen ⇌ ammonia.

After some time it is found that the composition of the mixture of nitrogen, hydrogen, and ammonia in the vessel does not change any

further. Which of the following best describes the situation inside the vessel?
A Hydrogen and nitrogen are reacting together more slowly than they were initially.
B Ammonia is decomposing into nitrogen and hydrogen.
C Reaction is taking place in both directions.
D Reaction is taking place in both directions at the same rate.
E All reaction has stopped.

4 In question 3, only one of the statements A, B, C, D, and E is completely incorrect. Which one?

5 When calcium carbonate is heated in an open evaporating dish it decomposes completely:

calcium carbonate(s) → calcium oxide(s) + carbon dioxide(g).

When calcium carbonate is heated in a closed container an equilibrium is set up:

calcium carbonate(s) ⇌ calcium oxide(s) + carbon dioxide(g).

Why is equilibrium set up in the closed container but not in the open dish?

Temperature °C	Volume percentage of ammonia in equilibrium mixture at pressures of			
	1 atmosphere	100 atmospheres	200 atmospheres	1000 atmospheres
200	15.33	80.6	85.8	98.3
300	2.18	52.1	62.8	92.6
400	0.44	25.1	36.3	79.8
500	0.129	10.4	17.6	57.5
600	0.049	4.47	8.25	31.4
700	0.0223	2.14	4.11	12.9
800	0.0117	1.15	2.24	—
900	0.0069	0.68	1.34	—
1000	0.0044	0.44	0.87	—

Figure 3.8

6 Figure 3.8 shows the influence of pressure and temperature on the equilibrium between nitrogen, hydrogen, and ammonia.

$$N_2 + 3H_2 \rightleftharpoons 2NH_3$$

(i) What is the effect of pressure on the equilibrium?
(ii) What is the effect of temperature on the equilibrium?

(iii) What would be the optimum (best) conditions of pressure and temperature to give the best yield of ammonia?

(iv) What could be used to accelerate the production of ammonia?

(v) A typical set of operating conditions for an industrial ammonia plant is 200 atmospheres pressure and 550°C. Using the figures in the table draw a graph and from it state what percentage of ammonia there is in the equilibrium mixture. (*London*)

For the teacher

Experiments 3.1, 3.4, 3.5, and 3.7 4M hydrochloric acid or 2M sulphuric acid. Zinc foil or granulated zinc. M copper(II) sulphate in experiment 3.7.

Experiments 3.2 and 3.3 0.03 M sodium thiosulphate(VI) and 2 M hydrochloric acid. 100 cm^3 conical flasks. It is useful for each working group to have two measuring cylinders, preferably of different sizes or, if they are the same size, labelled to help avoid mixing the solutions. Thorough rinsing of the flasks between experiments is essential to remove traces of hydrochloric acid.

Experiment 3.6 0.15 M sodium thiosulphate(VI), otherwise as in experiments 3.2 and 3.3.

Catalysis Cigarette ash on a sugar lump (page 63) is an effective demonstration.

The combustion of hydrogen in oxygen is catalysed by platinum. Before the lesson heat a tuft of platinised asbestos for a few seconds to ensure that it is quite dry, and keep it in a closed bottle. If it is held at a jet of hydrogen from a cylinder or over a gas jar of hydrogen the hydrogen will ignite.

Catalytic decomposition of potassium chlorate(V). Heat 3 tubes under identical conditions: one with potassium chlorate(V) only, one with copper(II) oxide only, and one with a mixture of the two. Measure the time taken to produce enough oxygen to ignite a glowing splint. The tubes should be clamped horizontally to minimise the danger of a piece of burning splint falling into hot potassium chlorate(V). The use of manganese(IV) oxide to catalyse the thermal decomposition of potassium chlorate(V) is potentially dangerous and should be avoided.

Experiment 3.8 100 cm^3 conical flasks. 50 cm^3 plastic syringes. 20-volume hydrogen peroxide. For (c) use *pyrolusite*, obtainable from R. F. D. Parkinson or Lythe Minerals, as lump manganese(IV) oxide. For (f) black pudding provides a convenient source of blood. Any substance within reason can be tried as a catalyst.

Experiment 3.9 Starch solution must be freshly made: mix 10 g of starch into a thin cream with cold water and pour into 1000 cm^3 of boiling water with stirring. Iodine solution: 0.001 M in 0.01 M potassium iodide. The reaction should be faster at 35–40°C, but at 95°C the enzyme is destroyed.

Section 3.2 It is important to put equation-writing into its proper perspective; too much emphasis has been placed on it in the past. Full chemical equations are certainly *not* quicker to write than word equations unless one is very experienced. At this level word equations are preferable except when one is interested in quantities, which is not often.

'Balancing' is a particularly arid exercise. Equations should only arise as the result of quantitative experimental work (one's own or, more often, someone else's). Unfortunately it is difficult to find a reaction which is readily susceptible to a full experimental treatment. In the example on page 69, for instance, the volume of hydrogen produced per mole of magnesium is easy to investigate using apparatus as in Figure 3.4 (page 65). The magnesium/hydrochloric acid ratio is not too difficult to establish, but the magnesium/magnesium chloride ratio is very tricky in practice because of the formation of a hydrate which hydrolyses to a basic chloride on dehydration. Unless *all* the quantitative relationships can be established experimentally the principle tends to be lost. It is sufficient that pupils should understand that it is possible to establish an equation, and should appreciate why in certain cases it is useful to do so.

Experiment 3.11 In this, as in experiments 3.12 and 3.14, the object is to experience reversibility (or equilibrium in the case of experiment 3.14): detailed understanding of the reactions involved is not necessary.

Experiment 3.12 M copper(II) sulphate, M sulphuric acid, and 2 M ammonia.

Experiment 3.13 M potassium iodide. At the end of the lesson iodine stains can be removed from fingers with 2 M sodium carbonate.

Experiment 3.14 0.02 M iron(III) chloride with 50 cm^3 concentrated hydrochloric acid per 1000 cm^3; 0.075 M potassium or ammonium thiocyanate. 100 cm^3 beakers.

4 Inside the atom

So far, we have thought of atoms as tiny solid spherical particles. From the time of Dalton (page 32), who first put forward an atomic theory, until the end of the nineteenth century, everyone thought that atoms were like this. Nothing was known of how atoms were made up.

4.1 The structure of atoms

Radioactivity

In 1896, Henri Bequerel discovered radioactivity, quite by accident. He noticed that some photographic plates, although completely wrapped in black light-proof paper, became fogged when left close to a type of rock called *pitchblende* (just as if they had been exposed to the light). Pitchblende is an ore of the metal uranium, and Bequerel realised that the uranium atoms must be producing some kind of radiation which passed through the black paper.

In 1899 Lord Rutherford found that this radiation was actually a combination of three different types which he named by the first three letters of the Greek alphabet: ALPHA (α), BETA (β), and GAMMA (γ). He studied the behaviour of the three types of radiation and was able to find out a lot about them.

ALPHA-RAYS are not very penetrating. They are absorbed by a thin sheet of paper. They consist of a stream of particles with a positive electrical charge, each of which has a mass of 4 atomic mass units.

BETA-RAYS are more penetrating. They will easily pass through paper, but they can be stopped by a sheet of lead only one millimetre thick. They consist of a stream of particles with a negative electrical charge called ELECTRONS, each of which has a mass of approximately $\frac{1}{2000}$ of an atomic mass unit.

GAMMA-RAYS are a highly penetrating form of radiation which will pass through lead sheet several centimetres thick. They have no electrical charge and no mass. They are a kind of invisible light ray, rather like X-rays, but with a shorter wavelength.

Figure 4.1   Ernest Rutherford, 1871–1937

The nucleus

The discovery that the alpha and beta radiations from uranium were composed of small electrically-charged particles led Rutherford to believe that atoms could not be solid as Dalton had thought. Atoms must themselves be made up of still smaller particles. In 1909 Hans Geiger and Ernest Marsden, two students studying with Rutherford, carried out a remarkable experiment which confirmed his belief and led to the modern idea of what an atom is like.

They directed a beam of alpha-particles at a very thin sheet of gold foil (Figure 4.2), only $\frac{1}{25000}$ cm thick. Almost all the particles passed straight through the gold foil, but about one in every 10 000 bounced back. Rutherford concluded from this experiment that atoms must consist mainly of empty space. He realised that an atom must be made up of a tiny positively-charged NUCLEUS, containing almost all of its mass, and one or more negatively-charged ELECTRONS moving about the nucleus but at some distance away. The electrical attraction between the positive nucleus and the negative electrons holds the atom together. In Geiger and Marsden's experiment most of the alpha-particles had passed through the space between the nucleus and the electrons. Only an alpha-particle going very near to the nucleus was bounced back as a result of the repulsive force between the positive charges on the alpha-particle and nucleus.

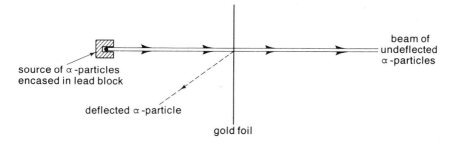

Figure 4.2

Inside the atom   83

To get some idea of the scale of size of the atom and its nucleus, imagine an atom magnified to a diameter of 10 metres. The nucleus and electrons would then be about the size of the full stops on this page. Alpha-particles, also about the size of full stops, would be far more likely to go through the space between the nucleus and the electrons than to go near enough to the nucleus to be bounced back.

Protons, neutrons, and electrons

In 1932 Sir James Chadwick found that the nucleus itself was made up of two kinds of particle: positively-charged PROTONS, each with a mass of one atomic unit, and NEUTRONS, also with a mass of one atomic unit, but without any electrical charge. *All kinds of atoms are made up of three types of still smaller particles: protons, neutrons, and electrons* (Figure 4.3).

		Mass (atomic mass units)	Charge
Nucleus	protons	1	+1
	neutrons	1	0
	electrons	$\frac{1}{2000}$	−1

Figure 4.3

The simplest atom, the hydrogen atom, has no neutrons. Its nucleus consists of a single proton, and it has one electron moving outside the nucleus. The next simplest is the helium atom, which has two protons and two neutrons in its nucleus, and two electrons moving outside. Then comes the lithium atom, with three protons and four neutrons, together with three electrons.

Figure 4.4 shows the structure of some other atoms and there are three important things to notice from it.

1 *An atom always has the same number of protons as of electrons.* A proton has one positive charge and an electron one negative charge. Atoms are electrically neutral; the number of protons and electrons must be the same, so that each positive charge is balanced by a negative charge.

2 *The atomic mass is very nearly the sum of the number of protons and neutrons.* Each proton and each neutron has a mass of one atomic mass unit. The mass of the electrons is so small that it contributes very little to the mass of the atom.

3 *The number of protons in an atom (equal to the number of electrons) is called the atomic number of the element.* The atomic number is the number of the element in the periodic classification (Figure 4.5), reading across from left to right, starting at the top.

When Mendeléev devised the periodic classification (page 19) nothing

	Number of:			Atomic Number	Atomic Mass
	protons	neutrons	electrons		
Hydrogen	1	0	1	1	1
Helium	2	2	2	2	4
Lithium	3	4	3	3	7
Carbon	6	6	6	6	12
Nitrogen	7	7	7	7	14
Oxygen	8	8	8	8	16
Sodium	11	12	11	11	23
Magnesium	12	12	12	12	24
Aluminium	13	14	13	13	27
Sulphur	16	16	16	16	32
Chlorine	17	18	17	17	35
Iron	26	30	26	26	56
Copper	29	35	29	29	64
Zinc	30	35	30	30	65
Silver	47	61	47	47	108
Iodine	53	74	53	53	127
Gold	79	118	79	79	197
Mercury	80	121	80	80	201
Lead	82	125	82	82	207
Radium	88	138	88	88	226
Uranium	92	146	92	92	238

Figure 4.4

was known about the structure of atoms. We can now see that his classification is directly related to the number of protons in atoms. Since the periodic classification groups elements according to their chemical properties it must be the number of protons and electrons which gives an element its characteristic properties.

Questions

1 Which of the following is the reason why atoms are electrically neutral?
A They contain protons and electrons.
B They contain a nucleus.
C They contain an equal number of protons and electrons.
D The electrons travel round the nucleus.
E They contain protons.   *(North West)*

1 H																	2 He
3 Li	4 Be											5 B	6 C	7 N	8 O	9 F	10 Ne
11 Na	12 Mg											13 Al	14 Si	15 P	16 S	17 Cl	18 Ar
19 K	20 Ca	21 Sc	22 Ti	23 V	24 Cr	25 Mn	26 Fe	27 Co	28 Ni	29 Cu	30 Zn	31 Ga	32 Ge	33 As	34 Se	35 Br	36 Kr
37 Rb	38 Sr	39 Y	40 Zr	41 Nb	42 Mo	43 Tc	44 Ru	45 Rh	46 Pd	47 Ag	48 Cd	49 In	50 Sn	51 Sb	52 Te	53 I	54 Xe
55 Cs	56 Ba	57 La	72 Hf	73 Ta	74 W	75 Re	76 Os	77 Ir	78 Pt	79 Au	80 Hg	81 Tl	82 Pb	83 Bi	84 Po	85 At	86 Rn
87 Fr	88 Ra	89 Ac	104 Rf	105 Ha													

58 Ce	59 Pr	60 Nd	61 Pm	62 Sm	63 Eu	64 Gd	65 Tb	66 Dy	67 Ho	68 Er	69 Tm	70 Yb	71 Lu
90 Th	91 Pa	92 U	93 Np	94 Pu	95 Am	96 Cm	97 Bk	98 Cf	99 Es	100 Fm	101 Md	102 No	103 Lw

Figure 4.5 The periodic classification

2 (a) In what important respect does a neutron differ from a proton?
(b) In what *two* important respects does an electron differ from a proton? (*East Anglia*)

3 What are the atomic numbers and atomic masses of elements whose atoms are made up as follows:
  (a) 4 protons, 5 neutrons, 4 electrons;
  (b) 14 protons, 14 neutrons, 14 electrons;
  (c) 19 protons, 20 neutrons, 19 electrons;
  (d) 46 protons, 60 neutrons, 46 electrons;
  (e) 93 protons, 144 neutrons, 93 electrons?

4 How many protons, neutrons, and electrons are there in atoms of the following elements?
  (a) Boron (atomic number 5, atomic mass 11).
  (b) Phosphorus (atomic number 15, atomic mass 31).
  (c) Argon (atomic number 18, atomic mass 40).
  (d) Tin (atomic number 50, atomic mass 119).
  (e) Plutonium (atomic number 94, atomic mass 242).

5 A section of the periodic table is shown in Figure 4.6.

I	II	III	IV	V	VI	VII	VIII
Li	Be	B	C	N	O	F	Ne
Na	Mg	Al	Si	P	S	Cl	Ar
K	Ca	Ga	Ge	As	Se	Br	Kr

Figure 4.6

If the atomic number of lithium (Li) is 3, what is the atomic number of phosphorus (P)? The atomic weights of lithium and fluorine (F) are 7 and 19, respectively. How many more neutrons has fluorine than lithium? (*East Anglia*)

6 Try to find out something about the discovery of the radioactive elements which have atomic numbers from 84 to 92.

7 Sir William Crookes, Sir Joseph Thomson, Henri Bequerel, Marie Curie, Wilhelm Röntgen, Lord Rutherford, Frederick Soddy, Hans Geiger, Charles Wilson, and Sir James Chadwick all made important contributions to modern atomic theory. Try to find out something about what some of them did and write about one or two which interest you most.

4.2 Isotopes

The number of neutrons in an atom does not have any effect on its chemical behaviour. Many elements have two or more different types of atom with the same number of protons and electrons but different numbers of neutrons. Chlorine, for example, always has 17 protons in the nucleus and 17 electrons outside. But there are two types of chlorine atom, one with 18 neutrons in the nucleus and one with 20. These are called ISOTOPES of chlorine. (The word *isotope* comes from a Greek word meaning *in the same place*: isotopes are in the same place in the periodic classification.)

Since the atomic mass of an element is the sum of the numbers of protons and neutrons, isotopes have different atomic masses. The two isotopes of chlorine have atomic masses of:

$$(17 + 18) = 35 \quad \text{(chlorine-35)},$$

and

$$(17 + 20) = 37 \quad \text{(chlorine-37)}.$$

The existence of isotopes was discovered by using a mass spectrometer (page 37). Modern mass spectrometers produce results in the form of a graph, and a sample of chlorine produces the graph shown in Figure 4.7. This shows that the masses of the isotopes are 35 and 37 and that their proportions are 75 per cent of chlorine-35 and 25 per cent of chlorine-37. These are the proportions in any sample of chlorine, either the gaseous element or any of its compounds. The average atomic mass of chlorine therefore is:

$$(75/100 \times 35) + (25/100 \times 37) = 35.5.$$

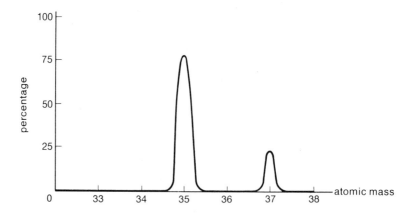

Figure 4.7

Some elements have as many as ten isotopes, others have only one. Some more examples are shown in Figure 4.8.

Atomic number	Element	Isotope	Number of:		Percentage in nature
			protons	neutrons	
1	Hydrogen	hydrogen-1	1	0	99.98
		hydrogen-2	1	1	0.02
		hydrogen-3*	1	2	a trace
6	Carbon	carbon-12	6	6	98.89
		carbon-13	6	7	1.11
		carbon-14*	6	8	a trace
17	Chlorine	chlorine-35	17	18	75.53
		chlorine-37	17	20	24.47
19	Potassium	potassium-39	19	20	93.22
		potassium-40*	19	21	0.01
		potassium-41	19	22	6.77
30	Zinc	zinc-64	30	34	48.90
		zinc-66	30	36	27.81
		zinc-67	30	37	4.11
		zinc-68	30	38	18.56
		zinc-70	30	40	0.62
92	Uranium	uranium-234*	92	142	0.01
		uranium-235*	92	143	0.71
		uranium-238*	92	146	99.28

Figure 4.8 Radioactive isotopes are identified by an asterisk

Questions

1 An atom of thallium has 81 electrons and 124 neutrons.
   (a) How many protons has it?
   (b) What would you expect the atomic mass to be?
   (c) In fact the exact atomic mass of thallium is 204.4. How do you explain this?
   (d) What is the atomic number of thallium?

2 What is meant by the terms *atomic number* and *atomic mass*?
   Explain why all the atoms of any *one* element have the same atomic number but may have different atomic masses.
   The element lithium can have atomic masses of 6 or 7 and its average atomic mass is 6.95. Calculate the percentage of each type of atom present.
   (*North West*)

3 Sir Joseph Thomson discovered that when neon was put in a mass spectrometer two lines appeared on the screen, very close together. One line, corresponding to an atomic mass of 20, was very bright. The other, corresponding to an atomic mass of 22, was much fainter. Can you explain this discovery?

4 Figure 4.9 shows an analysis of copper with a mass spectrometer. What is the average atomic mass of copper?

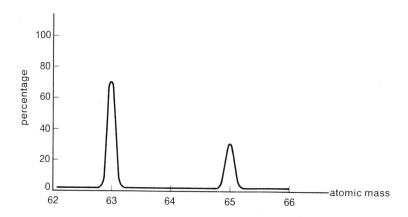

Figure 4.9

4.3 Radioactivity

When an atom has a large proportion of neutrons in its nucleus the forces which hold the nucleus together are weaker. An atom of this kind is unstable, and the nucleus may break up of its own accord. We say the atom is RADIOACTIVE, and the breaking-apart process is called RADIOACTIVE DECAY. The 90 elements found in nature have 300 isotopes altogether, of which about 50 are radioactive, though other radioactive isotopes have been made artificially. In Figure 4.8 (page 89) radioactive isotopes are marked with an asterisk (*).

There are two types of radioactive decay.

Alpha-decay

In alpha-decay an ALPHA-PARTICLE, consisting of two protons and two neutrons, is emitted from the nucleus. The α-particle has a mass of 4 atomic mass units, so when an atom undergoes α-decay its atomic mass falls by 4. Since two protons are lost the atomic number falls by 2, giving an atom of the element two places earlier in the periodic classification.

For example, the isotope of thorium (atomic number 90) with atomic mass 232 is radioactive. It decays by emission of an α-particle. The new atom produced has an atomic number of 88 (90 − 2) and an atomic mass of 228 (232 − 4). Element number 88 is radium, so the product of the radioactive decay of this thorium isotope is the isotope of radium with an atomic mass of 228:

$$\text{thorium-232} \rightarrow \text{radium-228} + \alpha\text{-particle.}$$

Beta-decay

In beta-decay a neutron in the radioactive atom changes into a proton and an electron, and the electron (or β-PARTICLE) is emitted from the nucleus:

	neutron	→	proton	+	electron.
mass:	1	=	1	+	0
charge:	0	=	(+1)	+	(−1)

The atomic mass of the atom is not changed in β-decay, since the electron lost has almost no mass. But a new proton is formed, so the atomic number increases by one, giving an atom of the element next highest in the periodic classification.

Radium-228 is an example of a radioactive isotope which decays by the emission of a β-particle (electron). The atomic number of radium is 88, so the new atom produced is element 89, which is actinium:

$$\text{radium-228} \rightarrow \text{actinium-228} + \beta\text{-particle.}$$

Mass into energy

Very accurate measurements with a mass spectrometer (page 37) show that the total mass of the products of radioactive decay is always slightly less than the mass of the original radioactive atom before decay. For example, the radioactive isotope of carbon, carbon-14, decays by emission of a β-particle into a non-radioactive isotope of nitrogen, nitrogen-14:

$$\text{carbon-14} \rightarrow \text{nitrogen-14} + \beta\text{-particle.}$$

The exact mass of an atom of carbon-14 is 14.0077 atomic mass units. The total mass of an atom of nitrogen-14 and a β-particle is 14.0075 atomic mass units. For every atom of carbon-14 which decays there is a loss of mass of 0.0002 (2×10^{-4}) atomic mass units. What happens to this lost mass? It cannot simply disappear.

A brilliant mathematician, Albert Einstein, had worked out the answer to this problem in 1916 even before the disappearance of mass in radioactive decay had been discovered. He said that under certain circumstances mass could be changed into energy, according to the equation:

$$E = mc^2.$$

E joules of energy are produced by the destruction of m kilograms of mass. c^2 is the square of the velocity of light ($c = 300\,000\,000$ metres per second). As c is so large, a very small loss of mass produces a huge amount of energy. The decay of 1 mole (14 g) of carbon-14 results in the destruction of only 0.0002 g of mass, but produces 15 000 000 000 joules of energy. This is as much energy as is obtained by burning half a tonne of coal.

It is this energy which shoots the alpha- or beta-particles out of the atom at high speeds. Often some of the energy is also produced in the form of gamma-rays (page 82), which may accompany α- or β-decay. The emission of gamma-rays does not involve any change in the nucleus since they have neither mass nor charge.

Measuring radioactivity

The emission of α- or β-particles is detected with an instrument called a GEIGER-MÜLLER COUNTER (or Geiger counter), invented by Hans Geiger and Hermann Müller. It consists of a thin-walled glass tube containing two wires and filled with a gas (Figure 4.10).

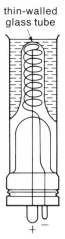

Figure 4.10

A voltage of between 300 and 500 volts is applied between the two wires. The gas cannot conduct electricity, so the current cannot pass between the wires. But when an α- or β-particle passes between the wires it collides with the gas molecules and disturbs their electrons. For a fraction of a second this makes the gas capable of conducting electricity and a short pulse of current passes between the wires.

This current may be fed into a loudspeaker so that every α- or β-particle passing through the tube produces a 'click'. The faster the clicking the more intense the radioactivity. Or, to measure the intensity of the radioactivity accurately, the current from the tube is fed into an electronic counter which counts the number of pulses of current caused by α- or β-particles passing through the tube.

The rate of radioactive decay

Different radioactive isotopes decay at different rates. For any particular isotope the rate of decay depends on the number of radioactive atoms present. The fewer radioactive atoms there are, the slower the rate of decay.

Suppose that a particular sample of a radioactive material contains 1 000 radioactive atoms, and that after one hour 500 of these have decayed. There are now only half as many radioactive atoms as there were at the beginning, so decay proceeds at half the original rate. At the end of the second hour 250 more atoms will have decayed. There will now be 250 radioactive atoms left, a quarter as many as at the beginning, so decay now proceeds at a quarter of the original rate. At the end of the third hour only 125 more atoms will have decayed, leaving 125 of the original radioactive atoms, and so on. Every one hour the number of radioactive atoms left is halved.

The relative rates at which different radioactive isotopes decay are measured by the time it takes for half the atoms to decay. In the imaginary example above, half the radioactive atoms decay in one hour. We would say that the HALF-LIFE of this isotope is one hour. The radioactive isotope of hydrogen, hydrogen-3, has a half-life of 12 years. After 12 years half of the atoms in a sample of hydrogen-3 will have decayed. After 24 years the radioactivity will have dropped to a quarter of the original amount, after 36 years to one-eighth, and so on.

Thorium-232	14 000 000 000 years
Uranium-238	5 500 000 000 years
Potassium-40	1 400 000 000 years
Uranium-235	700 000 000 years
Carbon-14	5 700 years
Hydrogen-3	12 years
Radium-228	7 years
Neptunium-239	$2\frac{1}{3}$ days
Uranium-239	24 minutes
Protactinium-234	68 seconds
Polonium-212	$\frac{3}{10\,000\,000}$ seconds

The half-lives of different radioactive isotopes vary from less than a millionth of a second to more than ten thousand million years. Some examples are shown in Figure 4.11.

Figure 4.11

Time of measurement	Number of counts in 1 minute
Start	43 200
After 1 day	35 400
After 2 days	27 800
After 3 days	21 600
After 4 days	17 700
After 5 days	13 900
After 6 days	10 800
After 7 days	8 900

Figure 4.12

The half-life is found by measuring the rate of radioactive decay at intervals with a Geiger-Müller counter. Figure 4.12 shows the results of an experiment in which measurements of the radioactivity of a sample of gold-198 were made every day for a week.

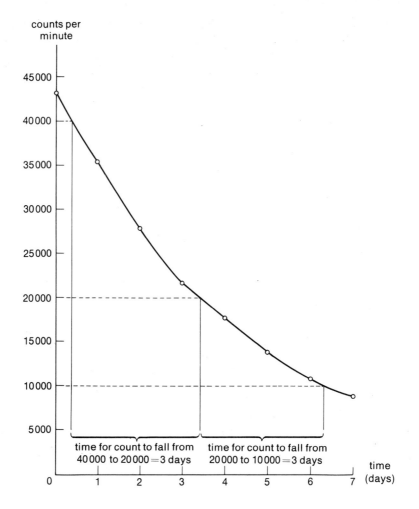

Figure 4.13

These results are plotted on a graph (Figure 4.13). The time for the rate of decay to drop by half (the half-life) can be read from this graph. The half-life of gold-198 turns out to be 3 days.

Questions

1 Complete the following equations for radioactive decay (you may need to use Figure 4.5, page 86).
    (a) uranium-238 → thorium-234 + ?
    (b) chlorine-38 → argon-38 + ?
    (c) astatine-210 → ? + α-particle
    (d) technetium-99 → ? + β-particle
    (e) ? → radium-223 + β-particle
    (f) ? → radon-222 + α-particle

2 Thorium-232 decays by emission of an α-particle to produce radium-228. Radium-228 is itself radioactive and decays by emission of a β-particle to produce actinium-228. This is also radioactive and there are seven more transformations before a stable non-radioactive isotope, lead-208, is obtained. Complete this series:

actinium-228 $\xrightarrow{\beta}$? $\xrightarrow{\alpha}$? $\xrightarrow{\alpha}$? $\xrightarrow{\alpha}$? $\xrightarrow{\alpha}$?

$\xrightarrow{\beta}$? $\xrightarrow{\beta}$? $\xrightarrow{\alpha}$ lead-208.

The α or β over each arrow shows which kind of decay is involved in each step. (Use Figure 4.5, page 86 if necessary.)

3 A sample of lead taken from a bottle of lead oxide in the laboratory is found to have an atomic mass of 207.2. A sample of lead extracted from rock containing thorium is found to have an atomic mass of exactly 208. Can you explain this? (Question 2 may help you to find the answer.)

4 Which of the following best explains why some isotopes are radioactive?
A They have too many protons.
B They have too many electrons.
C They have too many neutrons.
D They have too large a proportion of neutrons.
E They have too large a proportion of protons.

5 Einstein's equation is $E = mc^2$, where E is the energy produced in joules, m is the mass destroyed in kilograms, and $c = 3 \times 10^8$ metres per second.

(a) Calculate the energy in joules produced by the destruction of 1 gram of mass.
(b) When 1 tonne of coal is burnt 3×10^{10} joules of energy are produced. How many tonnes of coal would it take to produce the same amount of energy as is obtained by the destruction of 1 gram of mass?

6 The half-life of carbon-14 is 5700 years. A sample of carbon-14 gives a count-rate of 68 counts per minute on a Geiger-Müller counter. How long will it take for the count to drop to 17 counts per minute?

7 Uranium-239 has a half-life of 24 minutes. A solution of uranium-239 nitrate gives a count-rate of 8000 counts per minute on a Geiger-Müller counter. What will be the count-rate after (a) 48 minutes (b) 2 hours?

Inside the atom

8 An enriched form of a radioactive isotope was obtained in solution. The solution was placed in a Geiger-Müller liquid counter tube and the radioactive count was measured at regular time intervals. The results are given in Figure 4.14.

Time (minutes)	Count (counts per second)
0	400
30	150
60	54
90	20
120	7

Figure 4.14

(a) Plot a graph of the count rate against time, using the horizontal axis for time. Choose your own scale and mark the axes clearly.
(b) Deduce from your graph the count rate at the following times.
  (i) 50 minutes.
  (ii) 100 minutes.
(c) After what length of time from the beginning of the experiment was the count exactly
  (i) 100 counts per second;
  (ii) 50 counts per second?
(d) Evaluate the half-life of the radioactive element.
(e) What would the count rate be at 141 minutes? Explain how you arrive at your answer.
(f) If a solution of half the strength had been used what would the count rate have been after
  (i) 50 minutes;
  (ii) 100 minutes?
(g) What would the half life of the element have been, measured in this solution? Explain your answer.
(h) Explain the meaning of the term *isotope*. (London)

4.4 Nuclear fission

The uranium isotope of atomic mass 235, which forms less than 1 per cent of the natural mixture of isotopes in uranium, is able to capture a stray neutron. This disturbs the proton-neutron balance of the atom, and it breaks up. Two new atoms are formed, an atom of barium-144 and an atom of krypton-90. This is called a FISSION process. No protons are lost. The number of protons in the uranium atom is the same as the number in the barium and krypton atoms together. But the neutrons do not add up. Two spare neutrons are emitted:

uranium-235 + 1 neutron → barium-144 + krypton-90 + 2 neutrons
92 protons   56 protons   36 protons
143 neutrons   88 neutrons   54 neutrons

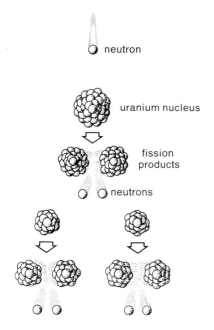

These two extra neutrons can cause the breakdown of two more uranium atoms, producing 4 neutrons. When these 4 collide with four more uranium atoms, 8 neutrons are produced, and so on (Figure 4.15). The number of atoms breaking up multiplies very rapidly until all the uranium-235 is used up. This is called a CHAIN REACTION.

In the fission of one uranium-235 atom there is a loss of mass of 0.2 atomic mass units. The fission of all the atoms in a mole of uranium-235 (235 g) would result in a mass loss of 0.2 g. According to Einstein's equation (page 91) this produces as much energy as is obtained by burning 600 tonnes of coal.

Figure 4.15

The atom bomb

In a block of uranium-235 not all of the neutrons produced by fission will have the chance of colliding with more uranium atoms to continue the chain reaction. Some escape through the sides of the block without colliding. If the block is larger than a certain size, called the CRITICAL MASS, the chain reaction produces neutrons inside the block faster than they can escape. The chain reaction then becomes an uncontrollable explosion, as a result of the enormous amount of energy suddenly released.

The earliest atom bombs (Figure 4.16) consisted of two blocks of the uranium-235 isotope, each smaller than the critical mass. The bomb was detonated by forcing these two blocks together by means of a small charge of ordinary explosive, producing a single block larger than the critical mass.

Inside the atom   97

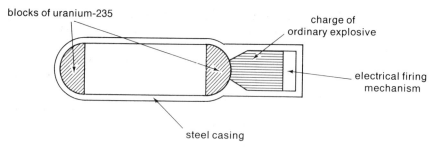

Figure 4.16

Nuclear reactors

In a reactor the same kind of chain fission process takes place as in the atom bomb, but it is carefully controlled so that it does not get out of hand. The bomb uses pure uranium-235 which is very expensive to separate from the much commoner isotope, uranium-238. Nuclear reactors use the natural mixture of isotopes or ENRICHED URANIUM, which contains a slightly increased proportion of the 235 isotope.

The neutrons produced in the fission of uranium-235 are travelling extremely fast. If they collide with more uranium-235 atoms the chain reaction is continued. But if they collide with the much commoner uranium-238 atoms, which is far more likely, they are absorbed. Uranium-238 is not capable of fission under these conditions, so that the chain reaction is interrupted. Slow neutrons, on the other hand, still cause fission when they collide with uranium-235 atoms but are *not* absorbed by uranium-238 atoms. So if the neutrons produced by fission

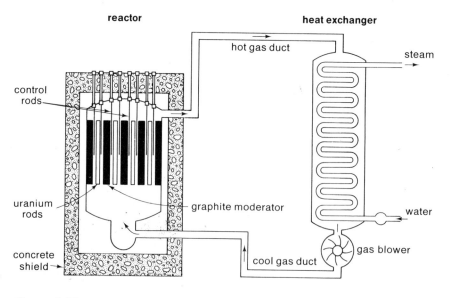

Figure 4.17

98 Chemistry today

can be quickly slowed down the chain reaction will not be interrupted. This is achieved by surrounding the uranium in a reactor with blocks of graphite (a form of carbon), which act as a MODERATOR (Figure 4.17).

So that the fission process does not get out of control, movable rods of an element such as boron, which readily absorbs neutrons, are placed in between the uranium rods. The chain reaction can be slowed down or speeded up by raising or lowering the boron rods. The whole reactor is shielded by lead and concrete to prevent the escape of radiation.

The energy produced by the fission process heats up the inside of the reactor. Most of the reactors now in use in Britain are cooled by circulating carbon dioxide gas around the graphite block. The heat produced by the reactor is transferred to the gas. The hot gas is then used to boil water, and the steam drives the turbines which generate electricity.

New elements from nuclear reactors

The invention of the nuclear reactor led to the discovery of some new elements which are not found in nature. Some of the neutrons produced by the fission of uranium-235 collide with uranium-238 nuclei before they can be slowed down by the moderator. A uranium-238 nucleus can capture a fast-moving neutron to produce another isotope, uranium-239:

uranium-238 + 1 neutron → uranium-239.

Like all uranium isotopes uranium-239 is radioactive. Unlike uranium-235 and uranium-238, which have very long half-lives (Figure 4.11, page 93), the half-life of uranium-239 is very short. It decays by the emission of a β-particle. This increases the atomic number by one (page 91) to produce element 93, neptunium:

uranium-239 → neptunium-239 + β-particle.

Neptunium is also radioactive, and decays by emission of a β-particle to produce another new element, plutonium (atomic number 94):

neptunium-239 → plutonium-239 + β-particle.

The discovery of these new elements as by-products of the nuclear reactor led scientists to look for ways of producing even more man-made elements. A device called a CYCLOTRON was developed, by means of which the atoms of heavy radioactive elements can be bombarded with much lighter atoms which have been accelerated to enormous speeds. Under these conditions the light and heavy nuclei join together to produce other new atoms. For example, uranium-238 atoms, when bombarded with carbon atoms, produce atoms of element number 98, californium:

uranium-238 + carbon-12 → californium-246 + 4 neutrons.
92 protons   6 protons   98 protons
146 neutrons 6 neutrons   148 neutrons

So far 15 artificial elements have been made.

Figure 4.18 The experimental fast breeder reactor at Dounreay, Scotland

The fast breeder reactor

The most modern type of nuclear reactor is the FAST BREEDER REACTOR. An experimental reactor of this type has been in operation at Dounreay in Scotland since 1959 (Figure 4.18).

The fuel is plutonium, instead of uranium. The plutonium-239 isotope can capture a stray neutron and undergo fission in a very similar way to uranium-235:

plutonium-239	+	1 neutron	→	lanthanum-140	+	rubidium-97	+	3 neutrons.
94 protons				57 protons		37 protons		
145 neutrons				83 neutrons		60 neutrons		

The three extra neutrons can collide with three more plutonium atoms, producing 9 more neutrons, and so on, causing a chain reaction just like the uranium-235 fission process. No moderator is required to slow down the neutrons. Plutonium will undergo fission with neutrons of any speed and there is no other isotope present to capture neutrons and interrupt the chain reaction.

The plutonium core of the reactor (Figure 4.19) is surrounded by natural uranium. Some of the neutrons from the fission of plutonium collide with atoms of uranium-238 and convert them into more plutonium:

$$\text{uranium-238} + 1 \text{ neutron} \rightarrow \text{uranium-239};$$
$$\text{uranium-239} \rightarrow \text{neptunium-239} + \beta\text{-particle};$$
$$\text{neptunium-239} \rightarrow \text{plutonium-239} + \beta\text{-particle}.$$

This type of reactor produces plutonium faster than it uses it up, hence the name *breeder reactor*. Its advantage is that it produces far more heat

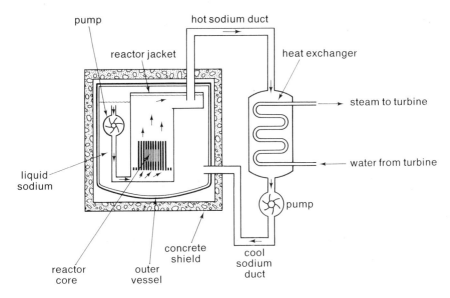

Figure 4.19

per tonne of uranium. In the early reactors which work on the fission of uranium-235 less than 1 per cent of the uranium fuel is used, the proportion which is the 235 isotope. The rest of the uranium, the 238 isotope, is wasted. By using fast breeder reactors all the uranium can be converted into plutonium, and all the plutonium can be fissioned to produce energy.

In fast breeder reactors liquid sodium is circulated around the reactor core to remove the heat produced, instead of carbon dioxide gas. The hot liquid sodium is used to generate steam to drive the turbines which generate electricity.

4.5 Nuclear fusion

If two light atoms can be made to collide extremely violently with one another, the nuclei can fuse together. For example, two atoms of the isotope of hydrogen which has one proton and one neutron in its nucleus (hydrogen-2, sometimes called *heavy hydrogen* or *deuterium*) will fuse together to produce one of the isotopes of helium (helium-3), together with an extra neutron:

hydrogen-2 + hydrogen-2 → helium-3 + 1 neutron.

The loss of mass in this process is 0.002 g per mole (2 g) of hydrogen-2 atoms, which produces as much energy as is obtained by burning over 5 tonnes of coal.

Once the fusion process has started the energy produced is great

Inside the atom 101

enough to make more fusions take place. The difficulty is getting fusion started.

Fusion in the sun

The sun produces its energy by nuclear fusion. It is mainly composed of hydrogen at a temperature of about 10 000 000°C (10^7°C). At this very high temperature the hydrogen atoms are moving so fast that their collisions are violent enough to cause fusion and to form helium atoms.

Some stars are more than ten times hotter than our sun (10^8°C). The even more vigorous collisions of nuclei in these stars make more complicated fusion processes possible. Helium nuclei, each with two protons and two neutrons, fuse to form atoms like carbon (6 protons and 6 neutrons), oxygen (8 protons and 8 neutrons), and even calcium (20 protons and 20 neutrons). Other elements, too, are formed by fusion of other light atoms. Astronomers believe that all the elements found in nature have been formed in this way.

The H-bomb

The hydrogen bomb consists of an atom bomb surrounded by a large quantity of hydrogen-2. The energy produced by the detonation of the atom bomb is sufficient to trigger off fusion between hydrogen-2 atoms, and the uncontrolled fusion process produces an even more massive explosion.

Energy from fusion

For more than 15 years experiments have been going on to try to develop a controlled fusion process to produce energy to make electricity. These experiments are still in their early stages, but there is little doubt that electricity from nuclear fusion will eventually become a practical reality. The most likely process is fusion of hydrogen-2 into helium. The great advantage of this would be that nuclear reactors could be run on hydrogen-2 as a fuel instead of using uranium. The supplies of uranium in the earth's crust will eventually run out but the supply of hydrogen-2 is almost unlimited. Every tonne of sea water contains 22 g of hydrogen-2 in the form of hydrogen-2 oxide (sometimes called *heavy water*). Cheap ways of separating it from the much commoner hydrogen-1 have already been found.

4.6 Radioisotopes

Although only a few of the elements found in nature have radioactive isotopes, it has now become possible to produce a RADIOISOTOPE of almost any element. A non-radioactive isotope is placed in a nuclear reactor where it is bombarded with neutrons. One or more neutrons may be

Figure 4.20 The experimental apparatus at the laboratory of the United Kingdom Atomic Energy Authority in Berkshire, used to investigate the controlled fusion of hydrogen-2

captured by the nucleus. This produces a new isotope which is often radioactive because of its excessive number of neutrons. Sodium, for example, occurs in nature as sodium-23 (11 protons, 12 neutrons). This is the only natural sodium isotope, and is not radioactive. When sodium-23 is bombarded with neutrons some sodium-24 is produced (11 protons, 13 neutrons):

$$\text{sodium-23} + 1 \text{ neutron} \rightarrow \text{sodium-24}.$$

This artificial isotope is radioactive.

Radioisotopes are finding more and more uses in research, industry, and medicine. They have four main properties.

They are easy to detect

A leak in a pipe or sewer can be traced by putting some sodium-24, in the form of sodium-24 chloride, into the water passing down the pipe. Radioactivity accumulates in the soil around the leak. The leak can then be located with a Geiger-Müller counter, even if the pipe runs a metre or more underground. This is a much easier way of finding the leak than digging up the whole pipe.

The movement of silt in a river can be followed by adding ground glass containing the radioisotope scandium-46 to the mud on the river bed. The progress of the silt up the river can then be followed from a boat with a counter. An understanding of how the silt is moving makes dredging more efficient.

The wear of an engine part such as a piston can be investigated by irradiating the whole piston in a nuclear reactor so that some of the metal atoms are converted to a radioisotope. As the piston is worn away radioactive atoms are carried into the oil. Measurement of the radioactivity of the oil shows the amount of wear on the piston. Information provided by this kind of study is used both to improve engine design and to develop better lubricants.

Radioisotopes can also be used to study the functioning of various parts of the human body. Diseases caused by the blood not circulating properly can be diagnosed by injecting sodium-24 chloride into the bloodstream and following its progress around the body with a special type of counter. Healthy brain tissue will not absorb certain compounds of the radioisotope technetium-99 from the blood: diseased cells in a brain tumour will do so. Brain tumours can be diagnosed at an early stage by injecting a compound of technetium-99 into the bloodstream and checking to see if its radioactivity appears in the brain. If there is a tumour its precise position can be located in this way, making surgery easier.

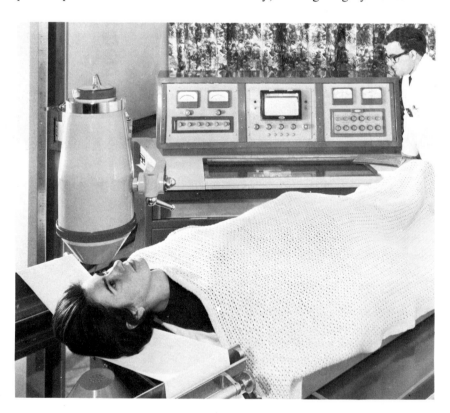

Figure 4.21 A special counter for tracing the movement of a radioisotope in the brain

Their radiation is penetrating

The amount of penetration of α-, β-, or γ-rays depends on the material involved and how thick it is. Radioisotopes can be used to check the thickness of materials such as paper, metal, and plastic, especially when they are manufactured in the form of continuous sheets. A radioisotope is placed below the sheet and a Geiger-Müller counter above. As long as the thickness of the sheet remains the same the counter registers the same reading. If the sheet becomes thinner the reading goes up, and if it becomes thicker the reading goes down. A change in the reading can be arranged to sound a warning, switch off the machine, or even feed information to a computer which will automatically correct the mistake the machine is making.

Radioisotopes can also be used to check whether containers on a production line have been correctly filled. The radioisotope is placed so that if the container is full no radiation passes through. If the container is empty or partly empty the radiation passes through to a counter which sets off an automatic mechanism to reject the faulty container.

Figure 4.22 A can which was not properly filled has been detected by a radioisotope source and counter

They produce energy

Radioactive elements were used to produce energy even before the cause of radioactivity was understood. The luminous spots on a watch or clock used to be made of a mixture of a uranium compound and a phosphorescent material. The energy released by the radioactive decay of the uranium caused the phosphorescent material to emit light. Nowadays it is more usual to make luminous paint from compounds of the radioisotope hydrogen-3.

If strontium-90 is enclosed in a sealed container and shielded with lead to prevent the escape of radiation, the energy released by radioactive decay is produced as heat. This can be converted into electricity to run a radio transmitter to guide aircraft or a light beacon to guide ships. This is an expensive way of producing energy, but has the advantage of being very reliable and needing no attention for five years or more.

One of the most recent uses of radioisotopes to produce energy is the *pacemaker*. This is a device implanted in the body, near the heart, which gives it a series of small electric shocks to keep it beating when the body's normal mechanism has broken down. The earliest pacemakers ran on an ordinary battery and had to be replaced by a surgical operation every two years. In July 1970 the first radioisotope pacemaker, powered by a plutonium isotope, was successfully implanted in a patient, and is expected to run for at least ten years without replacement.

They destroy living cells

Intense α-, β-, or γ-radiation can kill living cells. An intense dose of γ-rays from cobalt-60 is used to sterilise medical equipment such as syringes and surgical dressings by destroying bacteria. This is a quicker, cheaper, and more certain method than the old process of heating the equipment to a high temperature.

Radioisotopes are sometimes used to destroy cancer cells in a living body, without destroying the surrounding healthy cells. This is done either by implanting a small pellet of a radioisotope in the middle of the cancerous growth, or by carefully directing an intense beam of radiation at the spot required.

The fact that the radiation can kill living cells means that elaborate precautions need to be taken in handling radioisotopes. In using them for medical diagnosis the quantities must be carefully calculated so that no damage is done to the patient. Fortunately, Geiger-Müller counters can be made so sensitive that incredibly minute traces of a radioisotope can easily be detected. In cancer treatment the dose of radiation must be just right: large enough to destroy the cancer without killing too many of the surrounding healthy cells. For purposes such as testing leaks in pipes, where radioisotopes are being released into the environment, it is important to choose an isotope with as short a half-life as possible, so that after a few days it has almost completely decayed to a stable isotope.

At the laboratories in Amersham, where most of the radioisotopes are manufactured in Britain, there is special handling equipment designed so that the workers are completely protected from radiation. Atomic reactors and radioisotope sources used in industry must be shielded with lead or concrete thick enough to prevent any radiation escaping.

Figure 4.23 The first plutonium-powered pacemaker

Figure 4.24 Remote handling of radioisotopes

Questions

1 (a) Uranium is an example of a radioactive element. Explain the meaning of the term *radioactive element* and give two further examples.

(b) Name and describe briefly the two kinds of radioactive particles found in radioactive emissions.

(c) Radioactivity can be of great benefit to mankind, but it can also be a danger. Describe *two* different beneficial uses and *two* different dangers arising from its use. (*Wales*)

2 Describe how any one type of nuclear reactor works. Draw a sketch to help make your explanation clearer.

3 Explain the following:
(a) Nuclear fission is a chain reaction.
(b) A fast breeder reactor is the most efficient way of obtaining energy from uranium.
(c) It is possible that all of the elements found in nature have been formed from hydrogen.
(d) If nuclear fusion can be controlled it will provide an almost limitless supply of energy for mankind.

4 What is the difference between a natural radioisotope and an artificial radioisotope? How are artificial radioisotopes made?

Inside the atom 107

5 Radioisotopes can be used to study the movements of swarms of insects such as locusts which, in some parts of the world, do tremendous damage to crops. Understanding their movements is the first step towards controlling them. One isotope which has been used for this purpose is carbon-14.
  (a) What is an isotope?
  (b) What does the figure 14 in carbon-14 mean?
  (c) Why is the isotope carbon-14 radioactive?
  (d) How do you think carbon-14 is made?
  (e) How could the presence of carbon-14 in a locust be detected?
  (f) How do you think carbon-14 could be used to trace the movement of locusts?

6 Plutonium-238 has a half-life of 86 years. Caesium-137 has a half-life of 30 years. Bromine-82 has a half-life of 36 hours. Plutonium-238 decays by emission of α-particles, but caesium-137 and bromine-82 decay by emission of β-particles together with γ-rays. Which of these radioisotopes would be most suitable for the following jobs and why?
  (a) Checking the thickness of aluminium kitchen foil while it is being manufactured in a continuous sheet.
  (b) Detecting a leak in an oil pipeline.
  (c) Producing the power to heat an astronaut's spacesuit.

For the teacher

N.B. Administrative memorandum 1/65 from the Department of Education and Science states that *No instruction shall be given in the school or educational establishment involving the use of radioactive material other than a compound of potassium, thorium or uranium normally used as a chemical reagent ... unless the Secretary of State has given his approval.* The experimental work below does not go beyond this limitation, but if further work is contemplated the full memorandum should be consulted.

Exposure to radiation from natural thorium and uranium compounds presents no hazard but pupils should be aware of precautions to avoid ingestion or absorption through the skin. Disposable polythene gloves should be worn, all operations should be conducted over a tray lined with paper tissues, and mouth pipettes should not be used.

Bequerel's discovery is easy to demonstrate. Uranyl(VI) nitrate or thorium(IV) nitrate can be used. Alternatively, *pitchblende* may be obtained from R. F. D. Parkinson to provide a more realistic illustration. Any fast film or plates can be used; polaroid film is obviously ideal. At least 24 hours exposure to the radiation is necessary, preferably longer.

Another interesting experiment is to soak a photographic plate for

10 minutes in 0.02M thorium(IV) nitrate solution in a darkroom, wash with distilled water, and leave to dry. After development, examination with a microscope or microprojector will reveal star-like α-particle tracks. This is most effective using nuclear emulsion plates (Kodak or Ilford), which have a thicker emulsion than ordinary plates.

Detection of radiation Any form of counter is useful to give pupils first-hand experience of (a) the random nature of radioactive decay, (b) the existence of background radiation, and (c) the radioactivity of natural potassium, thorium, and uranium compounds. A Geiger-Müller liquid counter (e.g. Mullard MX142 tube with holder) together with a scaler (physics department?) is ideal so that measurement of a half-life (see below) is also possible.

The background radiation can be observed with tap water in the counter. Counting over several 2-minute periods emphasises the random nature of radioactive decay. A saturated solution of any potassium salt will show an appreciable increase over the background count due to the radioactive potassium-40 isotope. Dilute solutions of uranyl(VI) nitrate and thorium(IV) nitrate give a much larger count. It is advisable to try these after the potassium salt, as the liquid counter is difficult to decontaminate; rinsing with concentrated nitric acid is the most effective method. With the thorium or uranium salt it is useful, after counting the original solution, to dilute it $\times$ 2, $\times$ 10, etc and recount.

Determination of a half-life The half-life of protactinium-234, one of the products of the uranium-238 decay series is the simplest to measure. Solvent extraction of a solution of uranyl(VI) nitrate with ethyl ethanoate or pentyl ethanoate separates protactinium from uranium and all of its other decay products.

Dissolve 1 g of uranyl(VI) acetate in 3 cm^3 of water and add 7 cm^3 of concentrated hydrochloric acid. Transfer to a separating funnel and shake for half a minute with 10 cm^3 of ethyl or pentyl ethanoate. Reject the lower aqueous layer and transfer the organic layer to the liquid counter. Count for 10-second periods at 10-second intervals and plot a graph of counts per second against time in seconds. (Since the half-life is so short, counting must be begun very quickly after extraction to get a good result.)

For further school experiments in radioactivity see Faires, R. A. *Experiments in Radioactivity* (Methuen, 1970).

5 Atoms joining together

5.1 The effect of electricity on chemicals

Experiment 5.1   Which solids conduct electricity?

Use the apparatus shown in Figure 5.1 to find out which of the following substances conduct electricity: carbon, copper, copper sulphate, lead, lead bromide, lead iodide, polythene, potassium iodide, sodium chloride, sugar, sulphur, and zinc.

Hold the two wires firmly against each substance in turn and see if the bulb lights up. After testing each substance write down its name and whether or not it conducts electricity.

Which of the substances you have tested are elements and which are compounds? Which of the elements are metals and which are non-metals?

What kind of solids conduct electricity?

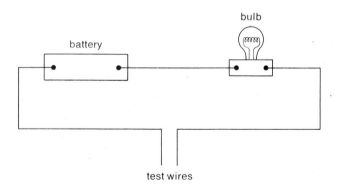

Figure 5.1

Experiment 5.2   Which liquids conduct electricity?

(a) Use the apparatus shown in Figure 5.1 to find out if any of the following liquids conduct electricity: alcohol, oil, paraffin, and distilled water.

Make a note of your results.

110   Chemistry today

(b) Now try melting some of the substances you investigated in experiment 5.1 and see if they conduct when they are liquid. You could try lead, lead bromide, lead iodide, polythene, potassium iodide, sugar, and sulphur. Melt each substance in turn in a crucible supported on a pipeclay triangle on a tripod and test it with the apparatus shown in Figure 5.1.

Melt sugar and sulphur very carefully so that they do not catch fire. Have a pair of tongs ready so that if they do you can put the crucible quickly under a tap.

If any of the molten substances do conduct, examine them closely while the electric current is passing through them to see if any changes appear to be taking place. After testing each substance make a note of the result.

Are there any substances which did not conduct electricity in the solid state but do when they are molten? If a molten substance conducts, what does the electric current do to it?

Experiment 5.3 Which solutions conduct electricity?

Use the apparatus shown in Figure 5.1 again. You can test solutions of alcohol, copper sulphate, potassium iodide, sodium chloride, and sugar in water. One-quarter-fill a beaker with each solution in turn, dip in the two wires, and see if the bulb lights up. If a solution does conduct examine it closely while the electric current is passing through it to see if any changes are taking place.

Make a list of the solutions which do conduct and a list of the solutions which do not. Describe any changes you have seen while a solution conducts.

Experiment 5.4 Investigating the effect of an electric current on molten lead bromide

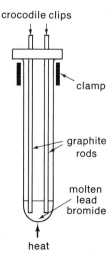

Put 5 cm depth of lead bromide into a large test tube. Clamp the tube and heat the lead bromide slowly until it has just melted. Then adjust the bunsen flame to keep the lead bromide *just* molten.

Fit a pair of carbon rods into the tube as shown in Figure 5.2 and connect them with crocodile clips to a battery.

Pass a current through the molten lead bromide for 5 minutes. Watch carefully what is happening inside the tube. Then disconnect the battery, remove the carbon rods, turn off the bunsen, and leave the tube to cool.

What does the gas look like that is produced from the lead bromide while the current is passing? What do you think this gas is? Which

Figure 5.2

Atoms joining together

carbon rod does it come from, the one connected to the positive terminal of the battery or the one connected to the negative terminal?

Can you see anything of interest in the bottom of the test tube at the end of the experiment? What do you think it is?

Experiment 5.5 Finding out what happens when a current is passed through some solutions of electrolytes

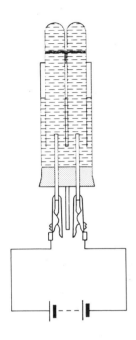

Figure 5.3

(a) Half-fill an electrolysis cell (Figure 5.3) with copper chloride solution. Fill the two small test tubes with copper chloride solution and invert them carefully over the carbon rods.

Connect the two carbon rods to a battery. Watch carefully what happens while the current is passing.

What is produced at the cathode (the negative carbon rod)?

What happened at the anode (the positive carbon rod)? What do you think the gas is? Remove the tube of gas and carry out a test to discover what it is. Refer to pages 430–31 if you do not know a suitable test.

(b) Empty and rinse out the electrolysis cell and test tubes. Refill them with zinc bromide solution and electrolyse it.

What is produced at the cathode this time?

What is produced at the anode?

(c) Repeat the experiment with sodium chloride solution.

What gas is produced at the anode? Carry out a test to check that you are correct.

Remove the tube of gas from the cathode. Bring a lighted splint up to the mouth of the tube. What gas is it? Is this what you would have expected to be produced at the cathode? Can you explain why you get this product?

(d) If you have time, repeat this experiment with potassium iodide solution, calcium chloride solution, dilute hydrochloric acid, and sodium hydroxide solution. In each case find out what is produced at the cathode and the anode and try to explain your results.

Conductors and non-conductors

Solids The only solids which conduct electricity well are the metal elements (and carbon, in the form of *graphite*). Non-metal elements (apart from carbon) do not conduct, nor do any solid compounds.

Liquids Some compounds which do not conduct electricity in the solid state do conduct when they are molten. These are always compounds of a metal with a non-metal, such as lead bromide, potassium iodide, and sodium chloride.

The metal elements also conduct electricity well when they are molten. Many other substances do not conduct in the liquid state, for example water, alcohol, molten sulphur, and molten sugar.

Solutions Some solutions of compounds in water do conduct electricity, though pure water is a bad conductor. These are mainly compounds of a metal and a non-metal, the same kind of compounds that conduct when they are molten. Examples are copper sulphate, potassium iodide, and sodium chloride.

Many other compounds, such as sugar and alcohol, do not conduct in solution in water.

Gases All gases are bad conductors of electricity. The individual particles (atoms or molecules) in a gas are too far apart to pass electric current from one to the other.

Substances which do not conduct in the solid state, but do conduct when they are molten or dissolved in water, are called ELECTROLYTES.
Substances which do not conduct under any conditions are called NON-ELECTROLYTES.

Why do metals conduct electricity?

Some of the electrons (usually one or two) in a metal atom are not very strongly attracted by the positive protons because they are rather far away from the nucleus. In a solid lump of metal these electrons move freely around between the metal atoms. A lump of metal is a GIANT STRUCTURE of millions of regularly-packed atoms. The structure is held together by the electrical forces of attraction between the positively-charged nuclei and the negatively-charged electrons moving between them.

An electric current is a stream of electrons. When a current is applied between the ends of a piece of metal the free electrons move from one end of the block to the other, carrying the current.

Non-metal atoms do not have any of these loosely-held electrons, so they have no means of conducting an electric current.

What happens when molten electrolytes conduct electricity?

Lead bromide, a compound of the elements lead and bromine, is an electrolyte. It will not conduct electricity in the solid state, but will conduct when molten.

The conduction of a current by molten lead bromide is very different from conduction by a metal. A metal is not changed by the passage of a current. It may get hot, like the bar in an electric fire, or it may glow and give out light, like the filament in a light bulb. But when the current is switched off it is still the same substance. Molten lead bromide, as well as conducting, is changed by the passage of the current.

The terminals by which the current enters and leaves the electrolyte are called ELECTRODES. As current passes through molten lead bromide a brown gas (bromine) is produced at the positive electrode (called the ANODE). At the negative electrode (called the CATHODE) a shiny bead of molten lead is formed. The current decomposes the compound, lead bromide, into its elements, lead and bromine. This process is called ELECTROLYSIS (*lysis* comes from a Greek word meaning *to loosen*: electrolysis means *loosening with electricity*).

Sodium chloride is another electrolyte. When molten it conducts electricity and it is decomposed by the current. Molten sodium is formed at the cathode. Chlorine gas is formed at the anode.

The electrolysis of molten electrolytes is often used in industry to split up compounds into their elements. The electrolysis of molten sodium chloride is used to manufacture both sodium and chlorine. Aluminium is manufactured by the electrolysis of aluminium oxide. Calcium and magnesium are made by electrolysing their molten chlorides.

What happens when solutions of electrolytes conduct electricity?

Solutions of electrolytes in water are also decomposed by the passage of an electric current. Figure 5.4 shows the products that are obtained at the anode and cathode when some solutions are electrolysed.

	At the cathode	At the anode
Copper chloride	Copper	Chlorine
Zinc bromide	Zinc	Bromine
Sodium chloride	Hydrogen	Chlorine
Potassium iodide	Hydrogen	Iodine
Calcium chloride	Hydrogen	Chlorine
Hydrochloric acid	Hydrogen	Chlorine
Sodium hydroxide	Hydrogen	Oxygen

Figure 5.4

The elements produced at the cathode are all metals, except for one non-metal, hydrogen.

Non-metal elements (except hydrogen) are produced at the anode.

Metals in Groups I and II of the periodic classification are never produced in the electrolysis of a solution. This is hardly surprising, since all these metals react with water. Hydrogen from the water is produced instead.

Questions

1 Figure 5.5 shows the behaviour of five substances, A, B, C, D and E towards an electric current.

	Does the solid substance conduct?	Does the substance conduct when molten?	Does a solution of the substance in water conduct?
A	No	Yes	Yes
B	Yes	Yes	(Insoluble in water)
C	No	Yes	(Insoluble in water)
D	No	No	No
E	No	(Decomposes before it melts)	Yes

Figure 5.5

(a) Which of the substances A, B, C, D and E are electrolytes?
(b) Which of the substances A, B, C, D and E are non-electrolytes?
(c) What sort of substance is B?

2 Which of the following is the best description of electrolysis?
A A physical change brought about by an electric current.
B The decomposition of water.
C The production of electricity.
D A process for the extraction of metals.
E A chemical change brought about by an electric current.

3 Fill the gaps in the following sentence to make correct chemical sense.
If an electric current is passed through molten _____ a gas called _____ is evolved at the anode and metallic _____ is set free at the cathode. (*Wales*)

4 Figure 5.6 shows an experiment to investigate the electrolysis of a solution of sodium chloride in water.

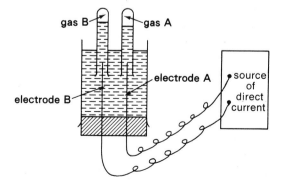

Figure 5.6

Atoms joining together

(a) When gas A is mixed with air and placed near a bunsen flame an explosion results. Name gas A and state the terminal of the direct current supply (positive or negative) to which the electrode A would be attached. Is this electrode the anode or cathode?

(b) Name gas B and give a chemical test you could use to help identify the gas.

(c) Electrode B must not be metallic but must be made of carbon. Suggest a reason for this. *(West Midlands)*

5 A variety of liquids was tested for their electrical conductivity. The results are set out in Figure 5.7.

Substance	Does it conduct?	Products	
		CATHODE	ANODE
A	No	—	—
B	Yes	Hydrogen	Chlorine
C	Yes	Shiny silvery metal	Brown vapour
D	No	—	—
E	Yes	Hydrogen	Oxygen
F	Yes	—	—
G	No	—	—
H	Yes	Hydrogen	Brown solution and black solid
I	Yes	Hydrogen	Oxygen
J	Yes	Pinkish metal	Oxygen

Figure 5.7

(a) Give the letters of TWO which are electrolytes.
(b) Give the letter of ONE which could be salt solution (sodium chloride).
(c) Give the letter of ONE which could be potassium iodide solution.
(d) Give the letter of ONE which could be mercury.
(e) Give the letter of ONE which could be sugar solution.
(f) Name a substance which could be C.
(g) Name a substance which could be J. *(London)*

6 Explain briefly why the colour of a copper sulphate solution slowly disappears during electrolysis. *(East Anglia)*

7 What differences are there between the ways in which (a) copper and (b) copper bromide conduct electricity?

2 The ionic theory

Current electricity was discovered by an Italian scientist, Luigi Galvani, in 1791. By the beginning of the nineteenth century, chemists, notably Humphry Davy (page 12), were using an electric current to decompose compounds into their elements. But for quite a long time no one was able to explain why some compounds were decomposed by a current whereas some did not even conduct electricity. Michael Faraday, in 1833, first put forward a theory to explain what happens when a current passes through molten or dissolved electrolytes.

1 When an electrolyte conducts, something must be carrying the current between the electrodes. Faraday suggested that electrolytes are made up of charged particles, which he called IONS. Sodium chloride, for instance, is made up of sodium ions and chlorine ions.

Non-electrolytes are not made up of ions, and therefore cannot conduct.

2 An electrolyte like sodium chloride is electrically neutral: if it was not you might get an electric shock when you touched it. Some of the ions must be positively charged and some negatively charged. The positive and negative charges balance each other out.

Metal elements and hydrogen are always produced at the cathode (negative electrode) during electrolysis. Metal ions and hydrogen ions must therefore be positive ions, to which Faraday gave the name CATIONS, because they are attracted towards the cathode.

Non-metal elements (except hydrogen) are always produced at the anode (positive electrode), so they must be negative ions. Negative ions are called ANIONS, because they are attracted to the anode.

In sodium chloride, the sodium ions are cations (positive) and the chlorine ions are anions (negative) (Figure 5.8).

3 A solid electrolyte will not conduct electricity because the ions are not free to move. They are held together in a regular structure by the strong attractive forces between the positive and negative ions. A crystal of an electrolyte is a giant structure of millions of ions arranged in a regular pattern (Figure 5.9).

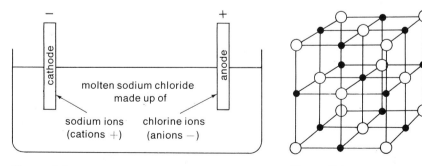

Figure 5.8   Figure 5.9

Atoms joining together

When a solid electrolyte is heated strongly enough the ions vibrate so vigorously that the forces between them are overcome. The solid melts and the ions are free to move: molten electrolytes conduct electricity.

In solution in water, the ions of an electrolyte are also free to move and to conduct the current.

4 During electrolysis, when the ions reach the electrodes they lose their charge. In the electrolysis of molten sodium chloride sodium ions give up their positive charge at the cathode and become molten sodium metal. Chlorine ions give up their negative charge at the anode and become chlorine gas.

How big are the charges on ions?

The *rate of flow* of an electric current is measured in AMPERES (amps). The *quantity* of electricity passed through an electrolyte during electrolysis is measured in COULOMBS. One coulomb of electricity flows when a current of one amp passes for one second.

Quantity of electricity   =   current   ×   time for which it flows
         coulombs   =   amps   ×   seconds

If molten sodium chloride is electrolysed with a current of 16 amps for 100 minutes, 23 grams of sodium are produced at the cathode.

The quantity of electricity passed through the molten sodium chloride is

16 × 6000 = 96 000 coulombs.   (100 minutes = 6000 seconds)

So in this experiment 23 g of sodium (1 mole) is produced by the passage of 96 000 coulombs of electricity. The ions have carried this amount of electricity through the electrolyte.

The charge carried by 1 mole of sodium ions is 96 000 coulombs.

The coulomb is rather a small unit for measuring quantity of electricity. In chemistry we prefer to use a larger unit called the FARADAY, chosen because it is the quantity of electricity carried by one mole of electrons (6.02×10^{23} electrons).

1 faraday = 96 000 coulombs.

The charge on 1 mole of sodium ions is one faraday. In chemical shorthand we represent 1 mole of sodium ions as Na^+, meaning 1 mole of sodium atoms (Na) carrying 1 faraday of positive charge (+).

The formula of sodium chloride is NaCl: one mole of sodium is combined with one mole of chlorine. Since sodium chloride is electrically neutral the mole of chlorine ions must carry one faraday of negative charge. 1 mole of chlorine ions is therefore represented Cl^-.

Explaining the electrolysis of molten sodium chloride

Sodium chloride is composed of sodium ions and chlorine ions:

$$NaCl = Na^+Cl^-.$$

At the cathode Sodium cations (positive ions) are attracted by the cathode (negative electrode). The cathode supplies electrons to neutralise the positive charge on the sodium ions and molten sodium is produced:

$$Na^+ + e \rightarrow Na(l).$$

(e represents a mole of electrons.)

At the anode Chlorine anions (negative ions) are attracted towards the anode (positive electrode). They give up their electrons to the anode and pair up as molecules of chlorine gas:

$$2Cl^- \rightarrow Cl_2(g) + 2e.$$

Another example

If molten lead bromide is electrolysed with a current of 2 amps for 32 minutes a bead of lead weighing 4.14 g is obtained at the cathode.

The quantity of electricity passed in this experiment is:

$2 \times 1920 = 3840$ coulombs.   (32 minutes is $32 \times 60 = 1920$ seconds)
4.14 g of lead are produced by the passage of 3 840 coulombs.
(To get rid of the decimals multiply by 100.)
414 g of lead are produced by the passage of 384 000 coulombs.
(Atomic mass of lead = 207)
$\frac{414}{207} = 2$ moles of lead are produced by the passage of 384 000 coulombs.
1 mole of lead is produced by the passage of $\frac{384000}{2} = 192 000$ coulombs.
(96 000 coulombs = 1 faraday.)
1 mole of lead is produced by the passage of $\frac{192000}{96000} = 2$ faradays.
1 mole of lead ions can be represented Pb^{2+}.

The formula of lead bromide is $PbBr_2$. To balance the 2 faradays of positive charge on 1 mole of lead ions there must be 2 faradays of negative charge shared between the 2 moles of bromine ions. 1 mole of bromine ions must therefore be represented *Br^-*.

Explaining the electrolysis of molten lead bromide

Lead bromide is composed of lead ions and bromine ions:

$$PbBr_2 = Pb^{2+}2Br^-.$$

At the cathode Lead cations are attracted towards the cathode, which supplies electrons to neutralise their positive charge:

$$Pb^{2+} + 2e \rightarrow Pb(l).$$

At the anode Bromine ions are attracted towards the anode, where they give up their electrons:

$$2Br^- \rightarrow Br_2(g) + 2e.$$

Experiment 5.6 Can we see ions moving?

Cut a strip of filter paper about 2 cm by 6 cm. Make it just damp with tap water and rest it on a white tile. By means of a crocodile clip on either end connect it to a 25 volt DC (direct current) supply.

Put one small crystal of potassium permanganate in the centre of the filter paper and leave it for 10 minutes.

Does the colour move towards the anode or the cathode? Which of the ions of potassium permanganate, the potassium ion or the permanganate ion, would you expect to be coloured?

Experiment 5.7 Measuring the charge on a copper ion

Half-fill a beaker with copper sulphate solution. Place two copper foil electrodes in the beaker and connect them to an ammeter, a 6 volt battery, a switch, and a variable resistance as shown in Figure 5.10.

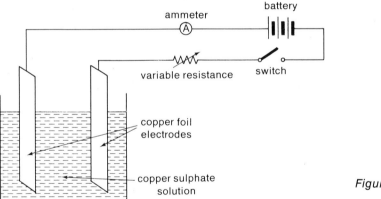

Figure 5.10

Adjust the variable resistance so that the current flowing through the circuit is 0.1 amp. (If a current as high as 0.1 amp cannot be obtained, increase the depth of copper sulphate solution in the beaker. If the ammeter needle is not steady there is a fault in your circuit: check that all the connections are secure.)

Take out both electrodes, clean them thoroughly with sandpaper, wash them, and dry them with paper tissues. Weigh each one separately and write down the weights, making sure you know which one is the cathode and which one is the anode.

Replace the electrodes in the beaker, switch on the current, and

120 Chemistry today

note the exact time. Allow the electrolysis to run for exactly 30 minutes. (You may find that the current changes during the experiment: if necessary, adjust the variable resistance to keep the current at exactly 0.1 amp.)

Remove the electrodes, rinse them under the tap, and dry them carefully with paper tissues. Reweigh them and write down the weights, again making sure that you remember which is the cathode and which is the anode.

What was the change in weight of (a) the anode (b) the cathode? Is there any connection between these two weights? Can you explain this?

A current of 0.1 amp flowed for 30 minutes (1 800 seconds). How many coulombs of electricity passed through the solution?

What weight of copper was lost from the anode as a result of the passage of this quantity of electricity? Calculate how much electricity would have been needed to make the anode lose 1 mole of copper (64 grams). How many faradays is this? What is the charge on 1 mole of copper ions?

Experiment 5.8 Measuring the charge on a silver ion

Set up the apparatus shown in Figure 5.10, but with silver nitrate solution in the beaker and silver electrodes in place of copper electrodes. Adjust the current to 0.1 amp.

Remove the silver anode, clean it with sandpaper, wash it, dry with paper tissues, and weigh it. Write down the weight.

Replace the anode in the beaker and switch on the current for exactly 30 minutes, adjusting the variable resistance if necessary to keep the current at 0.1 amp.

Remove the silver anode, wash it, dry carefully with paper tissues, and reweigh it. Write down the weight.

Calculate the charge on a mole of silver ions by the same method as in experiment 5.7.

Why is it not necessary to weigh both electrodes before and after the experiment? Why is it better to weigh the anode than the cathode?

Experiment 5.9 Another experiment to measure the charge on a copper ion

Set up the apparatus shown in Figure 5.10, but with a solution of sodium chloride in the beaker. Stand the beaker on a wire gauze on a tripod.

Heat the sodium chloride solution to 80°C, switch on the current, and adjust it to exactly 0.2 amp by means of the variable resistance.

Switch off the current, remove the copper anode, clean it with sandpaper, wash it, dry it, and weigh it. Write down the weight.

Replace the copper anode in the beaker and switch on the current for exactly 15 minutes, adjusting the variable resistance if necessary to keep the current at 0.2 amp.

Remove the copper anode, wash it thoroughly under the tap, dry it, and reweigh it. Write down the weight.

Calculate the charge on a mole of copper ions by the same method as in experiment 5.7. Do you get the same result as in experiment 5.7? If not, can you explain why?

The charges on ions

By experiments of this kind the amount of electricity carried by a mole of any ion can be found. *It is always one, two, or three faradays.* Figure 5.11 shows the charges on some important ions.

H^+	Mg^{2+}	Al^{3+}	Cl^-	O^{2-}	N^{3-}
Li^+	Ca^{2+}	Fe^{3+}	Br^-	S^{2-}	
Na^+	Pb^{2+}		I^-		
K^+	Zn^{2+}				
Cu^+	Fe^{2+}				
Ag^+	Cu^{2+}				

Figure 5.11

Two points are worth noticing from this table.
1 Elements in the same group of the periodic classification have the same charges on their ions.
2 Some of the transition elements, such as iron and copper, form more than one type of ion.

The formulas of compounds

Figure 5.11 is worth remembering because it makes it possible to work out the formulas of compounds, instead of having to do an experiment to find them, or having to remember them.

A compound is always electrically neutral. The positive charge on the cations must balance the negative charge on the anions. For example, a mole of potassium ions (K^+) carries 1 faraday of positive charge and a mole of iodine ions (I^-) carries 1 faraday of negative charge. The formula of potassium iodide must be KI, so that the positive and negative charges balance exactly.

Silver oxide To balance the two faradays of negative charge on a mole of oxygen ions (O^{2-}), 2 faradays of positive charge are needed. A mole of silver ions (Ag^+) carries 1 faraday of positive charge, so two moles of silver ions are needed for each mole of oxygen ions. The formula of silver oxide is therefore Ag_2O.

Aluminium oxide A mole of aluminium ions (Al^{3+}) carries 3 faradays of

positive charge. A mole of oxygen ions (O^{2-}) carries 2 faradays of negative charge. For the positive and negative charges to balance we must have:
2 moles of aluminium ions (6 faradays of positive charge) and
3 moles of oxygen ions (6 faradays of negative charge).
So the formula of aluminium oxide is Al_2O_3.

Elements forming more than one type of ion Copper forms two types of ion, one carrying 1 faraday of positive charge per mole of copper (Cu^+), and one carrying 2 faradays of positive charge per mole of copper (Cu^{2+}). In order to distinguish between them we call them:

                CuCl (Cu^+ and Cl^-)'   copper(I) chloride
and $CuCl_2$ (Cu^{2+} and $2Cl^-$)   copper(II) chloride.

Iron also forms two types of ion, Fe^{2+} and Fe^{3+}. The two iron oxides are called:

                FeO (Fe^{2+} and O^{2-})   iron(II) oxide
and Fe_2O_3 ($2Fe^{3+}$ and $3O^{2-}$) iron(III) oxide.

Questions

1 *Copy and fill in the gaps.* Molten potassium iodide conducts electricity. It is called an ＿＿＿＿. It is composed of potassium ＿＿＿＿ and iodine ＿＿＿＿. The potassium ＿＿ are ＿＿＿＿ charged and are called ＿＿＿＿. The iodine ＿＿＿ are ＿＿＿＿ charged and are called ＿＿＿＿. During electrolysis the potassium ＿＿＿ are attracted to the ＿＿＿＿ charged electrode which is called the ＿＿＿＿, and the iodine ＿＿ are attracted towards the ＿＿＿＿ charged electrode which is called the ＿＿.

2 What experimental evidence led Michael Faraday to put forward his ionic theory?

3 Which of the following moves at about the same speed as ions in electrolysis?
A Light   D A bicycle
B A supersonic plane   E A snail.
C A racing car

4 Which of the following is a possible value for the charge carried by one mole of ions of an element?
A 2 amps   D 2 faradays
B 2 volts   E 2 electrons.
C 2 coulombs

5 Molten sodium chloride was electrolysed with a current of $\frac{2}{3}$ amp for 20 minutes. 0.3 g of chlorine were produced at the anode.
   (a) For how many seconds did the current flow?
   (b) How many coulombs were passed in the experiment?
   (c) How many coulombs would have been needed to produce 1 mole (36 g) of chlorine?
   (d) How many faradays is this? (1 faraday = 96 000 coulombs)
   (e) Is the chlorine ion an anion or a cation?
   (f) What is the shorthand symbol for 1 mole of chlorine ions?

6 A solution of a gold compound was electrolysed with a current of 6 amps for 8 minutes. 1.97 g of gold were produced at the cathode. What is the charge on 1 mole of gold ions? (Atomic mass of gold = 197.)

7 An industrial electrolysis cell for the manufacture of calcium from molten calcium chloride uses a current of 10 000 amps and produces 7500 g of calcium per hour. Calculate the size of the charge on a calcium ion.

8 Two solutions of copper compounds, A and B, were electrolysed with the same current (0.32 amps) for the same time ($\frac{1}{2}$ hour), using copper electrodes. In solution A, 0.192 g of copper were lost from the anode. In solution B, 0.384 g of copper were lost from the anode. What are the charges on the copper ions in solutions A and B? (Atomic mass of copper = 64.)

9 Sodium is manufactured from common salt by the process of electrolysis.
   (a) In what state must the salt be to obtain sodium as one of the products?
   (b) What is the other product?
   (c) Name a suitable material for the electrode at which this by-product is discharged.
   (West Yorkshire)

10 (a) Describe the electrolysis of any fused salt, explaining the reactions which occur at the anode and cathode.
   (b) State *two* industrial uses of electrolysis.
   (c) Why is it that solid salts do not conduct an electric current?
   (Wales)

11 Complete the following equations which show what happens at electrodes during certain electrolyses.

(a) Al^{3+} + _____ → Al.
(b) $2O^{2-}$ → O_2 + _____.
(c) _____ + e → Ag.
(d) _____ → Cl_2 + 2e.
(e) $2H^+$ + 2e → _____.

12 Which of the electrode reactions in question 11 take place at the cathode and which take place at the anode?

13 Work out the formulas of the following compounds (use Figure 5.11, page 122, to look up the charges on the ions if you need to).

(a) lithium bromide
(b) calcium oxide
(c) potassium oxide
(d) zinc chloride
(e) aluminium chloride
(f) sodium nitride
(g) magnesium nitride
(h) copper(II) bromide
(i) copper(I) oxide
(j) iron(II) sulphide
(k) iron(III) sulphide
(l) water.

.3 Electroplating

If a solution of a copper compound, such as copper(II) sulphate, is electrolysed, copper ions are attracted towards the cathode. They collect electrons to neutralise their positive charges and a layer of metallic copper is deposited on the cathode:

$$Cu^{2+}(aq) + 2e \rightarrow Cu(s).$$

If the anode is made of carbon this process continues until all the copper ions in the electrolyte have been used up. But if the anode is made of copper its copper atoms lose electrons, forming more copper ions to replace the ones which have been removed at the cathode:

$$Cu(s) \rightarrow Cu^{2+}(aq) + 2e.$$

The effect is that copper dissolves from the anode as copper ions, at the same speed as copper ions at the cathode are deposited as copper atoms. This process is made use of in ELECTROPLATING.

Figure 5.12 Car bumpers being taken out of an electroplating bath

Atoms joining together

For copper-plating the object to be plated is made the cathode, with copper(II) sulphate solution as the electrolyte. The anode is a piece of pure copper. Electroplating can be carried out with other metals in the same way. For silver-plating the anode is a piece of pure silver and the electrolyte is a solution of a silver compound. For gold-plating a gold anode and a solution of a gold compound are used, and so on.

For electroplating to be successful the anode deposit must stick firmly, and to make sure of this the object to be plated must be perfectly clean. It is washed with a special solvent to remove grease, and then *pickled* in dilute acid. The cathode deposit must also be fine and smooth. The smoothest deposit is obtained by using a very low current, so that electrolysis proceeds very slowly. The choice of electrolyte also has an effect on the smoothness of the deposit. For copper- and nickel-plating, solutions of copper(II) sulphate and nickel sulphate are used, but for silver- and gold-plating, solutions of complex silver and gold cyanides must be used as electrolytes.

Experiment 5.10 Plating with nickel

Set up the circuit shown in Figure 5.13, with a piece of nickel foil as anode and a piece of copper foil as cathode (the object to be plated). The copper foil should first be scrubbed clean with steel wool moistened with sodium hydroxide solution, and then rinsed under the tap before connecting it into the circuit.

Switch on the current and electrolyse for 3 minutes. Then turn the cathode round the other way and electrolyse for another 3 minutes.

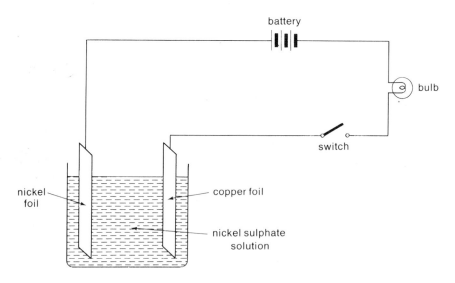

Figure 5.13

Can you explain what is happening during the passage of the current? Why is the cathode turned round half-way through?

What do you think would happen if you changed over the anode and cathode and passed the current again? If you have time you could try this and see if you were right.

Questions

1 Which of the following are examples of electrolysis, and which are not?
   (a) Producing light by passing electricity through a gas.
   (b) Producing hydrogen by passing electricity through water.
   (c) Electric welding.
   (d) Nickel plating.
   (e) Electric arc lamp. *(East Anglia)*

2 Explain the process known as electrolysis. Describe, with the aid of diagrams, the application of electrolysis to electroplating.

3 Explain what happens at the anode and what happens at the cathode during the electrolysis of:
   (a) copper(II) chloride solution using carbon electrodes
   (b) copper(II) chloride solution using copper electrodes.

.4 How ions are formed

A sodium atom has 11 protons in its nucleus and 11 electrons moving at very high speeds outside the nucleus. It is electrically neutral. A sodium ion is formed by the atom losing one of its electrons. It then has 11 protons and 10 electrons, giving it a surplus of one positive charge:

$$Na \rightarrow Na^+ + e.$$

A magnesium atom loses two electrons to form a magnesium ion:

$$Mg \rightarrow Mg^{2+} + 2e.$$

Cations are all formed as a result of atoms losing 1, 2, or 3 electrons.

A chlorine atom has 17 protons and 17 electrons. It forms an ion by gaining an extra electron. It then has 17 protons and 18 electrons, a surplus of one negative charge:

$$Cl + e \rightarrow Cl^-.$$

An oxygen atom gains two electrons to form an oxygen ion:

$$O + 2e \rightarrow O^{2-}.$$

Anions are formed as a result of atoms gaining 1, 2, or 3 electrons.

Ions and the periodic classification

A sodium atom *always* loses *one* electron, so that sodium ions always have one positive charge (Na^+). Sodium atoms never lose more than one electron to give an ion such as Na^{2+} or Na^{3+}, nor do they ever gain electrons to become anions such as Na^- or Na^{2-}. Chlorine atoms, on the other hand, always gain one electron each to become chlorine ions (Cl^-). Most atoms gain or lose a fixed number of electrons and can form only one type of ion, though atoms of some of the transition elements can form more than one type of ion (for example, Cu^+ and Cu^{2+}).

A theory to explain why atoms form particular types of ions was put forward by an American called Irving Langmuir in 1919. He suggested that the electrons in an atom are grouped together in sets, called SHELLS, and that complete shells are very stable. The noble gases (Group O of the periodic classification) never form ions, and Langmuir thought this must be because their electrons are all grouped in complete shells. The atomic numbers of the noble gases (the number of electrons in their atoms) are shown in Figure 5.14. *This suggests that complete shells consist of 2, 8, and 18 electrons.*

	Atomic number	Size of complete electron shells
Helium	2	2
Neon	10	2 and 8
Argon	18	2, 8, and 8
Krypton	36	2, 8, 18, and 8
Xenon	54	2, 8, 18, 18, and 8

Figure 5.14

I		II	III	IV
Hydrogen 1				
Lithium (2)1	Beryllium (2)2	Boron (2)3	Carbon (2)4	
Sodium (2)(8)1	Magnesium (2)(8)2	Aluminium (2)(8)3	Silicon (2)(8)4	
Potassium (2)(8)(8)1	Calcium (2)(8)(8)2	Gallium (2)(8)(18)3	Germanium (2)(8)(18)4	

Figure 5.15

The lithium atom (atomic number 3) has 3 electrons, a complete shell of 2, and 1 extra. The sodium atom (atomic number 11) has complete shells of 2 and 8 electrons, and 1 extra. The potassium atom (atomic number 19) has a complete shell of 2, two complete shells of 8, and 1 extra. In fact all of the Group I elements (the alkali metals) have one extra electron apart from complete shells. In forming ions this one extra electron is always lost, leaving complete shells only. All the alkali metals always form ions with one positive charge (Li^+, Na^+, K^+, Rb^+, Cs^+).

Figure 5.15 shows part of the periodic classification of elements with the groupings of electrons marked. Complete shells of electrons are shown in brackets.

The elements of Group II all have two electrons which are not part of a complete shell. They therefore all form ions with two positive charges by losing these two electrons (Be^{2+}, Mg^{2+}, Ca^{2+}, Sr^{2+}, Ba^{2+}). Similarly, the elements of Group III form ions with three positive charges (such as Al^{3+}). It is these 1, 2, or 3 easily-lost electrons which make it possible for metal elements to conduct electricity (page 113).

The elements of Group VII (the halogens) all have 7 electrons which are not part of complete shells. Rather than lose all of these 7 odd electrons, they find it much easier to gain one electron to complete another shell. They therefore form anions with one negative charge, for example F^- (with complete shells of 2 and 8 electrons), and Cl^- (with one complete shell of 2 electrons and two complete shells of 8).

In the same way the elements of Group VI form ions like O^{2-} and S^{2-} by gaining two electrons to complete another shell.

The transition elements are not shown in Figure 5.15. The arrangement of their electrons is rather more complicated, so that they can often form more than one type of ion (for example, copper can form the ions Cu^+ and Cu^{2+}, and iron can form the ions Fe^{2+} and Fe^{3+}).

				O
V	VI	VII		Helium (2)
Nitrogen (2)5	Oxygen (2)6	Fluorine (2)7		Neon (2)(8)
Phosphorus (2)(8)5	Sulphur (2)(8)6	Chlorine (2)(8)7		Argon (2)(8)(8)
Arsenic (2)(8)(18)5	Selenium (2)(8)(18)6	Bromine (2)(8)(18)7		Krypton (2)(8)(18)(8)

Atoms joining together

The formation of ionic compounds

Sodium atoms each need to lose one electron to form ions. Chlorine atoms each need to gain one electron. If sodium and chlorine atoms come into contact with one another a chemical reaction can occur in which one electron from each sodium atom is transferred to a chlorine atom. The compound sodium chloride, composed of sodium ions and chlorine ions, is formed:

$$Na(2, 8, 1) + Cl(2, 8, 7) \rightarrow Na^+(2, 8) + Cl^-(2, 8, 8).$$

(The grouping of the electrons is shown in brackets.)

Similarly, magnesium atoms can each give two electrons to oxygen atoms, forming magnesium oxide:

$$Mg(2,8,2) + O(2,6) \rightarrow Mg^{2+}(2,8) + O^{2-}(2,8).$$

Compounds which are composed of ions, formed by the transfer of electrons from one atom to another, are called ionic compounds or ELECTROVALENT COMPOUNDS.

The properties of electrovalent compounds

1 They are electrolytes, since they are made up of ions.
2 They have high melting points and boiling points. For instance, sodium chloride melts at 808°C and boils at 1465°C, and magnesium oxide melts at 2900°C and boils at 3600°C. A high temperature is needed to make the ions vibrate sufficiently vigorously to overcome the strong forces of electrical attraction holding them together. A crystal of an electrovalent compound is a giant structure of millions of positive and negative ions arranged in a regular pattern. Figure 5.9, page 117, is a diagram of a sodium chloride crystal. It has a cubic shape with alternating sodium ions and chlorine ions. Each sodium ion is held in place by the attraction of the six chlorine atoms surrounding it, and each chlorine ion is attracted by six sodium ions, with the result that a very strong network of forces holds the crystal together.
3 Many electrovalent compounds are soluble in water.

Questions

1 *Copy and fill in the gaps.*
   (a) When a chlorine atom gains an electron it becomes a particle called a _____ _____.
   (b) When a sodium atom loses an electron, it is converted into a _____ _____. *(London)*

2 Which of the following is the reason why an atom and its ion have almost the same mass?

A   There is no significant difference between an atom and its ion.
B   Electrons are almost without mass.
C   The nucleus of an atom is unaffected when atoms become ions.
D   Atoms and ions are too small to be seen.
E   Most atoms have isotopes.   (North West)

3 A certain atom normally loses one electron when it reacts. Which of the following will be the particle formed?
A   An ion with one negative charge.
B   An ion with one positive charge.
C   An isotope of the original atom.
D   An atom of a noble gas.
E   An atom of a gaseous element.   (North West)

4 Arrange the following elements in order of increasing number of outer electrons present in their atoms: carbon, sodium, neon, chlorine.   (East Anglia)

5 Here are some structures of atoms and ions:
A   11 protons, 12 neutrons, 11 electrons;
B   10 protons, 10 neutrons, 10 electrons;
C   17 protons, 18 neutrons, 18 electrons;
D   12 protons, 12 neutrons, 10 electrons;
E   16 protons, 16 neutrons, 16 electrons.
Which of A, B, C, D and E best fits each of the following:
   (a) A metal atom;
   (b) a positive ion;
   (c) a negative ion;
   (d) an inert gas atom?   (West Midlands)

6 (a) Complete Figure 5.16 concerning the structure of the atoms of elements for which the symbols for one mole are Na and Cl.

Symbol	Number of protons	Number and arrangement of electrons
Na		2, 8, 1,
Cl	17	

Figure 5.16

(b) The compound with the formula NaCl is an ionic crystalline substance. Copy and complete Figure 5.17 concerning the IONS formed from the atoms of the elements for which the symbols for one mole are Na and Cl.

Symbol for ion	Number of protons	Number and arrangement of electrons
		2, 8
Cl⁻		

Figure 5.17 (*West Midlands*)

7 Explain the transfers of electrons involved in forming the following electrovalent compounds from their elements: lithium fluoride, calcium oxide, sodium oxide, and aluminium fluoride.

8 Which of the following is true about most ionic substances?
A They are only slightly soluble in water.
B They are easily vaporised.
C They exist as liquids.
D They have high melting points.
E They are poor electrolytes. (*North West*)

9 How do you account for the fact that sodium chloride is a high melting point crystalline solid which is difficult to decompose by heating?

10 What can you say about the properties of the electrovalent compound lithium fluoride?

5.5 Non-electrolytes

Carbon's six electrons are arranged in a complete shell of 2, with 4 extra. An atom like this is in a difficult position. If it lost four electrons, leaving it with a complete shell of 2, it would form an ion with four positive charges (C^{4+}). If it gained four electrons to give complete shells of 2 and 8, it would form an ion with four negative charges (C^{4-}). A charge of 4+ or 4− is too large to be carried comfortably by a small atom. In fact it is very unusual for an atom to carry more than two charges, and charges of four or more are never found.

Elements like carbon form compounds by *sharing* electrons with other elements. In methane (CH_4) each carbon atom shares its 4 extra electrons with the single electrons from four hydrogen atoms. Figure 5.18 shows the way in which the electrons are distributed in a methane molecule.

Each of the shaded areas represents the space within which two electrons are moving. Each hydrogen nucleus therefore has a share in 2 electrons (a complete shell), and the carbon atom has a share in 8 electrons (also a complete shell).

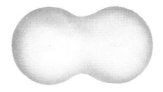

Figure 5.18 (left) The distribution of electrons in a methane molecule
Figure 5.19 (above) The distribution of electrons in a hydrogen molecule

Compounds which are formed by sharing electrons are called COVALENT COMPOUNDS. Some other examples of covalent compounds are water (H_2O), ammonia (NH_3), and carbon tetrachloride (CCl_4). The gaseous elements (hydrogen, oxygen, nitrogen, and the halogens) also form molecules by electron-sharing. In hydrogen gas, for example, there are no separate atoms because the hydrogen atom has only one electron, an incomplete shell, and is therefore unstable. Instead, hydrogen atoms are joined in pairs (Figure 5.19) so that each hydrogen nucleus has a share in 2 electrons, forming a complete shell.

The properties of covalent compounds

1 They are non-electrolytes, since they are not made up of ions.
2 They have low melting points and boiling points: most are gases or liquids at room temperature. For example, methane melts at $-182°C$ and boils at $-161°C$, and carbon tetrachloride melts at $-23°C$ and boils at 77°C. Although the atoms in each individual molecule are held together strongly by the attractive forces between the electrons and the nuclei, there are no strong forces holding one molecule to another: very little heat energy is needed to separate the molecules from one another.
3 Covalent compounds are mostly insoluble in water.

Experiment 5.11 Finding out whether compounds are electrovalent or covalent

You will be provided with a number of compounds without being told their names. Carry out the following experiments on each one in turn and from the results of your experiments decide whether each compound is electrovalent or covalent.

(a) Heat a spatula-full of the substance in a small test tube. Is it easy to melt, difficult to melt, or is it impossible to melt with a bunsen?
(b) Shake a spatula-full of the substance with half a test tube of water. Does it dissolve completely, dissolve partly, or not dissolve at all?

(c) If the substance does dissolve, test the solution with the apparatus shown in Figure 5.1, page 110, to see if it conducts electricity. Does it conduct well, slightly, or not at all?

Questions

1 (a) The elements sodium and carbon are both good conductors of electricity. The compounds of these two elements with chlorine are sodium chloride and carbon tetrachloride. How would you account for molten sodium chloride being a good conductor and carbon tetrachloride being a bad conductor?

(b) How do you account for sodium chloride being a solid with a high melting point and carbon tetrachloride being a liquid with a low boiling point? (London)

2 Consider the following series of elements in the periodic classification (the numbers are the atomic numbers of the elements):

$_{11}$Na $_{12}$Mg $_{13}$Al $_{14}$Si $_{15}$P $_{16}$S $_{17}$Cl $_{18}$Ar.

Explain simply why (a) sodium forms an ionic compound with chlorine, (b) phosphorus forms a covalent compound with chlorine, (c) argon is almost completely unreactive. (North West)

3 Explain why neon and argon gases are made up of separate atoms whereas hydrogen and chlorine are made up of two-atom molecules.

4 Figure 5.20 gives some information about some common substances.

Substance	Melting point °C	Boiling point °C	Electrical conductance of solid substance	Electrical conductance of liquid substance	Electrical conductance of solution of substance in water
P	−25	144	Poor	Poor	Insoluble
Q	−51	−35	Poor	Poor	Good
R	730	1 380	Poor	Good	Good
S	1 455	2 835	Good	Good	Insoluble
T	651	1 300	Poor	Good	Good
U	0	100	Poor	Poor	—
V	961	2 193	Good	Good	Insoluble
W	1 083	2 582	Good	Good	Insoluble
X	712	1 412	Poor	Good	Good

Figure 5.20

(a) Which of these substances are liquids at room temperature?
(b) Which of these substances is a gas at room temperature?
(c) Which of these substances has a structure made of ions?
(d) Which of these substances has a structure made up of covalent molecules?
(e) Which of these substances could be metals? (*London*)

For the teacher

Experiments 5.1, 5.2, and 5.3 6 volt DC supply. M solutions for experiment 5.3.

Experiment 5.4 20 cm carbon electrodes and electrode holders are available from most suppliers.

Experiment 5.5 Electrolysis cells are available ready-made from some suppliers but may be cheaply made from 75 mm lengths of 27–29 mm (outside diameter) glass tubing closed at one end with a rubber bung carrying two carbon electrodes. The cell is conveniently held by a spring clip on a small wooden stand. Gases are collected in 75 × 10 mm rimless test tubes, supported by a rubber band attached to a small piece of wood resting across the top of the cell. M solutions.

Determination of the charge on a mole of lead ions Connect a 12 volt DC supply in series with a switch, an ammeter (5 amp), and a 10 ohm rheostat (e.g. a wire-wound potentiometer from Radiospares). Lead(II) bromide can be melted in a crystallising dish or 250 cm^3 beaker to a depth of about $\frac{1}{2}$ cm. The carbon electrodes should be placed as far apart as possible to minimise recombination of lead and bromine.

Place the electrodes in position in the melt and allow time for remelting around them before switching on the current. Maintain the current at a fixed value between 2 and 4 amps and electrolyse for 10 minutes. Switch off, remove the electrodes, and while the electrolyte is still molten decant it carefully into another container, leaving the bead of lead behind. When the lead has solidified it can be prised away from the glass and residual lead(II) bromide can be scraped off before weighing.

Experiment 5.6 A supply of 20–25 volts DC is necessary to obtain a reasonable movement of the purple boundary within 10 minutes.

Many more sophisticated demonstrations of ionic movement have been devised. One of the best uses copper(II) chromate(VI), both of whose ions are coloured. One-third-fill a U-tube (at least 125 × 15 mm) with 2 M hydrochloric acid. Dissolve 15 g of copper(II) chromate(VI) in the minimum quantity of 2 M hydrochloric acid and then dissolve as much carbamide (urea) as possible in this solution to increase its density. By means of a pipette run this solution into the U-tube below the hydrochloric acid to form a separate layer. Fit carbon electrodes

Atoms joining together

into each limb of the U-tube and connect them to a 20–25 volt DC supply. After 10 minutes movement of the blue/green copper(II) and orange dichromate(VI) ions will be visible.

Experiment 5.7 5×3 cm copper foil electrodes can be supported in a 100 cm^3 beaker with an electrode holder consisting of a piece of wood, about $70 \times 25 \times 13$ mm, with a crocodile clip screwed on each side. 6 volt DC supply and 0–0.5 amp meter. The rheostat should be 100 ohms (e.g. wire-wound potentiometers from Radiospares). The electrolyte is 0.05 M copper(II) sulphate.

Experiment 5.8 As for experiment 5.7, but silver foil electrodes and 0.05 M silver nitrate. It is extremely difficult to obtain a cohesive cathode deposit, but once the cathode–anode weight equivalence has been established in experiment 5.7 only the loss in weight of the anode need be measured.

Experiment 5.9 As for experiment 5.7, but electrolyte 1.5 M with respect to sodium chloride and 0.02 M with respect to sodium hydroxide.

Experiment 5.10 0.125 M ammonium nickel sulphate as electrolyte. With a 6 volt DC supply and a 6.5 volt 0.3 amp bulb (Radiospares) the current density will be adequate for plating. Electrode foil holder as for experiment 5.7.

For zinc plating a suitable electrolyte is 0.2 M zinc sulphate with 10 g boric acid and 10 cm^3 M sulphuric acid per litre.

Structural models Ball-and-spoke models are available from Gallenkamp. A $4 \times 4 \times 4$ ion sodium chloride model (*vide* Figure 5.9, p. 117) can be constructed from 32 yellow (for sodium) and 32 green (for chlorine) spheres, each with 6 holes, and 144 6 cm springs. Other models, such as methane, ammonia, water, and the diatomic gases, can also be constructed from these models, or from polystyrene spheres (each pack of Griffin and George's *Grifzote* spheres is supplied with a booklet giving full instructions for making this type of model).

Experiment 5.11 Substances could include acetamide, lead chloride, naphthalene, potassium bromide, sodium carbonate, sodium chloride, starch, and sucrose.

6 Metals

More than three-quarters of the elements are metals, about 80 in all. Some, like iron, are very common; others, like platinum, very rare. Some, gold and copper for example, have been used by man since ancient times. Others, such as titanium and germanium, have only become important in the last 25 years.

Only about two-dozen metals are widely used, but it is hard to imagine life without them. At home you will find aluminium pots and pans, chromium-plated taps, germanium transistors in the television set, lead and copper pipes, tin-plated cans of food, tungsten filaments in light bulbs, a dustbin or water tank of iron galvanised with zinc, and perhaps a mercury-filled thermometer, a silver teaspoon, and a piece of gold jewellery. A whole range of items from bed-springs to cutlery are made of different types of steel, which is iron mixed with carbon, manganese, chromium, nickel, molybdenum, or vanadium.

Metals are not usually found as free elements in the earth, but are combined chemically with other elements. Before they are available for use a great deal of work is necessary. Geologists find the metal-bearing rocks, engineers mine them, and chemists extract the metals from them. Even when the pure metal elements have been won from the earth they are not always very useful. Pure iron, for instance, is not very strong and rusts very easily. Mixed with small quantities of other elements such as carbon and manganese it becomes steel, which is both stronger and more resistant to corrosion. A mixture of metals is called an ALLOY.

6.1 The chemical properties of metals

Experiment 6.1 Investigation of the reaction of metals with oxygen in air

You will be heating a number of metals in air to find out if they react with the oxygen and, if they do, how vigorously. After trying each metal, describe what happens.

(a) **Calcium** Using tongs, hold a piece of calcium in the hottest part of a fierce bunsen flame.

(b) **Copper** Using tongs, hold a piece of copper foil in the bunsen flame.

(c) **Iron** Using tongs, hold a tuft of iron wool in the bunsen flame.

(d) **Magnesium** Using tongs, hold a piece of magnesium ribbon in a fierce bunsen flame.

(e) **Zinc** Lay a piece of asbestos paper on a wire gauze standing on a tripod. Put a spatula-full of powdered zinc on the asbestos paper. Heat the zinc strongly from underneath and stir it gently with the spatula while it is being heated.

What is the order of reactivity of these metals towards oxygen, from the most reactive to the least reactive?

Experiment 6.2 Investigation of the reaction of metals with water

(a) Put a small piece of copper in a test tube and add a third of a tube of water. Is there any sign of a reaction? Repeat with small pieces of calcium, iron, magnesium, lead, and zinc. Which metals react with water and which do not?

(b) Heating often helps a reaction to go. Set up the apparatus shown in Figure 6.1 to find out if magnesium will react with steam. Heat the centre of the tube with a fierce bunsen flame until the magnesium starts to burn. Then very quickly move the flame to the bottom of the tube to boil the water. At the same time hold a lighted splint at the mouth of the glass tube.

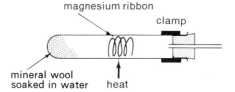

Figure 6.1

What happens at the mouth of the glass tube? What is it that is burning here? What has happened to the magnesium?

(c) Repeat experiment (b) with a strip of copper foil in place of the magnesium. Does copper react with steam?

(d) Repeat experiment (b) with a piece of zinc in place of magnesium. Does zinc react with steam?

(e) You may have seen the reactions of sodium and potassium with water. How do sodium and potassium compare with calcium in their reactivity?

What is the order of reactivity of the metals calcium, copper, magnesium, potassium, sodium, and zinc with water, from the most reactive to the least reactive?

Experiment 6.3 Investigation of the reaction of metals with dilute acids

(a) Arrange 6 test tubes in a rack. Place one piece of the metals calcium, copper, iron, lead, magnesium, and zinc in each tube (one metal to each tube). Label the tubes so that you will remember which is which. Now, as quickly as possible, one-third-fill each tube with dilute hydrochloric acid.

Judge which of the metals reacts fastest, which second fastest, and so on. Make a list of the 6 metals in order from the most reactive to the least reactive.

(b) Repeat the experiment with dilute sulphuric acid instead of dilute hydrochloric acid. What is the order of reactivity this time?

Experiment 6.4 Testing predictions about which metal will win in competitions for oxygen

(a) Judging from what you have learnt from experiments 6.1, 6.2, and 6.3 about the relative reactivity of zinc and lead, do you think there will be a reaction if zinc is heated with lead oxide? Will the zinc win the oxygen away from the lead, in which case a reaction will be seen, or will the zinc be unable to win the oxygen, in which case no reaction will be seen?

Test your prediction by trying it out. Mix together a spatula-full of powdered zinc with a spatula-full of lead oxide. Lay a piece of asbestos paper on a wire gauze standing on a tripod. Put the mixture of zinc and lead oxide on the asbestos paper and heat it from underneath with a fierce bunsen flame.

Are there any signs of a reaction taking place? If there are, write a word equation for the reaction.

(b) You are provided with the following metals and their oxides:

copper	copper oxide
iron	iron oxide
lead	lead oxide
zinc	zinc oxide.

Choose *three* pairs of a metal and the oxide of another metal. Choose one pair which you think should react vigorously together, one pair which you think should react together only slightly, and one pair which you think should not react at all. Test your predictions by heating each of the pairs separately, using the same method as in experiment (a).

Experiment 6.5 Investigating how easily compounds of metals are decomposed

(a) Put one spatula-full of sodium carbonate in a small test tube.

Using a test-tube holder heat the tube in a fierce bunsen flame for about two minutes. Does the sodium carbonate seem to have decomposed?

Repeat the experiment with copper carbonate instead of sodium carbonate. Is there any sign of decomposition this time? What do you think the substance left in the tube is?

(b) Try the same experiment again, first with sodium nitrate and then with copper nitrate. Hold a glowing splint near the mouth of each tube while you are heating it.

Describe what you noticed while the two nitrates were being heated. Which one decomposed more easily?

Which are more stable towards heat: compounds of very reactive metals or compounds of very unreactive metals?

The activity series

Figure 6.2 is a summary of the results of experiments to investigate the reactivity of metals towards substances such as oxygen, water, and dilute acids.

Some metals are much more reactive than others. Metals that are reactive to one substance are reactive to others also, and those which are unreactive to one substance are unreactive to all. The list of metals in Figure 6.2, from the most reactive at the top in order of decreasing reactivity to the least reactive at the bottom, is known as the ACTIVITY SERIES. The activity series is a great help in understanding the chemistry of metals.

1 If two metals are competing for oxygen, the series will tell us which one should win. For example, if zinc is heated with iron oxide a reaction occurs and the more active metal takes the oxygen:

 zinc + iron oxide → zinc oxide + iron.

If magnesium is heated with calcium oxide there is no reaction because the less active metal cannot take the oxygen:

 magnesium + calcium oxide → no reaction.

2 Metals high in the list are more easily corroded than those lower down. Potassium and sodium are so reactive that they have to be stored under oil to protect them from the atmosphere. Copper, silver, and gold are so resistant to corrosion that they have been used for thousands of years for coins and ornaments.

3 Metals at the top of the activity series react to form compounds more easily than those at the bottom. On the other hand, compounds of metals at the top of the list, once formed, are not decomposed as easily as compounds of metals at the bottom. The position of a metal in the activity series therefore gives us some idea of how easy or how difficult it will be to extract it from its compounds which are found in the earth.

Metal	Reaction with oxygen (in air)	Reaction with water	Reaction with dilute acid
Potassium	■	■ React with cold water to form hydrogen and the metal hydroxide	■
Sodium	■	■	■
Calcium	■ Burn in air to form the metal oxide	■	■
Magnesium	■	■ Do not react with cold water. React with steam to form hydrogen and the metal oxide	■ React to form hydrogen and a metal salt
Aluminium	■	■	■
Zinc	■	■	■
Iron	■	■	■
Tin	■ Do not burn. React slowly to form the metal oxide	■	■
Lead	■	■	■
Copper	■	■	■
Mercury	■	■ Do not react with water or steam	■
Silver	■ Do not combine easily with oxygen at all	■	■ Do not react
Gold	■	■	■
Platinum	■	■	■

Figure 6.2

Questions

1 Elements have different affinities for one another. A list of some elements with DECREASING affinity for oxygen is: sodium, calcium, magnesium, zinc, iron, lead, hydrogen, copper, gold.

    (a) (i) Which element will react most readily with oxygen?

        (ii) Which element is most likely to be found in the native (uncombined) state?

    (b) (i) What would you expect to happen if magnesium was heated with copper oxide?

        (ii) Write a word equation for this reaction.

    (c) The above series of elements can be regarded as an activity series. Use this idea to explain why

        (i) iron will liberate hydrogen from dilute hydrochloric or sulphuric acids whilst copper will not;

(ii) zinc will displace lead from lead nitrate solution whilst gold will not. *(North West)*

2 (a) Figure 6.3 represents an apparatus which may be used to study the reaction of magnesium with steam.

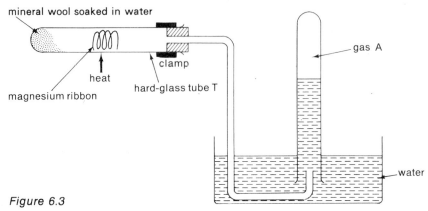

Figure 6.3

(i) Name the gas A.
(ii) How would you test for this gas?
(iii) Why would the first test tube of gas collected be unsuitable for testing?
(iv) Name the compound of magnesium left in tube T after the reaction.
(v) What would happen if the apparatus was left set up as in the diagram after the heating process was over?
(b) (i) Name a metal which will react more readily with water than will magnesium.
(ii) Name a metal which will react less readily with water than will magnesium. *(North West)*

3 The following is a series of metal elements placed in order of reactivity. X and Y are imaginary elements (you are *not* asked to suggest what elements they might be):

K (most reactive), X, Ca, Mg, Zn, Y, Fe, Cu (least reactive).

Answer the following questions using your knowledge of the chemistry of the metals listed.

(i) Would you expect metal X to liberate hydrogen from cold water? Give a reason for your answer.
(ii) Would you expect metal Y to liberate hydrogen from cold water? Give a reason for your answer.
(iii) Would metal Y liberate hydrogen from dilute hydrochloric acid? Give a reason for your answer.
(iv) Explain what you would expect to see if a piece of Y was

dipped into a solution of copper nitrate for a few minutes and then withdrawn and examined. *(West Midlands)*

4 (a) When potassium metal is placed in water, hydrogen is evolved which spontaneously bursts into flame. With sodium the hydrogen does not burn spontaneously unless the movement of the sodium is restricted.

(i) What does this tell you about the relative reactivities of sodium and potassium?
(ii) How can you *safely* restrict the movement of sodium?
(iii) What colour does potassium give to the hydrogen flame?
(iv) What colour does sodium give to the hydrogen flame?

(b) Rubidium is one below potassium in Group I of the periodic classification. From your studies of the periodic classification and of the activity series answer the following questions.

(i) What do you think would happen to a piece of rubidium placed in water?
(ii) Would you expect rubidium chloride to be soluble or insoluble in water? State your reason. *(London)*

5 A metal X is displaced from a solution of one of its salts by a metal Y, and a metal Z displaces Y from a solution of one of its salts. The activity of the metals decreases in which of the following orders?
A   X, Y, Z.
B   Y, X, Z.
C   Z, X, Y.
D   Y, Z, X.
E   Z, Y, X.   *(North West)*

6 Explain why a metal such as copper can occur uncombined, whereas a metal such as sodium never occurs in the free state. Give *one* other example of each type. *(North West)*

7 The following is a list of some metals arranged in an order of activity—the most reactive being the first.

    Sodium, aluminium, iron, tin, lead, copper, mercury, gold.

Use this list to answer the following questions.
(a) Which statements are correct?
(i) Tin reduces aluminium oxide.
(ii) Copper does not react with hydrochloric acid therefore mercury will not.
(iii) Mercury appears native, so does gold.
(b) (i) Name a metal in the list which would displace lead from lead nitrate solution.
(ii) Where would magnesium fit in the above list?
(West Yorkshire)

8 (a) Study the results in Figure 6.4 of the reactions of six metals towards water. Place them in their order of chemical reactivity (most active first).

Metal	Reaction
Calcium	Hydrogen evolved slowly in cold water.
Copper	Hydrogen not evolved at red heat with steam.
Iron	Hydrogen evolved at red heat with steam.
Magnesium	Hydrogen evolved when heated with steam.
Potassium	Hydrogen evolved rapidly in cold water and ignites spontaneously.
Sodium	Hydrogen evolved rapidly in cold water.

Figure 6.4

(b) The Bronze Age came before the Iron Age and gold was known in ancient cultures. From your study of the activity series explain why they occurred in this order.

(c) Titanium is a metal of great importance in space probes. It is the eighth most abundant element in the earth's crust. Why, in our scientific age, do you think that this metal has not been more fully used? *(London)*

6.2 The occurrence of metals

When the earth was being formed the elements had plenty of opportunity to combine with one another. The form in which they are now found depends on their reactivity (their position in the activity series). Figure 6.5 shows the commonest important ore of the metals you are studying. Ore is the geologist's name for a type of rock from which a metal can be extracted.

The four most reactive metals are extracted from the sea (in the case of calcium, sea creatures have converted calcium compounds into calcium carbonate for their shells and over millions of years have provided us with deposits of *limestone, chalk,* and *marble*). The metals of medium reactivity are found in the earth in the form of their oxides and sulphides. The five least reactive metals are found NATIVE (as the free uncombined elements). The metals in Figure 6.5 are arranged in the order of the activity series. The position of a metal in the series is a guide to the form in which it is found in the earth.

Some metals are much more plentiful than others. The commonest is aluminium, which makes up $7\frac{1}{2}$ per cent of the earth's crust. Iron is the next most common, making up nearly 5 per cent of the crust. The four most active metals (potassium, sodium, calcium, and magnesium) are all quite

Metal	Main form in nature	Minor forms
Potassium	Sea water (potassium chloride, KCl)	
Sodium	Sea water (sodium chloride, NaCl)	
Calcium	*Limestone* (calcium carbonate, $CaCO_3$)	*Chalk, marble* ($CaCO_3$)
Magnesium	Sea water (magnesium chloride, $MgCl_2$)	*Dolomite* ($MgCO_3.CaCO_3$)
Aluminum	*Bauxite* (aluminium oxide, Al_2O_3)	
Zinc	*Zincblende* (zinc sulphide, ZnS)	*Calamine* ($ZnCO_3$)
Iron	*Haematite* (iron oxide, Fe_2O_3)	*Magnetite* (Fe_3O_4) *Pyrites* (FeS_2)
Tin	*Tinstone* (tin oxide, SnO_2)	
Lead	*Galena* (lead sulphide, PbS)	
Copper	*Copper pyrites* (copper iron sulphide, $CuFeS_2$)	*Malachite* ($CuCO_3$) Native
Mercury	*Cinnabar* (mercury sulphide, HgS)	Native
Silver	Impurity in *galena* (silver sulphide, Ag_2S)	Native
Gold	Native	
Platinum	Native	

Figure 6.5

common, and make up about 10 per cent altogether. All the others together (zinc, tin, lead, copper, silver, mercury, gold, and platinum) make up only $\frac{3}{100}$ per cent of the earth's crust.

When the earth was first formed the compounds of all the elements were evenly spread out. Two main processes, occurring slowly over millions of years, have concentrated the metal ores. While the earth was

Metals

still very hot some of the metal sulphides vaporised. The vapour rose towards the surface and condensed in cool cracks in the upper crust. As a result of this process metal ores are often found in VEINS between layers of rock. Figure 6.6 shows a vein of *copper pyrites*, broken by a FAULT caused by earthquake action after the vein was formed.

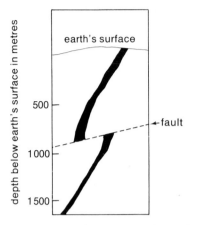

Figure 6.6 A cross-section of the earth's crust at Anaconda, in the state of Montana, USA, which shows a vein of copper pyrites

Later, when the earth had cooled, the concentration process continued with the action of wind and water. Soluble compounds of metals like potassium, sodium, calcium, and magnesium were washed into the sea. Wind and running water gradually broke up rocks containing insoluble metal compounds. The lighter particles of rock were washed away and the denser particles of metal ore were concentrated in pockets.

If the metal compounds had remained evenly spread out the task of extracting them would almost certainly have been impossible. Even with an element as common as iron, 20 tonnes of rock would have had to be processed to obtain one tonne of iron, and in the case of copper one tonne of rock would have produced only about 70 grams of the metal. Fortunately, as a result of the natural processes of concentration, ore deposits with much higher concentrations are found. For example, iron ore with 50 per cent iron or copper ore with 2 per cent copper are quite common.

Questions

1 Which of the following is the best description of an ore?
A   A purified metal.
B   The oxide of a metal.
C   A mixture of rock and metal.
D   A mixture of rock and a metal compound.
E   A mixture of rock and a metal oxide.

2 Some metal ores are very attractive in appearance. Describe two or three metal ores that you have seen.

3 Molybdenum is found as an ore called *molybdenite*, containing molybdenum sulphide. Palladium is found native. Titanium is found as an ore called *rutile*, which contains titanium oxide. What do you think is the order of these three metals in the activity series (give the most reactive first)?

4 Try to find out something about the main ores of chromium, cobalt, manganese, nickel, tungsten, and vanadium.

5 It is estimated that $2\frac{1}{2}$ per cent of the earth's crust is sodium, and $\frac{1}{10000}$ per cent of the crust is molybdenum. Yet in certain parts of the world there are huge deposits of almost 100 per cent pure sodium chloride, and smaller deposits of rock containing as much as 10 per cent of molybdenum sulphide are found. Explain how these concentrated deposits of sodium and molybdenum compounds might have been formed.

3 The extraction of metals

Concentration of the ore

Some ores contain a very high proportion of the metal compound. *Haematite*, for instance, may contain as much as 70 per cent of iron oxide. Ores like this are pure enough to be fed directly into a furnace to extract the metal. But many metal ores are not nearly so rich. *Galena* may contain only 4 per cent of lead sulphide and *copper pyrites* as little as 2 per cent of copper sulphide. The rest of these ores is useless rock. They must be concentrated before the metal can be extracted.

The ore is first broken up into small pieces. This is sometimes done underground in the mine and sometimes after the ore has been brought to the surface. The small pieces are then ground to a fine powder in BALL MILLS (Figure 6.7). The mill is a steel container containing tonnes of loose steel balls which grind the ore to powder as the container rotates.

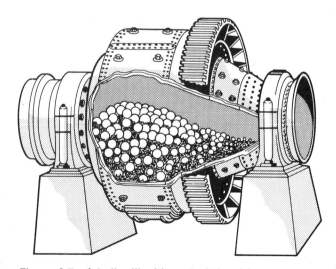

Figure 6.7 A ball mill with part of the side cut away to show the steel balls which crush the ore to powder as the mill rotates

Metals 147

The method most often used nowadays to concentrate the ore is FROTH FLOTATION. The crushed ore is fed into tanks of water containing a chemical frothing agent. Air is blown through so that the whole mixture froths up.

The rock particles become soaked with water and sink to the bottom of the tank, but the metal suphide particles, into which the water cannot soak, are carried to the top of the tank by the air bubbles and can be skimmed off and dried. This ore concentrate contains between 50 per cent and 75 per cent of the metal sulphide.

Roasting of sulphides to oxides

For metals which occur as sulphide ores the next stage in extraction is to convert the metal sulphide into the metal oxide. This process is known as ROASTING. The concentrated sulphide ore is heated in air, which converts the sulphur in the ore to sulphur dioxide gas (which is used in the manufacture of sulphuric acid, page 236). For example, with *galena concentrate*:

lead sulphide + oxygen → lead oxide + sulphur dioxide

$2PbS(s) + 3O_2(g) \rightarrow 2PbO(s) + 2SO_2(g)$.

Roasting is carried out on a large scale in a REVERBERATORY FURNACE (Figure 6.8). The sulphide ore is heated in the hearth by blowing very hot air across the top.

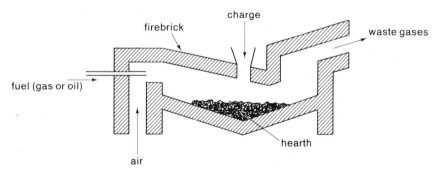

Figure 6.8 *A reverberatory furnace for roasting sulphide ores*

A more modern method is FLASH ROASTING, in which the finely powdered ore concentrate is blown into a furnace together with very hot air. The air can make better contact with the ore particles, and the reaction takes place much more quickly than in the reverberatory furnace. Extra heating is unnecessary because the reaction between the metal sulphide and oxygen, like so many chemical reactions, produces heat, and the heat produced is sufficient to keep the reaction going.

Experiment 6.6 The roasting of a sulphide

Set up the apparatus shown in Figure 6.9. Put one spatula-full of lead sulphide in the tube and position a piece of damp blue litmus paper as shown. Connect the glass tube to a filter pump so that a stream of air is drawn through the apparatus.

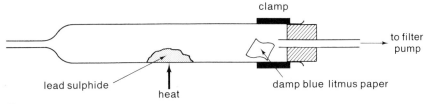

Figure 6.9

Heat the lead sulphide strongly for at least 5 minutes.
   What happens to the litmus paper? What has caused this?
   Describe any changes you observe in the lead sulphide. What will the lead sulphide be changed into if the heating is continued for a long enough time?

Experiment 6.7 Which metal oxides are reduced by town gas?

(a) Set up the apparatus shown in Figure 6.10 with a spatula-full of copper oxide in the middle of the tube.

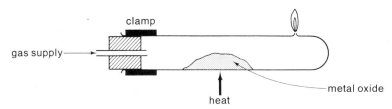

Figure 6.10

Turn on the gas supply to the tube, count slowly up to five, then light the gas at the hole in the tube. Adjust the gas tap to give a flame not more than 3 cm high.
   Heat the copper oxide gently, with a *small* bunsen flame. Describe what you see happening in the tube.
   Has the copper oxide been reduced? Write a word equation to show the reaction that has taken place.
   (b) Repeat the experiment with magnesium oxide in place of copper oxide. Is the magnesium oxide reduced?

Metals 149

Experiment 6.8 Which metal oxides are reduced by carbon?

(a) Mix one spatula-full of lead oxide with two spatulas-full of powdered carbon on a piece of asbestos paper. Lay the paper on a gauze standing on a tripod.

Heat the mixture from above with a fierce bunsen flame. When no further changes appear to be taking place, allow the paper to cool and examine it carefully.

Describe what you see on the asbestos paper. Write a word equation to represent the reaction that has taken place.

(b) Repeat the experiment, using aluminium oxide in place of lead oxide. Has any metallic aluminium been formed?

(c) Mix one spatula-full of iron oxide with two spatulas-full of powdered carbon in a test tube. Holding the tube in tongs, heat it with a fierce bunsen flame for at least two minutes.

Did you notice any change in the appearance of the mixture?

Tip half of the mixture from the tube on to an asbestos square and test it with a magnet. What happens? Why?

When the tube is cool add enough dilute sulphuric acid to the rest of the mixture to half-fill the tube. Examine the tube carefully. What could the gas be that is being given off?

Write a word equation to show what happened when the iron oxide and carbon were heated together.

Reduction of oxides to metals

REDUCTION is the removal of oxygen from a compound. This is the final stage in the extraction process. Some metal oxides, such as lead oxide, can easily be reduced with hydrogen or carbon:

$$\text{lead oxide} + \text{hydrogen} \rightarrow \text{lead} + \text{water}$$
$$PbO(s) + H_2(g) \rightarrow Pb(s) + H_2O(g);$$
$$\text{lead oxide} + \text{carbon} \rightarrow \text{lead} + \text{carbon monoxide}$$
$$PbO(s) + C(s) \rightarrow Pb(s) + CO(g).$$

Others, such as aluminium oxide, cannot be reduced by hydrogen or carbon. The oxides of metals low in the activity series can be reduced but oxides of metals towards the top of the series cannot. If the oxide is heated with hydrogen or carbon using a bunsen burner:

calcium oxide
magnesium oxide
aluminium oxide   } are not reduced to the metal
zinc oxide

iron oxide
tin oxide
lead oxide   } are reduced to the metal.
copper oxide

Whether or not the metals can be obtained by using these reducing agents depends on the temperature used. At higher temperatures than can be obtained with a bunsen burner oxides of metals higher than iron in the activity series can also be reduced. With carbon the temperatures required for reduction are shown in Figure 6.11.

Metal oxide	Temperature required for reduction
Calcium oxide	2 100°C
Magnesium oxide	1 600°C
Aluminium oxide	2 100°C
Zinc oxide	900°C
Iron oxide	700°C
Tin oxide	500°C
Lead oxide	400°C
Copper oxide	100°C

Figure 6.11

These are the minimum temperatures. Considerably higher temperatures are required for rapid reduction.

In industrial furnaces temperatures of at least 1200°C can be reached fairly cheaply. Metals from zinc downwards in the activity series can therefore be extracted by the reduction of their oxides with carbon. For metals above zinc the cost of reaching the enormous temperatures required is too high. Vast quantities of fuel would be needed and furnaces which can withstand very high temperatures are expensive to build. Reduction with hydrogen generally requires even higher temperatures.

Metals above zinc in the activity series are usually obtained on an industrial scale by electrolysis. A compound of the metal is melted and an electric current is passed through it. Although electricity is expensive electrolysis is cheaper than chemical reduction for the extraction of these metals. The method chosen for the extraction of a metal, like so many other properties of metals, depends on its position in the activity series:

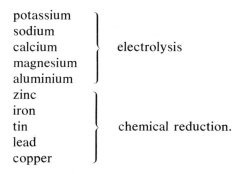

Questions

1 Native copper is found in Canada. What is *native copper*? Copper pyrites (which we can take as Cu_2S) is a source of the metal. The metal can be obtained by roasting the ore in air. Explain what is meant by *roasting*. Write a word equation for the reaction. What important chemical compound could be made from the by-product of the reaction? (London)

2 The following metals are written in order of reactivity, calcium being the most reactive:

        calcium, aluminium, zinc, iron, copper, silver.

(a) Which of the metals would be most easily extracted from its oxide, using hydrogen gas as the reducing agent? Give a reason for your answer.
(b) Explain what is meant by a 'reducing agent'.
(c) What would the hydrogen be converted into when it reduced the metal oxide?
(d) Draw a diagram of a simple laboratory apparatus which you could use to illustrate the reaction described in (a).
(West Midlands)

3 If a little lead oxide is heated on a block of carbon, small silvery globules appear.
(a) Of what substance are the globules composed?
(b) What other substance is formed in this reaction?
(c) What chemical change has the lead oxide undergone?
(d) What does this experiment indicate about the position of lead in the activity series? (Wales)

4 A geologist discovers a vein of *zincblende*, estimated to contain over a million tonnes of the ore. Analysis shows that the ore contains 5 per cent of zinc sulphide.
(a) Outline briefly *all* the stages necessary to win pure zinc from the rock.
(b) Calculate the approximate weight of zinc that might be obtained from the vein. (The formula of zinc sulphide is ZnS. Atomic mass of zinc = 65, atomic mass of sulphur = 32.)

6.4 Reduction of a metal oxide with carbon: the extraction of iron

The most important ores of iron are oxides: *haematite* (Fe_2O_3), and *magnetite* (Fe_3O_4). British iron ore contains between 20 per cent and 50 per cent of iron oxide; the rest of the ore is useless sandy rock. The iron can be extracted directly: concentration is unnecessary.

Reduction of iron oxide with carbon, in the form of *coke*, is carried out in a BLAST FURNACE (Figure 6.12). It gets its name from the blast of hot air which is blown through a number of holes at the base called TUYÈRES (pronounced *tweers*). The furnace is about 30 metres high and 8 metres in diameter, and is made of thick steel plates lined with firebrick. A single furnace can produce up to 5000 tonnes of iron per day.

A mixture of *iron ore, coke,* and *limestone* is fed into the top of the furnace. The exact composition of the CHARGE depends on the quality of the iron ore, but on average the production of 1000 tonnes of iron requires: 2500 tonnes of iron ore, 650 tonnes of coke, and 75 tonnes of limestone.

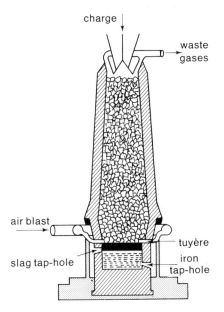

Figure 6.12 A blast furnace

The air which is blown through the *tuyères* is heated to a temperature of 900°C or more. At this temperature the coke burns fiercely, producing carbon dioxide and a great deal of heat:

$$\text{carbon} + \text{oxygen} \rightarrow \text{carbon dioxide} + \text{heat}$$
$$C(s) + O_2(g) \rightarrow CO_2(g).$$

The carbon dioxide is then converted into carbon monoxide:

$$\text{carbon dioxide} + \text{carbon} \rightarrow \text{carbon monoxide}$$
$$CO_2(g) + C(s) \rightarrow 2CO(g).$$

The temperature at the base of the furnace reaches 1800°C and the iron oxide is reduced to iron, partly by the coke and partly by the carbon monoxide:

$$\text{iron oxide} + \text{carbon} \rightarrow \text{iron} + \text{carbon monoxide}$$
$$Fe_2O_3(s) + 3C(s) \rightarrow 2Fe(l) + 3CO(g);$$

and

$$\text{iron oxide} + \text{carbon monoxide} \rightarrow \text{iron} + \text{carbon dioxide}$$
$$Fe_2O_3(s) + 3CO(g) \rightarrow 2Fe(l) + 3CO_2(g).$$

This temperature is well above the temperature at which iron melts, and molten iron trickles down to the bottom of the furnace.

The limestone reacts with the sandy impurities in the iron ore to form SLAG, which is also molten at this high temperature. Molten slag is less dense than molten iron and the two do not mix. They form separate layers at the bottom of the furnace. When the level of slag has almost reached the *tuyères* the two layers are tapped off separately through holes in the firebrick lining. After tapping the holes are plugged up with clay until another batch of iron and slag has collected.

Fresh charge is constantly fed in to keep the furnace full so that the whole process is continuous. Blast furnaces are kept in operation 24 hours a day until the firebrick lining wears out. As much as two million tonnes of iron can be made before the furnace has to be shut down to replace the lining.

The mixture of gases which comes out of the top of the furnace is at a temperature of about 250°C, and consists mainly of nitrogen and carbon dioxide with about 25 per cent of carbon monoxide. These waste gases go to special stoves where the carbon monoxide is burnt:

$$\text{carbon monoxide} + \text{oxygen} \rightarrow \text{carbon dioxide} + \text{heat}$$
$$2CO(g) + O_2(g) \rightarrow 2CO_2(g).$$

The heat of the gases and the heat produced by burning the carbon monoxide is used to generate electricity for light and power in the ironworks and for heating the air blast.

Low-grade British iron ores may produce more slag than iron. Slag is used for making cement, concrete (page 348) and light-weight building materials, as well as for road-making. A great deal, however, cannot be used up and is dumped to produce the slag-heaps which are an all too familiar sight in certain parts of the country.

Iron from the blast furnace is not very pure. It may contain up to 4 per cent of carbon and 2 per cent of silicon, as well as smaller quantities of sulphur, phosphorus, and manganese. It has to be purified to make steel (page 155). In some plants the iron is tapped into huge ladles in which it is transported, still molten, to steelworks on the same site. In other plants the molten iron is poured into moulds called PIGS, in which it solidifies. These moulds give impure iron its familiar name of PIG IRON.

The extraction of other metals by reduction of their oxides

LEAD occurs principally as *galena*, a sulphide ore. The ore is concentrated by froth flotation, roasted to produce the oxide, and the oxide is fed into a blast furnace together with coke and limestone.

ZINC occurs as *zincblende*, also a sulphide ore, which is concentrated and then converted into zinc oxide just like lead ore. Nowadays a blast furnace is usually used to reduce the zinc oxide to zinc. The only difference is that zinc boils at a rather low temperature (908°C). Instead of collecting as a liquid at the bottom of the furnace, like iron and lead, the zinc boils and comes out of the top of the furnace. A spray of molten lead is used to condense the zinc vapour out of the waste gases.

TIN occurs as *tinstone*, an oxide ore. After concentration, the ore is mixed with coke and limestone and heated in a reverberatory furnace (Figure 6.8, page 148) to reduce the tin oxide to tin.

COPPER occurs as *copper pyrites*, an ore containing copper and iron sulphides. After concentration by froth flotation, the ore is roasted with limestone in a reverberatory furnace. A molten mixture of copper and iron sulphides called MATTE is produced, and the impurities form a slag. The matte is transferred to a converter rather similar to a Bessemer converter (page 156), in which air is blown through the molten mass. The more active metal, iron, is converted to its oxide which floats to the top and can be skimmed off. The sulphur is burned out as sulphur dioxide. Copper is so low in the activity series that at the temperature of the furnace the copper sulphide is decomposed directly to copper. Copper from the converter is not very pure and usually requires purification by electrolysis before use (page 162).

6.5 Iron into steel

Pig iron from the blast furnace contains up to 4 per cent of carbon, 2 per cent of silicon, and smaller quantities of other elements, especially phosphorus, sulphur, and manganese. The main difference between pig iron and steel is that steel is more pure. The carbon content must be reduced to between $\frac{1}{10}$ per cent and $1\frac{1}{2}$ per cent, and the concentration of the other impurities must also be cut down to produce high-quality steel.

The impurities are removed by OXIDATION. Oxygen is supplied either from the air or from iron oxide. Carbon and sulphur are converted into carbon monoxide and sulphur dioxide, which escape in the waste gases from the furnace:

$$\text{carbon} + \text{oxygen} \rightarrow \text{carbon monoxide}$$
$$2C(s) + O_2(g) \rightarrow 2CO(g);$$

and
$$\text{sulphur} + \text{oxygen} \rightarrow \text{sulphur dioxide}$$
$$S(s) + O_2(g) \rightarrow SO_2(g).$$

Phosphorus and silicon are oxidised to phosphorus pentoxide and silicon dioxide:

$$\text{phosphorus} + \text{oxygen} \rightarrow \text{phosphorus pentoxide}$$
$$4P(s) + 5O_2(g) \rightarrow 2P_2O_5(s);$$

and
$$\text{silicon} + \text{oxygen} \rightarrow \text{silicon dioxide}$$
$$Si(s) + O_2(g) \rightarrow SiO_2(s).$$

Calcium oxide (*lime*) is added to react with these solid oxides, forming calcium phosphate and calcium silicate, which float to the surface of the iron as slag:

$$\text{calcium oxide} + \text{phosphorus oxide} \rightarrow \text{calcium phosphate}$$
$$3CaO(s) + P_2O_5(s) \rightarrow Ca_3(PO_4)_2(l);$$

and calcium oxide + silicon oxide → calcium silicate
$$CaO(s) + SiO_2(s) \rightarrow CaSiO_3(l).$$

There are several different types of furnace in which iron is converted into steel.

The Bessemer converter

This is the oldest method, invented in 1856 by Sir Henry Bessemer, and was the first to be used for the large-scale production of steel. Fifteen per cent of British steel is still made in the Bessemer or similar converters. The converter is a large brick-lined steel vessel, mounted on bearings about which it can be rotated (Figure 6.13).

The mouth of the converter is tipped sideways and it is charged with up to 50 tonnes of molten pig iron, together with some lime. The converter is then turned into the upright position and a blast of hot air is blown through the iron from underneath. For a few minutes nothing is seen. Then, as the reaction between the carbon in the iron and the oxygen in the air gets under way, great heat is produced and a huge sheet of flame leaps from the mouth of the converter. The BLOW lasts for about 20 minutes, by which time all the carbon and silicon have been oxidised. The air blast is continued for a few more minutes and brown smoke pours out as the sulphur and phosphorus are oxidised out of the iron. The foreman stops this AFTER-BLOW when he judges purification is complete. This process removes all the carbon from the pig iron, so a calculated quantity of ferro-manganese (a mixture of iron, manganese, and carbon) is added to produce the type of steel required. The converter is then tipped sideways to pour the molten steel into a ladle in which it is transported for casting into ingots.

Figure 6.13 A blow from a Bessemer converter

The Siemens open-hearth furnace

Three-quarters of British steel is made by this method, which was invented by Sir William Siemens in 1866. The furnace is a shallow bath with a roof over the top, built of steel with a firebrick lining (Figure 6.14). It is charged with molten or solid iron (a good deal of scrap iron is re-used in this way), together with lime and iron oxide. An open-hearth furnace may take a charge of up to 500 tonnes. A mixture of liquid fuel and gas is pumped to burners at one end and the flame is deflected on to the surface of the metal by the roof. The impurities are oxidised by the iron oxide and combine with the lime to form slag.

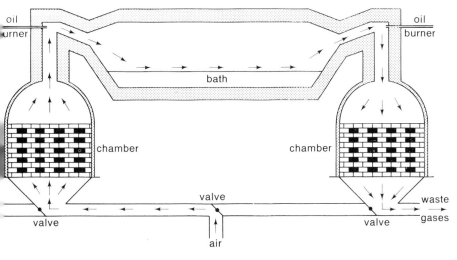

Figure 6.14 The Siemens open-hearth furnace

The waste gases heat up a chamber containing bricks. Every 20 minutes or so the direction of operation of the furnace is reversed so that the brickwork heats up the incoming air; this reduces heat wastage. Samples of metal are taken at intervals to check on the progress of purification, which generally takes between eight and twelve hours. Final additions of carbon and manganese are made before the steel is tapped.

Electric furnaces

In the Bessemer and open-hearth processes it is difficult to control accurately the purity of the steel produced. The main problem is that gases, especially nitrogen from the air, dissolve in the steel to some extent, and bubbles of gas weaken the final product. These processes are therefore not suitable for the production of special high quality steels (for example, for tools and ball bearings). Although electricity for heating is very expensive, electric furnaces are more easily controlled and all gases can be excluded. About 10 per cent of British steel, mainly high quality steel for special purposes, is produced in various types of electric furnace. Figure 6.15 shows one type, the ARC FURNACE, in which the charge is heated by an electric arc struck between carbon electrodes and the surface of the metal.

Metals   157

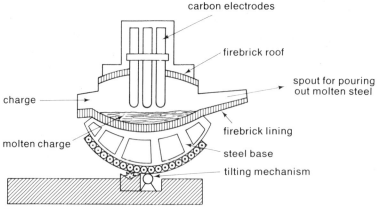

Figure 6.15 An arc furnace

Modern developments in steelmaking

The problem of nitrogen from air dissolving in steel has been mentioned. One way in which this problem has been overcome in recent years is by making steel in a converter, similar to the Bessemer converter, but using pure oxygen instead of air. Rather than blow air through the molten iron from underneath, these modern converters use oxygen, which is blown on to the surface of the iron from above.

An even more revolutionary process, developed in the 1960s, is SPRAY STEELMAKING (Figure 6.16). Molten iron from the blast furnace falls through the furnace in a stream which meets blasts of oxygen and powdered lime. By the time the spray has fallen to the bottom of the furnace the impurities have been oxidised out and the slag and refined iron collect in separate layers.

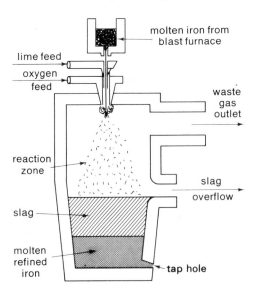

Figure 6.16 Furnace for spray steelmaking

Questions

1 (a) Name one of the chief ores from which iron is obtained.
   (b) As well as iron oxide and coke, another chemical (X) is introduced into the blast furnace. This compound assists in helping to remove impurities as 'slag'. Give the common name and chemical name of compound X.
   (c) The following represents a summary of the reactions which occur in the blast furnace leading to the production of molten iron:

   oxygen + coke → gas Y
   gas Y + coke → gas Z
   gas Z + iron oxide → gas Y + iron.

   Identify the gases Y and Z. Explain why gas Z is said to act as a reducing agent in the last step in the list.
   (d) What is the essential purpose of any process which converts the iron from the blast furnace into steel? *(West Midlands)*

2 In the blast furnace iron is obtained by the reduction of the iron ore.
   (a) Give *either* the chemical name *or* the common name of the iron ore used in the blast furnace.
   (b) What does *reduction* mean?
   (c) What material is added to the blast furnace to bring about this reduction?
   (d) Why is the furnace called a *blast* furnace?
   (e) Why is limestone added to the furnace? *(East Anglia)*

3 The equation for the reaction between iron(III) oxide and carbon monoxide can be written as follows:

$$Fe_2O_3 + 3CO \rightarrow 2Fe + 3CO_2.$$

The atomic masses of the elements concerned are Fe = 56, O = 16, C = 12. Show all your working in answering the following.
   (a) Find the weight of one mole of iron(III) oxide.
   (b) How much iron could be obtained from 40 tonnes of iron(III) oxide?
   (c) How much carbon monoxide would be required to reduce the 40 tonnes of iron(III) oxide? *(West Midlands)*

4 Describe *briefly* the manufacture of any metal other than iron which involves the use of *coke* as a reducing agent.

5 Describe any one method of steelmaking which you find most interesting.

6 Try to find out something about one of the other methods of

steelmaking not described in this book, for example the Linz-Donawicz (LD) converter, the Kaldo converter, or the high-frequency induction furnace.

6.6 Extraction of a metal by electrolysis: aluminium

Most of the aluminium in the earth's crust is combined with silicon and oxygen in rocks from which extraction is too expensive. The most useful ore of aluminium is bauxite, which contains between 50 and 70 per cent of aluminium oxide; the main impurity is iron oxide. The ore is crushed and purified by heating under pressure with concentrated sodium hydroxide solution. The aluminium oxide dissolves as sodium aluminate:

aluminium oxide + sodium hydroxide → sodium aluminate + water

$$Al_2O_3(s) + 2NaOH(aq) \rightarrow 2NaAlO_2(aq) + H_2O(l).$$

The iron oxide does not dissolve and is filtered off. The solution of sodium aluminate is cooled, which reverses the reaction so that pure aluminium oxide is slowly precipitated. The precipitate is filtered off and dried by heating. The sodium hydroxide solution can be used again.

Aluminium is extracted by electrolysis of the pure aluminium oxide. Compounds can be electrolysed either molten or in solution. Unfortunately the melting point of aluminium oxide is extremely high (over 2 000°C) and it does not dissolve in any of the usual laboratory or industrial solvents. It was not until 1886 that extraction of aluminium on a large scale became possible with the discovery that aluminium oxide will dissolve in molten *cryolite*. (Cryolite is sodium aluminium fluoride, Na_3AlF_6, a rare mineral found in Greenland.)

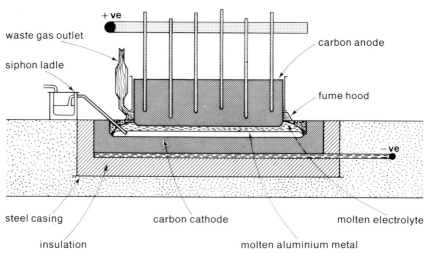

Figure 6.17 Electrolytic cell for the production of aluminium

Figure 6.17 shows the type of electrolysis cell used, a steel container with carbon electrodes. The electrolyte is a solution of 5 per cent of aluminium oxide in molten cryolite. A current of 100 000 amps is used at about 5 volts. The resistance of the electrolyte to this large current provides enough heat to melt the cryolite.

Aluminium oxide is composed of ions:

$$Al_2O_3 = 2Al^{3+}3O^{2-}.$$

When the current is passed, aluminium ions move to the cathode where they collect electrons and form aluminium atoms:

(*at the cathode*) $\quad Al^{3+} + 3e \rightarrow Al(l).$

Molten aluminium sinks to the bottom of the cell and is sucked out into a ladle by a siphon. Oxygen ions move to the anode, to which they give up their electrons, forming oxygen gas:

(*at the anode*) $\quad 2O^{2-} \rightarrow O_2(g) + 4e.$

This oxygen combines with the carbon anode to form carbon dioxide:

$$\text{carbon} + \text{oxygen} \rightarrow \text{carbon dioxide}$$
$$C(s) + O_2(g) \rightarrow CO_2(g).$$

A single cell produces half a tonne of aluminium per day, using up a quarter of a tonne of carbon anode in the process. Two hundred or more cells are operated in series.

Extraction of other metals by electrolysis

The source of SODIUM is sodium chloride (*salt*) from the sea. A molten mixture of sodium chloride with 60 per cent of calcium chloride is electrolysed. Chlorine gas is produced at the anode. Molten sodium floats to the top of the electrolyte at the cathode and is siphoned off (page 406).

CALCIUM is extracted from calcium chloride, a by-product of the Solvay process (page 417). Molten calcium chloride is electrolysed and solid calcium is deposited on a steel cathode.

MAGNESIUM is extracted from sea water. Calcium hydroxide (*slaked lime*) is added to sea water to precipitate solid magnesium hydroxide, which is filtered off:

magnesium chloride + calcium hydroxide →
       magnesium hydroxide + calcium chloride
$\quad MgCl_2(aq) \;+\; Ca(OH)_2(s) \;\rightarrow\; Mg(OH)_2(s) \;+\; CaCl_2(aq).$

The magnesium hydroxide is dissolved in hydrochloric acid to produce a solution of magnesium chloride:

magnesium hydroxide + hydrochloric acid →
       magnesium chloride + water
$\quad Mg(OH)_2(s) \;+\; 2HCl(aq) \;\rightarrow\; MgCl_2(aq) \;+\; 2H_2O(l).$

The solution is evaporated to give pure magnesium chloride, which is mixed with sodium chloride, melted, and electrolysed. Molten magnesium floats to the surface and is removed with ladles.

Experiment 6.9 The purification of a metal by electrolysis

Set up the apparatus shown in Figure 6.18. Take care (a) that the brass screw is connected to the positive terminal and the copper wire to the negative terminal and (b) that the crocodile clips do *not* dip into the dilute sulphuric acid.

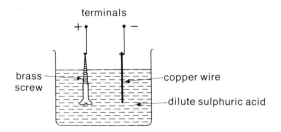

Figure 6.18

Allow the current to pass for at least five minutes (longer if you have time). Then remove the copper wire and the brass screw and examine them carefully.

What has happened to the part of the brass screw that was dipping into the sulphuric acid? What has happened to the part of the copper wire that was dipping into the sulphuric acid?

If the electrolysis was continued for long enough the brass screw would completely disappear. Brass is a mixture of two metals: copper and zinc. Where does the copper from the brass go? Where does the zinc go?

The purification of copper by electrolysis

More copper is used in the manufacture of wire than for any other purpose. Pure copper is an extremely good conductor of electricity but even small traces of impurities reduce the conductivity a great deal. Crude copper must therefore be refined to at least 99.5 per cent purity and for some specialised purposes to 99.95 per cent purity. The method used is electrolytic refining.

The crude copper is cast into large slabs for use as anodes in the electrolytic tank. The cathodes are thin sheets of pure copper and the electrolyte is a solution of copper(II) sulphate in dilute sulphuric acid. Each tank holds 30 to 40 anodes and a similar number of cathodes. A large tank house may contain more than 1 000 of these tanks.

When the current is passed copper atoms in the anode lose electrons to form copper ions:

(*at the anode*) $\quad\quad\quad Cu(s) \rightarrow Cu^{2+} + 2e.$

Figure 6.19 A tank house for the electrolytic purification of copper. A batch of pure copper cathodes is being lowered into one of the tanks.

These copper ions are attracted to the cathode, where they get their electrons back:

(*at the cathode*) $\qquad Cu^{2+} + 2e \rightarrow Cu(s)$.

The overall effect is that copper gradually dissolves from the anodes and is deposited on the cathodes. It takes about a fortnight for all the copper in the anodes, which usually weigh about 150 kilograms each, to be transferred.

Impurities which are higher than copper in the activity series, such as iron, also dissolve from the anode but are not redeposited on the cathode. They accumulate in solution in the electrolyte. Impurities which are lower than copper in the activity series do not dissolve at all. They fall to the bottom of the tank as ANODE SLUDGE. The anode sludge often contains small quantities of silver, gold, platinum, selenium, and other precious elements which were present in the original copper ore. The sludge is a valuable source of these elements and is collected and purified.

Questions

1 Naturally-occurring aluminium compounds are very common yet the metal has been extracted on a commercial scale only within the last hundred years. Why?

What would happen if aluminium oxide and carbon were heated strongly? Explain. (*London*)

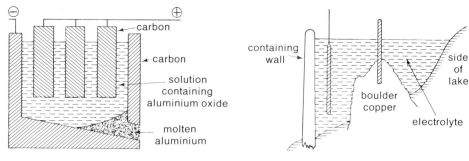

Figure 6.20 *Figure 6.21*

2 Aluminium is made industrially by passing an electric current through a solution of aluminium oxide in molten cryolite at a temperature above 700°C (Figure 6.20). Carbon electrodes are placed in the vessel which is lined with carbon. Molten aluminium collects on the floor of the vessel. The electrodes placed in the vessel gradually become smaller and need to be replaced. It is found that a current of 26.5 amperes must be passed for 30 hours (i.e. 30 faradays) to obtain 270 grams of aluminium.

(a) Why is carbon used for electrodes?

(b) Aluminium collects at the cathode. What is the sign of the charge on the aluminium ion?

(c) What other ion is present in the solution besides aluminium?

(d) Give the name of the electrode at which it is discharged.

(e) What is produced by the discharge of this ion?

(f) Why do you think the carbon electrodes placed in the vessel become smaller and need replacing occasionally?

(g) What can you say about the melting point of aluminium?

(h) The material in the vessel is kept molten but no heat is apparently supplied to it. Suggest an explanation for this.

(i) If a solution of an aluminium salt in water is electrolysed, no aluminium is formed at the cathode. What do you think *is* formed at the cathode? Suggest a reason for this.

(j) The weight of aluminium deposited from an excess of aluminium oxide depends on two factors. What are they?

(k) How many moles of aluminium have been produced? (Atomic mass = 27)

(l) If 30 faradays of electrical charge have been used up, how many faradays are required to discharge 1 mole of aluminium?

(m) The symbol for 1 mole of aluminium atoms is Al. What is the symbol for 1 mole of aluminium ions? (*London*)

3 If you were given a piece of impure copper, a piece of pure copper, some copper sulphate solution, and a source of direct current, together with any other apparatus you require, draw a diagram to

show how you would obtain pure copper from the piece of impure copper.

Write the following labels on your diagram: +ve and −ve terminals of the source of current, anode, cathode, pure copper, impure copper, copper sulphate solution. (*West Midlands*)

4 Copper occurs native on the shores of Lake Superior as large 'boulders'. The impure boulder copper is extracted by building a containing wall around the metal and using the metal as one electrode in an electrolytic cell (Figure 6.21).

(a) What is the name given to this method of extraction?
(b) What type of electrical current would be required?
(c) Which electrode in the cell would the boulder copper be?
(d) What would the other electrode be made of?
(e) Suggest a suitable solution for use as an electrolyte.
(f) Explain, with equations where possible, how the pure copper is obtained. (*London*)

6.7 The structure of metals

Experiment 6.10 Growing metal crystals

Three-quarters-fill a test tube with lead nitrate solution. Hang a piece of zinc foil over the edge of the tube so that it dips into the solution as shown in Figure 6.22.

After about half an hour (longer, if possible) examine the zinc foil. What do you see? What are the crystals on the zinc foil?

Transfer some of the crystals to a slide and examine them under a microscope. Make a drawing of what you see.

Metal crystals

Metals are CRYSTALLINE: they are made up of crystals. But metal crystals are not often large enough to be seen easily. Galvanised iron is made by

Figure 6.22

Figure 6.23 *The surface of galvanised iron*

dipping iron articles into a bath of molten zinc. The zinc crystallises as it solidifies around the iron and the crystals in this case are often large and easily visible. Lead crystals can be grown from a lead nitrate solution by pushing the lead out with a more active metal such as zinc:

$$\text{lead nitrate} + \text{zinc} \rightarrow \text{zinc nitrate} + \text{lead}.$$

Usually, however, metal crystals are so small that they can only be seen with a microscope. To examine them the metal is first polished absolutely smooth. If the smooth surface is viewed through a microscope the crystals still cannot be seen because they fit together so neatly that the joins between them are invisible. The crystals become visible if the polished surface is etched. This involves dipping it into a suitable chemical, usually an acid, which attacks the metal. The edges of the crystals are attacked fastest, producing slight grooves between the crystals which are clearly visible under the microscope. The effect of etching is illustrated in Figures 6.24 and 6.25.

Figure 6.24(a) Metal surface before etching (b) Metal surface after etching

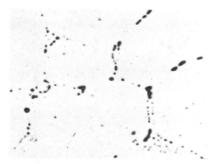 

Figure 6.25(a) Photograph (magnification × 200) of the surface of a piece of polished copper before etching   (b) Photograph (magnification × 200) of the surface of a piece of polished copper after etching

The effect of grain-size on the properties of metals

Metal crystals are called GRAINS. The size of the crystals, the GRAIN-SIZE, is very important in determining the properties of the metal. Metals with a large grain-size are very MALLEABLE and DUCTILE; the large grains can slide over one another when the metal is hammered into shape or pulled into wires. But the weakest points in metals are the boundaries between the grains, so when the grains are large the metal can break easily.

Metals with a small grain-size are much less malleable and ductile. A lot of small grains arranged in an irregular way will not easily slip past one another. On the other hand, they are much stronger because the irregularity of the grain boundaries prevents easy breakage.

The effects of grain-size can be summed up as follows.
Large grain-size: weak, very malleable and ductile.
Small grain-size: strong, not very malleable or ductile.

Alteration of the grain-size in metals

For most engineering purposes a metal must be strong, so a small grain-size is usually required. If you have grown crystals of substances like copper sulphate and alum, you will know that the more slowly the crystals are allowed to grow the larger they become. Metal crystals are similar. If a metal is heated to a temperature a little below its melting point and then cooled very quickly by plunging it into cold water (QUENCHING) very small crystal grains are formed (Figure 6.26). If, however, a metal is allowed to cool very slowly from a high temperature (ANNEALING) the grains have time to grow much larger (Figure 6.27). Controlled heating and cooling can be used in this way to alter grain-size. This is known as HEAT TREATMENT of metals.

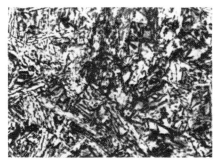
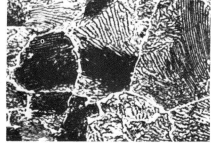

Figure 6.26 Photograph (magnification × 200) of the etched surface of steel which has been cooled rapidly (quenched)

Figure 6.27 Photograph (magnification × 200) of the etched surface of the same type of steel which has been cooled slowly (annealed)

Any machining operation, such as rolling ingots to produce metal sheet or rod, distorts the crystal grains. Their irregularity makes the metal stronger, though less ductile and malleable. As a result of this WORK-HARDENING, operations such as wire-drawing become progressively more difficult as the metal grains become distorted, and eventually ductility may be reduced so much that the metal breaks. At certain stages in wire-drawing, to prevent breakage, the metal grains must be reformed by heat treatment.

Metals

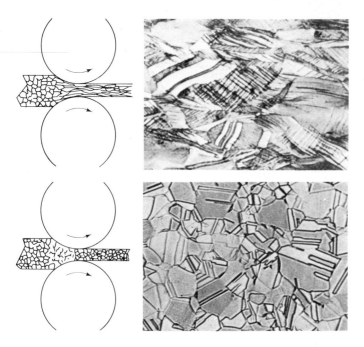

Figure 6.28
(a) Distortion of grains by cold-working
(b) Recrystallisation of grains in hot-working

Figure 6.29
(a) Photograph (magnification × 200) of the etched surface of bronze after cold-working
(b) Photograph (magnification × 200) of the etched surface of the same bronze after hot-working

The two processes of machining and heat treatment may be combined by machining the metal at a high temperature so that re-crystallisation takes place continuously. In HOT-WORKING the grains do not become distorted. Figures 6.28 and 6.29 illustrate the effect of cold-working and hot-working on the grain structure.

Lead is exceptional in its behaviour. It has the ability to recrystallise even at room temperature. Lead will therefore remain soft and malleable even during cold-working, a property which makes it useful for plumbing, roofing, etc.

The grain structure of alloys

Examination of grain structure with a microscope also provides useful information about alloys. Figures 6.30 and 6.31 show the grain structure of two alloys. In the photograph of gunmetal separate grains of two metals can be seen (Figure 6.30). This is perhaps what might be expected as a result of crystals of the two metals growing quite separately.

In the photograph of brass, on the other hand, the grains all appear the same. Each grain is a mixture of the two metals (Figure 6.31). Many important alloys are like this; they are called SOLID SOLUTIONS. Just as in a solution of salt in water the salt particles cannot be seen, so in a solid solution particles of the individual metals cannot be seen, even with a microscope. To understand what a solid solution is like it is necessary to find out how the atoms of metals are arranged within the grains.

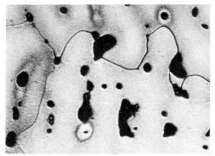 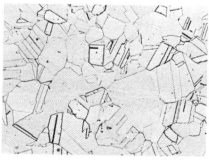

Figure 6.30 The etched surface of gunmetal, an alloy in which two different types of grain can be seen

Figure 6.31 The etched surface of brass, an alloy in which only one type of grain can be seen

X-ray diffraction

The fact that crystals have a regular shape led scientists to believe that the atoms within a crystal are most probably arranged in a regular pattern. But since atoms are too small to be seen, even with the most powerful microscope, it seemed for a long time that it would be impossible to find out exactly what the regular arrangement is like. Nevertheless, in 1912 Sir Lawrence Bragg developed a method of working out the arrangement of atoms in a crystal, using a technique called X-RAY DIFFRACTION.

When a small light source is viewed through a lot of very small holes a pattern of light and dark spots is seen. The light is said to be DIFFRACTED, and the pattern of spots is called a DIFFRACTION PATTERN. You can see this effect by looking at a torch bulb through a piece of terylene net curtain or a handkerchief. The cloth has a lot of very small holes between the threads and the shape of the diffraction pattern is directly related to the position of the holes in the cloth. If the cloth is turned the diffraction pattern turns. If the cloth is stretched the diffraction pattern changes shape. The distance between the light spots in the diffraction pattern depends on the distance between the threads in the cloth: if the threads are further apart the light spots are closer together. By photographing the diffraction pattern and measuring the distance between the light spots it is possible to work out how the threads in the cloth are arranged and how far apart they are.

Since atoms are roughly spherical in shape there are small holes between the atoms in a crystal. If these holes are made to produce a diffraction pattern it is possible to work out how the atoms are arranged and how far apart they are. To get a diffraction pattern the wavelength of the light used must be about the same as the size of the holes. It so happens that the distance between the threads in a piece of cloth is about the same as the wavelength of ordinary visible light (about 500 nanometres). Atoms are much closer together, so light of a much shorter wavelength is needed. The wavelength of X-rays (about 1 nanometre) is just right. A beam of X-rays produces a diffraction pattern when it is directed at a crystal (Figure 6.32). Since X-rays are invisible the pattern has to be photographed using special film. The patterns are very similar to the patterns obtained from a piece of cloth with visible light.

Using X-ray diffraction the sizes of atoms have been measured accurately and the way they are arranged has been discovered for many different substances, including metals and alloys.

The arrangement of atoms in alloys

Figure 6.33 shows how a few of the atoms in a gold crystal are arranged. If a small proportion of silver is alloyed with the gold the structure of the metal is not affected. The silver atoms simply replace a few of the gold atoms in the regular pattern (Figure 6.34). This can happen quite easily because silver atoms and gold atoms are exactly the same size (both 0.29 nm in diameter). Gold-silver alloy, which is called *electrum*, is a solid solution. Each grain is a mixed crystal of gold and silver.

More or less the same situation is found with another much more important alloy, *brass*. The atoms of the two metals in brass, copper and zinc, are not quite the same size:

diameter of copper atom = 0.26 nm

diameter of zinc atom = 0.27 nm.

However, the size difference is sufficiently small for zinc atoms to replace copper atoms in a copper grain without distorting the regularity of the pattern very much. Brasses are solid solutions and can be made with any proportions of copper and zinc. The grain structure of the brass in Figure

Figure 6.32 The X-ray diffraction pattern from a metal crystal

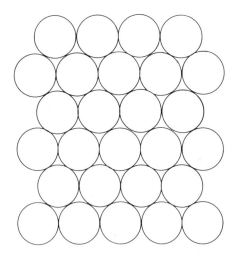

Figure 6.33

170 Chemistry today

6.31 (page 169) shows that this alloy is the solid solution type. Only one kind of grain can be seen.

If the atoms of the two metals of an alloy are very different in size solid solutions cannot be produced. The atoms of one metal cannot fit into the regularly-packed arrangement of the other type of atoms without considerable distortion (Figure 6.35). In these alloys the crystals of the two metals grow quite separately and two types of grain can be seen under the microscope. *Gunmetal*, an alloy consisting of 85 per cent copper with 5 per cent each of zinc, tin, and lead, is an example of this type. Lead atoms are much larger than copper atoms:

$$\text{diameter of copper atom} = 0.26 \text{ nm}$$
$$\text{diameter of lead atom} = 0.35 \text{ nm}.$$

The lead crystallises out in separate grains; the zinc and tin form a solid solution with the copper. Figure 6.30 (page 169) shows the etched surface of gunmetal: the dark lead grains are easily visible.

When atoms of the alloying element are exceptionally small some of the smaller atoms can fit into the spaces between the larger atoms without distorting their regular arrangement too much (Figure 6.36). This type of alloying, in which some of the gaps between the atoms of the pure metal are filled in, produces a much stronger material. An example is *beryllium–copper*:

$$\text{diameter of copper atom} = 0.26 \text{ nm}$$
$$\text{diameter of beryllium atom} = 0.22 \text{ nm}.$$

This is an alloy of copper with up to 2 per cent of beryllium, and is the hardest of all copper alloys.

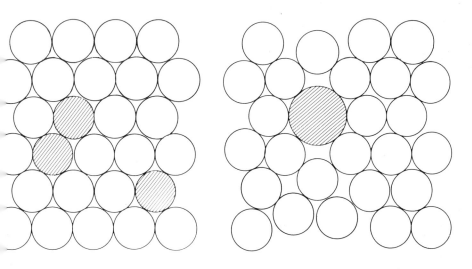

Figure 6.34   Figure 6.35

Figure 6.36 A model of an alloy of two metals, one with considerably larger atoms than the other. The regular arrangement of the larger atoms is not distorted very much.

The effect of alloying on properties

The properties of an alloy obviously depend on the properties of the pure metals which it contains, and are usually a combination of them. Pure metals are mostly of limited usefulness. A strong metal may have the disadvantage of being brittle or a malleable metal may have the disadvantage of being weak. Alloying of a strong brittle metal with a weak malleable one, however, may produce a material which is both strong and malleable. The addition of about 10 per cent of antimony, a hard and brittle metal, to lead is an example of this. The alloy produced is only a little less malleable than lead, but is very much stronger.

In one important respect the properties of an alloy are not a simple mixture of the properties of its pure metals. The melting point of pure tin is 232°C and that of pure lead is 327°C. But *solder*, composed of 62 per cent tin and 38 per cent lead, melts at 183°C. In an ordinary solution of a solid in a liquid the same thing happens. Water freezes at 0°C but a saturated solution of salt in water freezes at -23°C: it is for this reason that salt is put on icy roads. The use of antifreeze in a car radiator depends on the same principle: *a mixture always freezes or melts at a lower temperature than the pure substances.* Since alloys are solid solutions it is not surprising that they behave in the same way: an alloy usually has a lower melting point than the pure metals of which it is made. A remarkable example of this is *Wood's metal*, which is composed of 50 per cent bismuth, 25 per cent lead, $12\frac{1}{2}$ per cent tin, and $12\frac{1}{2}$ per cent cadmium. It melts at the remarkably low temperature of 70°C and is used in safety valves to protect against fire. A water jet is blocked by a piece of the alloy, and if the temperature rises above 70°C the alloy melts and water is automatically sprayed on the fire.

Experiment 6.11 The properties of solder, an alloy of tin and lead

(a) *The melting point of solder*. Clamp a hard-glass boiling tube containing 5 cm depth of medicinal paraffin vertically. Put a piece of solder in the paraffin. Heat the tube slowly, stirring the contents *gently* with a 360°C thermometer. Watch the solder carefully and note the temperature at which it melts.

Take care! The tube is now very hot. Do not try to dismantle the apparatus until the temperature has dropped below 50°C.

The melting point of pure tin is 232°C and the melting point of pure lead is 327°C. Why is your result for the melting point of solder surprising?

(b) *The hardness of solder*. Strap a ball bearing to a block of solder as shown in Figure 6.37.

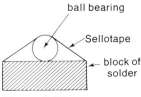

Place the block of solder with the ball bearing attached to it on the bench. Drop a one kilogram weight squarely on to the ball bearing from a height of exactly $\frac{1}{2}$ metre. Do the same with a block of tin and with a block of lead.

Figure 6.37

Measure the diameter of the dent in each of the three metal blocks in millimetres. Make a note of your results.

The composition of solder is about 2 parts of tin to 1 part of lead. Are your results as you would expect?

Experiment 6.12 The heat treatment of steel

(a) Hold a steel knitting needle in tongs and heat the centre to redness with a fierce bunsen flame. Then lay it on an asbestos square to cool.

When it is cool enough to hold, try to bend it. Does it bend? If so, how easily?

(b) Heat another needle as before. When it is red hot, put it very quickly under running water from the tap. Try to bend it. What happens?

The heat treatment of steel

The effect of heat treatment of steel is remarkable.
1 When it is heated to a temperature of 800°C or more and then cooled

Metals 173

quickly by plunging into cold water (QUENCHING) it becomes hard and brittle.

2 When it is heated to 800°C and then allowed to cool very slowly (ANNEALING) it becomes soft and malleable.

3 If, after quenching, it is heated to between 300°C and 700°C and then allowed to cool slowly in air (TEMPERING) it becomes tough and springy. With various combinations of quenching, annealing, and tempering and variation of the temperatures and rates of cooling, steel with almost any required properties can be produced: for springs, saw blades, razor blades, knives, drill-bits, etc.

Steel is iron containing between $\frac{1}{10}$ per cent and $1\frac{1}{2}$ per cent of carbon. To understand the remarkable effects of heat treatment it is necessary to think about the way in which the carbon and iron atoms in steel are arranged. Carbon atoms are much smaller than iron atoms:

$$\text{diameter of iron atom} = 0.25 \text{ nm}$$
$$\text{diameter of carbon atom} = 0.15 \text{ nm.}$$

A solid solution of carbon in iron can be produced in which carbon atoms fill up a few of the gaps between the iron atoms without much distortion of the pattern (Figure 6.36, page 172). At higher temperatures more carbon can be fitted in than at lower temperatures. Carbon is more soluble in iron at higher temperatures, just as sugar is more soluble in hot water than in cold.

When the solid solution of carbon in iron is cooled slowly from above 800°C some of the carbon remains in grains of a solid solution with iron called *ferrite*. Some of the carbon cannot be held in solution and separates as grains of *cementite*, a sort of compound of iron and carbon with the composition Fe_3C. Annealed steel consists of layers of ferrite and cementite grains. The large grain-size (Figure 6.38) produced by the slow cooling makes it very malleable, though rather weak.

If steel is cooled very rapidly from above 800°C to room temperature

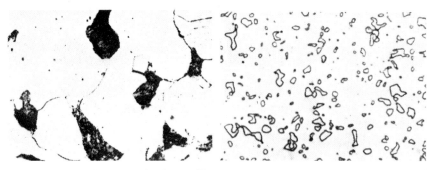

Figure 6.38 Photograph (magnification × 200) of the etched surface of annealed steel. The cementite shows up as dark grains.

Figure 6.39 Photograph (magnification × 200) of the etched surface of tempered steel. The small grains outlined in black are the cementite.

the iron and carbon atoms do not have time to organise themselves into these two different types of grain. Quenched steel consists of very tiny grains with a very distorted pattern of mixed-up iron and carbon atoms (Figure 6.39). As usual, the small grain-size produces a hard and brittle metal.

If quenched steel is tempered by heating it to a temperature of 300–700°C and cooling slowly, the distorted crystals can reorganise themselves into separate ferrite and cementite grains, though the grain-size is not as large as in annealed steel. The combination of the soft and malleable ferrite grains mixed up with the hard and brittle cementite grains gives a very tough steel which is both strong and malleable.

Questions

1 (a) What is a metal *grain*?
  (b) What is the effect of the *grain-size* on the properties of a metal?
  (c) How can the grain-size of a metal be altered?

2 Metal grains are too small, and fit together too closely, to be seen by the naked eye. Explain how the grain structure of a metal can be made visible.

3 What is meant by *work-hardening*? How can it be avoided?

4 What information about metals is provided by the technique of X-ray diffraction?

5 The aluminium atom has a diameter of 0.29 nm. Three other metals, X, Y, and Z, have atomic diameters of:

    X: 0.16 nm   Y: 0.29 nm   Z: 0.39 nm.

  (a) Which of the three metals X, Y, or Z, will form an alloy with aluminium which is weaker than pure aluminium?
  (b) Which of the three metals X, Y, or Z will form an alloy with aluminium which is stronger than pure aluminium?
  (c) Which of the three metals X, Y, or Z will form a solid solution with aluminium?

6 (a) What is an alloy?
  (b) What non-metal is frequently added to iron?
  (c) What good effect can this have?
  (d) What property of iron is impaired by this addition?

                                                        (*London*)

7 What is the effect on the properties of steel of quenching, annealing, and tempering? Explain how the arrangement of the atoms in steel is affected by these three processes.

6.8 Corrosion and its prevention

Corrosion is the slow destruction of metals as a result of chemical attack by air or water. Pure air contains oxygen, carbon dioxide, and water vapour, which can all react with some metals. Air polluted with smoke from factory chimneys contains acid gases such as sulphur dioxide which corrode metals even faster. Pure water does not react easily with many metals, but sea water, water contaminated by sewage, or impure water poured into rivers by factories can be very corrosive.

Some metals are much more easily corroded than others. The position of a metal in the activity series (page 141) is an indication of how easily it will be corroded. Metals at the top of the series like sodium and calcium are very quickly attacked by oxygen and water and are no use for structural purposes. Metals at the bottom like gold and platinum are almost completely resistant to corrosion, but they are unfortunately much too expensive for general use. The important structural metals (aluminium, iron, lead, and copper) are in the middle of the activity series. Lead and copper are much less easily corroded than aluminium or iron, but all of these metals are attacked by water and air to some extent unless they are protected in some way.

Protective coatings

Corrosion can be prevented by coating a metal with a protective layer. Paints and varnishes can be used, but have the disadvantage that a new coat has to be applied from time to time. Metal roofs are often coated with bitumen (tar) and wires may be covered with a plastic sheath to protect them against corrosion. A metal like iron which is quite easily corroded may be protected with a thin film of a metal such as tin or chromium which is less easily corroded.

Some metals build up their own protective layer. Copper is often used for roofing. When it is exposed to the atmosphere it reacts slowly with oxygen, water, and carbon dioxide, and a thin layer of a mixture of copper(II) carbonate and copper(II) hydroxide, called *verdigris*, is produced, which is responsible for the attractive turquoise colour of copper roofs. This protects the copper from further corrosion.

Aluminium is high in the activity series and reacts with water and oxygen quite easily, but once the outer surface has been converted to aluminium oxide no more corrosion is possible. The thin layer of oxide sticks firmly to the metal and protects it from further attack by water or oxygen. The natural oxide film can be thickened to increase corrosion resistance by ANODISING. The aluminium article is made the anode in an electrolytic bath with dilute sulphuric acid as electrolyte. At the anode, aluminium atoms give up electrons and react with water, forming aluminium oxide and hydrogen ions:

$$2Al(s) + 3H_2O(l) \rightarrow Al_2O_3(s) + 6H^+(aq) + 6e.$$

The hydrogen ions are attracted towards the cathode, where they receive electrons and become molecules of hydrogen gas:

$$6H^+(aq) + 6e \rightarrow 3H_2(g).$$

The anodised film can be dyed to produce the attractive finish seen, for example, on some saucepan lids.

Experiment 6.13 Anodising and dyeing aluminium

(a) *Cleaning* Any small piece of aluminium or small aluminium object can be used. Wipe it well with a piece of cloth dipped in carbon tetrachloride. **From now on do not handle the object: use tongs.**

Dip the object in a beaker of sodium hydroxide solution until vigorous effervescence occurs. Rinse it under the tap, dip it into a beaker of dilute nitric acid, and rinse again.

What is cleaned off with carbon tetrachloride? Why must the object not be touched? What is the gas given off when it is dipped into sodium hydroxide solution? Why is it dipped in sodium hydroxide solution? Why is it then dipped in dilute nitric acid?

(b) *Anodising* Wrap a piece of wire around the object and set up the apparatus shown in Figure 6.40.

The object must be connected to the positive terminal. The sheet of aluminium foil, which is attached to the negative terminal, should completely surround the object, **but must not touch it.**

Pass the current for at least 15 minutes (longer if you have time), then remove the object and rinse it.

(c) *Dyeing* Half-fill a beaker with dye solution and warm it to between 65°C and 70°C. Put the object into the dye for 3 minutes.

(d) *Sealing* Half-fill a beaker with sealing solution and heat it to boiling. Immerse the object in the boiling sealing solution for 15 minutes.

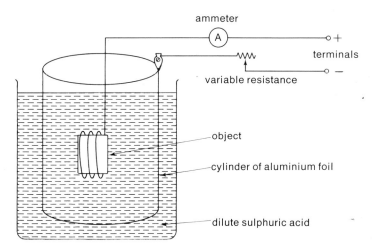

Figure 6.40

Corrosion-resistant alloys

Alloys may be more resistant to corrosion than the pure metals of which they are made. Steel rusts less easily than iron, and special steels containing various proportions of chromium, nickel, or copper (*stainless steel*) have a very high resistance to corrosion. An alloy of 70 per cent copper and 30 per cent nickel is used for the parts of ships' engines that are in constant contact with sea water. This alloy, *cupronickel*, is much less easily corroded by sea water than either pure copper or pure nickel.

Chemical protection against corrosion

The corrosion of iron, known as RUSTING, is a very unusual process. Iron is not corroded at all by either pure oxygen or pure water. Yet in the presence of oxygen and water together it rusts rapidly. This corrosion is electrical in origin. You may have seen how an electric current can be produced by dipping, for example, a copper rod and a zinc rod into dilute sulphuric acid. This arrangement is called an electric CELL. The electricity is produced from the chemical reaction as the zinc dissolves in the acid:

zinc + sulphuric acid → zinc sulphate + hydrogen + electricity.

The copper, which is lower than zinc in the activity series, is not affected.

Iron usually contains impurities such as carbon. In the presence of water a tiny electric cell is set up between an iron particle and a carbon particle. The more reactive element, iron, dissolves:

iron + water → iron oxide (*rust*) + hydrogen + electricity.

The reaction stops almost immediately because the iron particle soon becomes surrounded by hydrogen bubbles which prevent the electric current flowing. If oxygen is present as well as water, the hydrogen combines with the oxygen (forming water) and the iron continues to rust.

Car bumpers (chromium-plated iron) and tin cans (tin-plated iron) are not very resistant to rusting. As long as the chromium or tin covering is perfect, water and oxygen cannot get at the iron to rust it, but once the plating is scratched a cell can be set up between the chromium and the iron or the tin and the iron. Since iron is higher than chromium and tin in the activity series the iron is rusted away, although the chromium or tin remains unattacked. A car bumper can become a lump of rust surrounded by a thin skin of chromium! Plating iron with a metal which lies below it in the activity series such as chromium, nickel, tin, or silver is not a very good protection against corrosion if there is any danger of the plating being scratched.

In GALVANISING, iron is coated with zinc, which is higher than iron in the activity series. If the zinc coating is scratched through a cell can be set up. This time, however, it is the zinc which is attacked instead of the iron. Only when all the zinc has been corroded away will the iron start to rust.

Galvanising can hold up the rusting of iron for quite a long time and is much more useful than plating with metals such as tin or chromium.

This is called SACRIFICIAL PROTECTION: the iron is protected as a result of the zinc being destroyed. Sacrificial protection can be applied in a slightly different way to protect metal structures like oil pipelines against corrosion. It is not necessary to coat the pipes with zinc. Instead, zinc rods are buried in the ground at intervals and connected by wires to the pipe. The zinc rod and the pipe make up the cell and the zinc rods are corroded instead of the pipe. The only servicing that the pipeline requires is replacement of the zinc rods when they are almost used up. This is a very much cheaper method of protecting large installations like pipelines, piers, and the hulls of ships (Figure 6.41) than making them of stainless steel or galvanising or painting them.

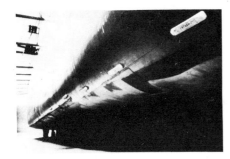

Figure 6.41 Zinc blocks attached to the steel hull of a ship for sacrificial protection against corrosion

Experiment 6.14   Sacrificial protection

Set up three beakers as shown in Figure 6.42.
In (1) stand a 7-cm nail.
In (2) dip a 7-cm nail connected by a wire to a piece of zinc foil.
In (3) dip a 7-cm nail connected by a wire to a piece of tin foil.

Put the three beakers on one side for at least 24 hours. At the end of this time examine the three nails for corrosion.

Which of the three nails has rusted the least? Why has this nail rusted less than the others?

Which of the three nails has rusted the most? Why has this nail rusted more than the others?

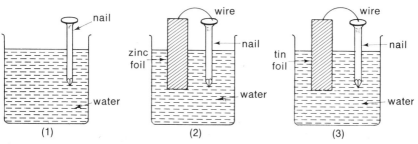

Figure 6.42

Questions

1 In which of the tubes A, B, C, D and E in Figure 6.43 would the iron nail remain unchanged if left for some considerable time?

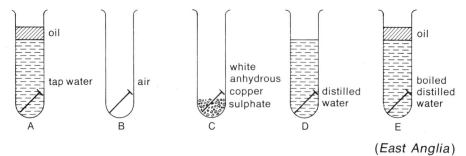

(East Anglia)

Figure 6.43

2 Iron will react with water in the presence of air to form rust.
(a) What is rust?
(b) Why is the rusting of iron and steel a serious problem?
(East Anglia)

3 How is iron or steel prevented from rusting when used for (a) the moving parts of a bicycle (b) cutlery? *(East Anglia)*

4 How is rusting prevented on (a) a car bumper (b) a food can?
(West Yorkshire)

5 Zinc has a greater affinity for oxygen than iron and yet zinc is used to protect iron from corrosion. Explain carefully why this is so.
(North West)

6 How are small aluminium objects protected against corrosion? What is the advantage of this method?

7 Corrosion is one of the most costly problems common to all kinds of industry. It is a problem in all forms of transportation, in large structures, small highly-specialised precision equipment, and is a problem domestically, too.

What chemically is corrosion? Give examples of where you have come across it. How can a chemical knowledge of a particular corrosion problem help to prevent it or arrest it? Anti-corrosion measures which deal with one problem may not be satisfactory for another—why? *(London)*

9 The uses of metals

Some metals have special properties which make them ideal for particular purposes. For example, copper conducts electricity very well (wiring); lead is soft and easy to work (plumbing); aluminium is very light (aircraft engineering). Often, however, there is a choice between several different metals to do the same job. You can find water pipes made of lead, copper, or iron. In a case like this the final choice depends on a number of factors such as which is cheapest, which corrodes least and will require replacement least often, and which is easiest to work with and will be cheapest to install.

The price of metals depends on how abundant they are and how easy they are to extract. Figure 6.44 shows the average prices of some metals in the United Kingdom during 1974. It also shows the total quantities of metals used for all purposes, and is an indication of the relative importance of the metals you are studying.

	Price (£) per tonne (1974)	Total number of tonnes used in Britain in 1973
Aluminium	315	490 000
Zinc	550	305 000
Iron	40	16 000 000
Tin	3 800	16 000
Lead	230	360 000
Copper	960	700 000

Figure 6.44

Iron

Iron direct from the blast furnace contains between 2 and 4 per cent of carbon. In this form it is known as CAST IRON. It is very fluid when molten and good for making intricate castings for car cylinder blocks, lamp-posts, drain pipes, manhole covers, etc. The large carbon content, however, makes cast iron brittle and of limited usefulness.

WROUGHT IRON is a much purer form of iron, from which almost all the carbon has been removed. It is very malleable and can be worked into decorative articles such as gates and ornaments.

Steel, which is by far the most useful form of iron, has less carbon than cast iron. There are two main types of carbon steel. SOFT STEEL contains less than $\frac{1}{4}$ per cent of carbon. It is used for making wire (from cables for suspension bridges to wire for cutting into nails and rivets) as well as for general engineering purposes. HARD STEEL contains between $\frac{1}{2}$ per cent and $1\frac{1}{2}$ per cent of carbon. Its properties can be altered by heat treatment (page 173) to make it suitable for a wide variety of items from springs and saw blades to hammers and chisels.

Figure 6.45 The Severn suspension bridge; 4500 tonnes of steel wire were used to make the suspension cables

Although carbon steels can be given all sorts of different properties by heat treatment, problems arise when very large articles are required which are impossible to heat and cool uniformly (the inside will always heat up and cool down more slowly than the outside). For large objects the properties of the steel must be altered by alloying rather than by heat treatment. Steel for bridges, heavy guns, car engines, and cutting tools is produced by adding various elements. Silicon strengthens steel, manganese toughens it, molybdenum helps it to withstand high temperatures, and chromium improves its resistance to corrosion. Stainless steel, for anything from architectural purposes to surgical instruments, as well as for household articles such as sinks, draining boards, and pressure cookers, contains up to 20 per cent of chromium and 10 per cent of nickel. For special uses other elements may also be included in alloy steels. High-speed cutting tools for lathes contain up to 20 per cent of tungsten, and a small proportion of boron produces exceptionally hard steel for turbine blades.

Aluminium

When the statue of *Eros* in Picadilly Circus, cast from pure aluminium, was put up in 1893, the metal was still a rarity. Now it is the second most important metal, but the statue illustrates one of its most important properties. In spite of aluminium's high position in the activity series it is very resistant to corrosion, as a result of the formation of a thin oxide film on the surface (page 176).

Of all the important metals used for structural purposes, aluminium is the least dense, which makes it particularly useful wherever weight must be saved. The bodywork of trains, buses, lorries, boats, and especially aeroplanes is now frequently constructed of aluminium or aluminium alloys. Even when lightness is not essential, as in building, the attractive appearance of aluminium makes it a popular choice.

Figure 6.46   The Silver Streak, *constructed in 1920, the first British all-metal aeroplane, consisting mainly of aluminium and aluminium alloys*

Aluminium conducts electricity about two-thirds as well as copper, but its density is less than one-third that of copper. Aluminium wire is therefore a better conductor than copper on a weight-for-weight basis. Almost the whole of the British overhead electrical grid system uses aluminium wire, strengthened with a central core of steel. Aluminium also conducts heat well and is used in the manufacture of low-price saucepans, frying pans, and kettles.

Treatment of aluminium with acid or alkali gives it a very bright surface which reflects light and heat well. This makes it useful in building to help keep the temperature inside constant in hot and cold weather. It is also used to reflect heat from the outside of space capsules and light in the reflector bowls of telescopes and car headlamps.

Pure aluminium can be rolled into foil as little as $\frac{1}{1000}$ cm in thickness. This foil, often called *silver paper* (though it contains no silver), finds more and more uses every year: wrappers for sweets and cigarettes, milk bottle tops, kitchen foil, and in condensers for radio and television sets.

Pure aluminium is rather weak but can be strengthened by alloying. Copper, magnesium, and silicon are frequently added, in varying proportions up to about 10 per cent in all, though other metals may be included as well. Some of these aluminium alloys, called *duralumin* alloys, possess the remarkable property of AGE-HARDENING; they get harder and stronger over a period of four or five days after they are made. This makes the alloy easy to work immediately after manufacture, but very strong in use. Another group of aluminium alloys, containing nickel and iron or nickel and cobalt, has the useful property of being even more powerfully magnetic than pure iron.

Figure 6.47 shows the proportions of the total British aluminium production used for various purposes.

Sheet and plate 40%	Castings 25%	Wire 10%	Foil 5%	Tube 5%	Other uses 15%

Figure 6.47

Copper

Copper is a soft, ductile metal which can very easily be drawn into wires. It is also the second best of all substances in conducting electricity (silver is slightly better). The electrical engineering industry is by far the biggest copper consumer. A big transformer may contain more than 40 tonnes of copper in the form of heavy gauge wire and contacts (Figure 6.48). A telephone exchange requires miles of fine gauge copper wire. Copper for electrical conductors needs to be extremely pure. Even minute traces of impurities very greatly reduce its conductivity.

Another useful property of copper is its resistance to corrosion, which makes it valuable for piping and roofing. Copper piping, both for water and for central heating systems, has become increasingly popular in recent years.

The resistance to corrosion and the good heat conductivity of copper make it a potentially valuable engineering material, but its softness is a disadvantage. It can be strengthened by alloying: with up to 40 per cent of zinc it makes *brass*, up to 10 per cent of tin *bronze*, and up to 10 per cent of aluminium *aluminium bronze*. These alloys are used in many fields: radiators for cars and commercial vehicles, whisky stills (see Figure 9.18 on page 262), and especially in shipbuilding to withstand the corrosive effect of sea water on boilers, condensers, propellers, and so on.

Our coinage is made from various copper alloys. The *copper* coins are 97 per cent copper with $\frac{1}{2}$ per cent tin and $2\frac{1}{2}$ per cent zinc; the *silver* coins are 75 per cent copper with 25 per cent nickel.

Figure 6.48 Copper wire and connecters in a 30 000 kilowatt generator

Figure 6.49 shows the proportions of the total British copper production used for various purposes.

Electrical industry (mainly wire) 45%	Engineering (shipbuilding, heat exchanges, etc.) 35%	Building industry (piping, roofing) 15%	Other uses 5%

Figure 6.49

Lead

Lead is used as sheathing for underground and underwater telephone and power cables. It is very resistant to corrosion and so soft and malleable

Figure 6.50 Part of the lead roof of Sidney Sussex College, Cambridge

that the cable remains pliable. Lead's resistance to corrosion and its property of being easily shaped also makes it a useful material in building, for roofing and piping.

Its exceptionally high density leads to its use in shielding against radioactive rays in atomic power stations and other atomic reactors. It is also used to protect the operators of X-ray machines.

Lead is such a soft, weak metal that for many applications it needs to be hardened by alloying with a small proportion of antimony. Car batteries contain plates of lead mixed with about 10 per cent of antimony. The lead alloy conducts the electricity but is not corroded by the sulphuric acid in the battery. Lead-antimony alloys are also used in engine bearings.

The exceptionally low melting point of lead is used to advantage in lead-tin alloys for solders. Most newspapers are printed with type made from an alloy called *type-metal* (74 per cent lead, 10 per cent tin, 16 per cent antimony) which has a low melting point and has the property of shrinking very little on casting so that it reproduces the intricate shapes of the letters very accurately.

Figure 6.51 shows the proportions of the total British lead production used for various purposes.

Batteries 25%	Cables 20%	Sheet and pipe 15%	Solder 5%	Other alloys 5%	Other uses 30%

Figure 6.51

186 Chemistry today

Figure 6.52 A large girder being lifted out of a galvanising bath

Zinc

The most important use of zinc is in protective coatings for iron and steel. The zinc coating can be applied in a number of ways depending on the size of the object to be coated. In HOT-DIP GALVANISING the object to be treated is dipped in a bath of molten zinc (Figure 6.52). For articles which are too large to be dipped into a galvanising bath molten zinc is sprayed on. Very small items which require a fine finish, such as nuts and bolts, can be coated electrolytically or the zinc can be applied by SHERARDISING, in which the articles are heated with zinc dust and sand in a slowly-rotating drum.

Zinc coatings usually range in thickness from $\frac{1}{500}$ cm to $\frac{1}{50}$ cm. Unlike tin or chromium, zinc does not form a simple coating on iron and steel. The zinc penetrates the iron to form an iron-zinc alloy at the surface of the articles. Galvanising is really a combination of alloying and plating. Most galvanised iron is used in the building and construction industries. The steel structure of the Forth Road Bridge is zinc-sprayed, the steel cables are hot-dip galvanised, and even the nuts and bolts are sherardised.

You can see pure zinc sheet on buildings in the form of roofing, gutters, and rainwater pipes, and as the outer casing of torch and transistor batteries.

Zinc alloys are also important. *Brass* is the alloy with copper (page 184). An alloy of zinc with 4 per cent aluminium is used for PRESSURE DIE-CASTING, in which the alloy is injected under pressure into a steel die, or mould. In this process thousands of identical castings can be made at a high production rate. Pressure die-casting is used to make a wide range of builders' hardware such as locks, door handles, and bathroom fittings, as well as for a wide variety of engineering components which need to be of accurate dimensions. In a car, for example, the door handles, carburettor, petrol pump, and many other parts may all be zinc alloy die-castings, sometimes plated with chromium to give them an attractive finish. You have probably owned scale model toys which were pressure die-cast.

Figure 6.53 shows the proportion of the total British zinc production used for various purposes.

Protective coatings 30%	Brass 30%	Die-casting 20%	Sheet 5%	Other uses 15%

Figure 6.53

Tin

Tin is a rather weak, soft metal. It is used mainly for making tinplate (steel with a thin coating of tin), which combines the strength and rigidity of steel with the attractive appearance and resistance to corrosion of tin.

Steel sheet is coated with tin either by electroplating (the most frequently used method) or by HOT-DIPPING, in which the steel strip is passed through a bath of molten tin. The thickness of the tin coating varies between $\frac{1}{3000}$ cm and $\frac{1}{500}$ cm. The thicker coatings are usually applied by the hot-dip method.

Tin is not poisonous and 90 per cent of all the tinplate made is used for tinning food. Tinplate is joined together to make food tins by soldering. *Solder* is an alloy of tin and lead: the tin sticks the edges of the metal together, the lead lowers the melting point. The strongest solder contains about 65 per cent of tin and 35 per cent of lead, though other proportions are used for special purposes. Solder can, of course, be used for joining many other metals, as well as tinplate.

Tin is also used in other alloys: with antimony and copper for making engine bearings and with copper as *bronze* (page 184).

Figure 6.54 shows the proportions of the total British tin production used for various purposes.

Tinplate 50%	Bearing metal 15%	Bronze 10%	Solder 10%	Other uses 15%

Figure 6.54

Metals for the future

In 1948 the total world production of TITANIUM was only 3 tonnes; tens of thousands of tonnes are now manufactured each year and production is still increasing rapidly. Titanium is a common element in the earth's crust (about $\frac{1}{2}$ per cent) but it is very expensive to extract and at present costs about £2000 per tonne. Titanium alloys are light and exceptionally strong, as well as resisting corrosion even at a temperature of 500°C. These properties make them more and more useful in the construction of supersonic aeroplanes, missiles, rockets, and space capsules.

TANTALUM is almost as resistant to corrosion as platinum, but considerably cheaper. It is being extracted on an increasing scale for use in the chemical engineering industry for reaction vessels which have to withstand the effects of concentrated acids and alkalis. It is also used in surgery for bone splints, screws, and skull plates which need to be left in the body.

NIOBIUM (called *columbium* in the United States) is another metal which, though still expensive (about £100000 per tonne), is becoming cheaper each year as more uses are found for it and it is extracted on a larger scale. It is a very strong metal and keeps its strength almost up to the very high temperature at which it melts (2500°C). The pure metal or its alloys are used in situations where this property is an advantage, for example in spacecraft and nuclear reactors.

Titanium, tantalum, and niobium are just three of the less common metals for which new uses are being found. Each year new inventions and new developments in technology present chemists and metallurgists with the problem of finding metals and alloys with special properties.

Questions

1 Aluminium's most useful property is its low density. Copper's is that it is an excellent conductor of electricity. What are the most useful properties of iron, lead, titanium, and zinc?

2 Water pipes are sometimes made of copper, sometimes of iron, and sometimes of lead. What are the advantages and disadvantages of using each of these metals for this purpose?

3 Name the different metals for the following uses:
 (a) in accumulators;
 (b) for galvanising iron;
 (c) as 'silver' paper. *(West Yorkshire)*

4 (a) Name an alloy which contains aluminium, and state one other metal present in the alloy. Give a use of the alloy you have named.
 (b) State three properties of aluminium which make it suitable for use in cooking utensils. *(West Midlands)*

5 Until 1920 silver coins of the realm were made of 'standard silver'—92.5 per cent silver, 7.5 per cent copper; then they became

50:50 silver and copper. In 1946 'silver' coins became cupro-nickel, 50 per cent copper 50 per cent nickel.

(a) Why is a pre-1920 silver coin valuable?

(b) You probably have some 'coppers' in your pocket or purse. What are they made of? *(London)*

6 Find out as much as you can about the use of any one metal which you find particularly interesting (not necessarily one of those mentioned in this chapter). You could either write about it or build up a pictorial chart, using drawings or pictures cut out of magazines.

For the teacher

Experiment 6.1 Calcium may be placed out of sequence, as its tenacious oxide film makes it difficult to ignite.

Experiment 6.2 A suitable mineral wool for use in this, and similar experiments is that sold under the trade-name *Rocksil*. **The use of asbestos wool in school laboratories should be avoided** (see DES *Safety in science laboratories*, Safety Series Pamphlet No. 2, 2nd edition, HMSO, 1976). In (b) the ribbon must be heated *strongly* until it starts to burn. The bunsen must then be transferred *quickly* to the asbestos wool. The experiment is best done by pairs of pupils so that one can manipulate the bunsen while the other holds the splint ready to ignite the hydrogen. Experiment (d) should be successful, but in (c) no hydrogen is formed.

The experiment can be extended by a demonstration of the reaction of sodium and potassium with water. Using tweezers, drop freshly-cut 2 mm cubes of the metals into a dish of water. A safety screen should be placed between the dish and the class. Pupils should **not** be allowed to react sodium and potassium with water themselves.

Experiment 6.3 2M hydrochloric acid and M sulphuric acid.

Experiment 6.4 The object of this experiment is *to test predictions*. Pupils should now have the activity series clear in their minds and must be encouraged to say (or write) what they think is going to happen *before* carrying out the experiments. **N.B. Combinations of metal and metal oxide other than those suggested should NOT be attempted by pupils; the reactions may be extremely violent.**

The teacher may wish to demonstrate a more violent reaction. Mix one spatula-full of magnesium powder with one of copper(II) oxide on a piece of asbestos paper resting on a tripod and gauze. Push a lighted bunsen underneath and stand well back. A safety screen should be placed between the experiment and the class. The result could be compared with that obtained from heating copper with magnesium oxide. A further possibility for extension is to moderate the reaction by adding some magnesium oxide to the mixture of magnesium powder and copper(II) oxide. There is then sufficient residue to heat with 2M hydrochloric acid and filter to obtain the copper.

Experiment 6.5 As an extension the teacher could demonstrate the thermal decomposition of silver nitrate or mercury(II) nitrate, where the metal is the final product.

Section 6.2 (pages 144–7) A good selection of ore samples adds interest to this topic. Free samples may be obtained as follows: copper pyrites (Copper Development Association), galena (The Long Rake Spar Co.), tinstone (Williams, Harvey & Co. Ltd), zincblende (The British Metal Corporation), and rock salt (I.C.I.). Limestone, chalk, and marble can be purchased from Griffin and George, and a wide selection of other specimens are available from Lythe Minerals or from R. F. D. Parkinson. The following alternative names are used: rock salt (halite), zincblende (sphalerite), calamine (smithsonite), tinstone (cassiterite), copper pyrites (chalcopyrite).

Experiment 6.6 Nuffield combustion tubes, as shown in Figure 6.9, may be obtained from Griffin and George. Alternatively, a 150 × 25 mm hard-glass tube may be used as shown in Figure 6.55.

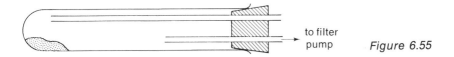

Figure 6.55

Experiment 6.7 The method of making reduction tubes is described on page 57. (a) gives copper very quickly; (b) is unsuccessful. The experiment can be extended to other oxides, but only lead(II) oxide is likely to give a positive result, and has the disadvantage that the lead fuses into the glass and ruins the tube.

Experiment 6.8 Lead(II) oxide and iron(III) oxide. 2M sulphuric acid. (a) is successful; (b) is not; (c) gives a magnetic product but some pupils may realise that this is not conclusive. Many samples of iron(III) oxide are magnetic and partial reduction to Fe_3O_4 would also give a magnetic product. Evolution of hydrogen with sulphuric acid is conclusive, but unfortunately the rate of production of gas is usually too slow for ignition to be possible.

Experiment 6.9 4 cm or longer brass screws and 22 s.w.g. copper wire. The electrolyte is 0.1 M copper(II) sulphate acidified with sulphuric acid. 12 volt DC supply.

Section 6.7 (pages 165–75) This section moves rather outside of the scope of the average CSE, or for that matter GCE syllabus. It justifies itself by its relevance to everyday experience (often overlapping work done in the metalwork department), as a particularly clear and straightforward example of the explanation of macroscopic properties in microstructural terms, and by the fact that most pupils enjoy it.

The author has found the following publications helpful.
Street, A. and Alexander, W. *Metals in the Service of Man* (Penguin, 1962)
Bailey, A. R. *A Textbook of Metallurgy* (Macmillan, 1960)
Nuffield Chemistry: The Sample Scheme Stage III (Longmans/Penguin, 1967)
A Guide to the Structure and Properties of Steel (British Steel Corporation; free)

Experiment 6.10 0.1 M lead(II) nitrate or lead(II) acetate. It is useful to have one or two samples of galvanised iron, on which large zinc grains are clearly visible, for passing round the class.

Etching It may be considered worthwhile to demonstrate the etching of a metal to make the grain structure visible. This is too tricky and time-consuming for a class experiment, but if most of the work is done before the lesson a demonstration of the final stage will help to make photomicrographs more comprehensible. Between 10 g and 20 g of plumber's solder can be used; it is readily obtainable from ironmongers or builder's merchants (multicore electrician's wire solder is not suitable as it contains flux). Alternatively a mixture of 35 per cent lead foil and 65 per cent granulated tin is melted in a crucible under powdered charcoal to prevent oxidation and cast into a small block using a mould of Plaster of Paris. One surface of the casting is smoothed by rubbing on a sheet of coarse emery cloth laid on a flat surface (a sheet of glass or metal). Successively finer grades of emery cloth are then used, down to the finest available. This is best carried out under water, and the casting should be rotated through 90° between each reduction in grade of paper. Finally the specimen is polished with a paste of magnesium oxide and water. The surface should now appear completely smooth and free from scratches when examined under a microscope at a magnification of ×100. Etching can be demonstrated by dipping the casting into a mixture of equal parts of concentrated nitric and acetic acids, rinsing with running

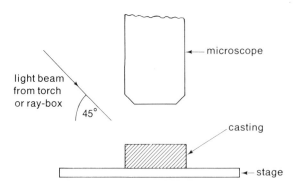

Figure 6.56

water, and drying with a paper tissue. Etch for half a minute and repeat as necessary until the desired effect is achieved. The grain structure can be examined under a microscope at a magnification of $\times 100$ with illumination arranged as shown in Figure 6.56.

Work-hardening Pupils can experience work-hardening for themselves by bending a piece of copper foil (15 × 5 cm) backwards and forwards until it snaps, and comparing this with the behaviour of lead foil.

X-ray diffraction An analogy with the diffraction of visible light aids understanding. Pupils can view a point source of light (such as a torch bulb) through a handkerchief or (better) a piece of *terylene* net curtain. A diffraction pattern is seen which rotates when the handkerchief is rotated and changes shape when the handkerchief is stretched. Pupils should realise from this that the diffraction pattern is related to the pattern of the threads in the cloth, which is visible with the naked eye or with a hand lens. A set of Nuffield diffraction grids makes it possible to extend this idea to a pattern which is too small to be visible. There are three types of grid with different patterns of dots. The pupils should first familiarise themselves with the diffraction patterns produced by each type, viewing the bulb with the grids on which the dot pattern is marked. They can then use the unmarked grids and by looking at the diffraction pattern deduce what the dot pattern must be, checking the accuracy of their deduction by looking at the grid under a microscope with a $\times 4$ objective.

The structure of alloys A number of useful demonstrations can be performed using polystyrene spheres in a wooden tray, about $\frac{1}{2}$ metre square with a 4 cm rim on three sides. To illustrate the way in which metal atoms are packed the tray is held at an angle of about 30° from the horizontal and about 100 white 38-mm spheres are rolled down. Most will pack automatically into a regular hexagonal structure but some dislocations and grain boundaries should be visible. The demonstration can be repeated with a mixture of 50 white and 50 coloured 38-mm spheres to simulate an alloy of the electrum or brass type. The possibility of the mixing of atoms within each grain will be apparent. If a mixture of 100 38-mm spheres with five or six 51-mm spheres is rolled down the distortion produced by the larger 'atoms' is evident and it can be pointed out that in an alloy of this type the two components prefer to gather in separate grains.

The fitting in of small atoms can be demonstrated with two triangles of side $25\frac{1}{2}$ cm and depth 3 cm, each with two sides made of wood and the third of perspex. A pyramid of 51-mm spheres can be built in one of the triangles with 15 spheres on the base, 10 in the second layer, 6 in the third, 3 in the fourth, and one on top. In the other triangle the pyramid can be built again, but it will be found possible to include a few coloured 19-mm spheres as the construction proceeds without distorting the structure.

Experiment 6.11 For (a) a small piece of plumber's solder, or a piece of lead-tin alloy prepared as on page 192. For (b), $1 \times 1 \times \frac{1}{2}$ cm blocks of tin, lead, and solder (or lead-tin alloy) are required, cast as on page 192. $\frac{1}{2}$ cm or 1 cm ball bearing.

Experiment 6.13 For cleaning: M sodium hydroxide and M nitric acid. The electrolyte for anodising is 3.5 M sulphuric acid. 12 volt DC supply with a 0–1 amp DC meter and a 100 ohm rheostat (e.g. wire-wound potentiometer from Radiospares). The current density is fairly critical; 1.0–1.5 amp per 100 cm^2 of total surface area of the object is suitable. A lower current does not produce an adequate anodic film in 15 minutes; a higher current gives a granular film which does not accept dye well. The dye solution contains 10 g of black 2 Y dye (available from Aluminium Federation) in 1 litre of water with 10 drops of concentrated acetic acid. The sealing solution contains 5 g of nickel(II) acetate and 5 g of boric acid per litre. The quality of the finish is not very good; the brilliance of professional finishes results from chemical or electrochemical brightening processes which are not practicable in the school laboratory.

Sacrificial protection A quick demonstration of a chemical cell is helpful (this is returned to in more detail in chapter 10). Connect a 10×2 cm strip of copper foil to the positive terminal of a DC voltmeter (between 1 V and 5 V full scale deflection) and a similar strip of zinc foil to the negative terminal. When the two foils are dipped into a beaker of M sulphuric acid about 1 volt will be recorded.

Experiment 6.14 The 7-cm nails and 7×1 cm zinc and tin foils can be supported upright by clipping to the sides of 100 cm^3 beakers with the crocodile clips on the end of the wires.

7 Acids

ACIDS are an important group of chemical compounds. Some of them have a pleasantly sour taste. Oranges and lemons (citrus fruits) contain *citric acid* (which is also used in making acid drop sweets and sherbet), apples contain *malic acid*, and vinegar is a solution of *acetic acid*. In fact the word *acid* comes from a Latin word *acidus*, meaning *sour*. Other acids are poisonous, such as *carbolic acid* (also called *phenol*), and *prussic acid*. Others are highly corrosive, such as sulphuric acid, used in car batteries, and nitric acid, used in the manufacture of fertilisers and explosives.

BASES are substances that react with acids. Bases which are soluble in water are also sometimes known as ALKALIS. The reaction of an acid with a base is called NEUTRALISATION. Some plants will not grow in soil which contains too much acid. *Lime* is used to neutralise the acidity of soil.

Figure 7.1 *Neutralisation on a large scale. Lime, a base, is spread on soil which contains too much acid to grow good crops.*

Your stomach contains acid which helps in the digestion of food, but too much acid causes indigestion. Stomach powders contain a base, usually *sodium bicarbonate*, to neutralise the extra acid. Acid left in your mouth after a meal slowly corrodes your teeth. Toothpaste contains a base to neutralise acid in your mouth. Some bases, like sodium bicarbonate, are quite harmless; others, like lime, are poisonous and corrosive.

7.1 Testing for acids and bases

All acids have a sour taste, but it would certainly not be safe to use taste as a test for acids. Fortunately, acids have another important property: they change the colour of certain dyes. Many of the substances responsible for the colours of flowers and vegetables are affected by acids. Most of these are not much use for laboratory work because they are slowly destroyed by the oxygen in the air. But one of them, called *litmus*, which is obtained from a type of moss, has been used by chemists as a test for acids and bases for several hundred years. Litmus is turned red by acids and blue by bases. Nowadays various man-made dyes are used instead of litmus. Some of the commoner ones are shown in Figure 7.2. Dyes which change colour in the presence of acids and alkalis are called INDICATORS.

Indicator	Colour with acids	Colour with bases
Litmus	Red	Blue
Methyl orange	Red	Yellow
Phenolphthalein	Colourless	Red

Figure 7.2

Simple indicators will tell us whether a substance is an acid, a base, or neither, but they will not tell us how strong an acid or base is. A more complicated indicator, called UNIVERSAL INDICATOR, which is made by mixing together several simple indicators, will do this. It changes through a whole range of different colours according to the strength of the acid or base.

The strength of acids or bases is measured on a scale from 0 to 14, called the *pH scale* (Figure 7.3). Universal indicator is supplied by the manufacturers along with a colour chart which gives the pH number corresponding to each colour.

pH 7 is NEUTRAL (neither acidic nor basic).

Acids have a pH of less than 7; the *smaller* the pH number the stronger the acid.

Bases have a pH of more than 7; the *larger* the pH number the stronger the base.

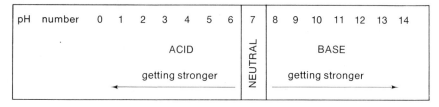

Figure 7.3 The pH scale

Experiment 7.1 Ways of detecting acidity

(a) Use beetroot juice, red cabbage juice, or some other plant extract to test each of the substances provided. If the substance is a liquid pour 1 cm depth into a test tube and add 5 drops of plant extract from a teat pipette. If the substance is a solid, dissolve a spatula-full in 1 cm depth of water in a test tube and add 5 drops of plant extract.

Make a list of the substances you have tested and write down the colour that each gives with the plant extract. Can you divide the substances into two groups according to the colour they give? What colour do the acids give and what colour do the bases give?

(b) Pour 1 cm depth of dilute hydrochloric acid into a test tube and add 5 drops of the coloured plant extract. Notice the colour. Now add some sodium hydroxide solution (a base), a few drops at a time. What do you think has happened to the acid? Can you guess what will happen if you add some more acid? Try it and see if you were right.

(c) The coloured plant extract can be used to find out if a substance is an acid or a base. There are better indicators which will tell you how acidic or how basic a substance is. Test the substances you used in experiment (a) with universal indicator paper.

If the substance is a liquid dip a glass rod into it and touch the end of the rod on to a piece of universal indicator paper. If the substance is a solid, dissolve a spatula-full in 1 cm of water in a test tube and then use a glass rod to transfer a drop of your solution on to a piece of universal indicator paper.

For each substance use the colour chart to find its pH number. Make a table with three columns headed 'substance', 'colour of universal indicator', and 'pH number' and list the results of your experiment.

Experiment 7.2 Curing acidity

(a) Measure 10 cm^3 of vinegar into a beaker, using a measuring cylinder. Add 10 cm^3 of water. Find the pH of your solution by using a glass rod to transfer a drop on to a piece of universal indicator paper. Make a note of the pH.

Add a spatula-full of lime to the vinegar. After stirring thoroughly, find the pH as before, and write it down. Continue adding spatulas-full of lime, stirring and measuring the pH after each addition, until you have added 6 spatulas-full altogether.

Use your results to draw a graph, planning your graph paper as in Figure 7.4.

Write a brief explanation of why the pH changes in the way it does as more and more lime is added.

(b) Measure 10 cm^3 of vinegar and 10 cm^3 of water into a beaker. Find the temperature of the solution and make a note of it.

Add a spatula-full of lime to the vinegar, stir thoroughly, and measure the temperature again. Do not forget to write it down.

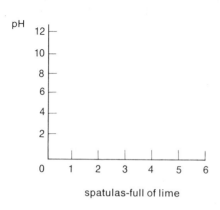

Figure 7.4

Continue adding spatulas-full of lime, stirring and measuring the temperature after each addition, until you have added 6 spatulas-full altogether.

Use your results to draw a graph, with temperature on the vertical axis and number of spatulas-full of lime on the horizontal axis.

Why does the temperature rise when lime is added to vinegar? Why does the temperature stop rising after a certain number of spatulas-full of lime have been added?

Questions

1 In the process of neutralisation which of the following happens?
A   Ions are converted to neutral atoms.
B   Neutrons are involved in radioactivity.
C   An acid reacts with a base.
D   A halogen reacts with hydrogen.
E   An acidic oxide is dissolved in water.   (*East Anglia*)

2 A base that is soluble in water is called _____.   (*London*)

3 Describe a neutralisation process which may take place (a) in the soil, (b) in the use of certain medicines. Mention the substances involved in each process.   (*East Anglia*)

4 Some acid from a car battery has been spilt on the garage floor. What TWO materials which you are likely to find in the kitchen could

you use to deal with this accident? Give reasons for your choice.
(South East)

5 (a) State any *two* facts that are true of *all* acids.
(b) Name *two* coloured substances (indicators) that will tell us whether an unknown liquid is alkaline or acidic. *(Wales)*

6 You suspect that the flowers in your window-box are not growing well because the soil is too acid. Which of the following should you add to the soil?
A Sugar
B Tea-leaves
C Fertiliser
D Baking powder
E Salt

7 Acid drops, lemons, and vinegar all have a sour taste.
(a) What pH value would you expect them to have?
(b) Which of the following substances would be most effective for removing the sour taste from lemon juice?
A Sugar
B A substance with pH 7
C A substance with pH below 6
D A substance with pH above 8
E Water

8 When acid is added to alkali which one of the following happens?
A Energy is released.
B There is no energy change.
C Hydrogen is evolved.
D Energy is absorbed.
E Oxygen is evolved.

7.2 Neutralisation: making salts from acids

When an acid and a base react together and neutralise each other a new type of compound is formed, called a SALT, together with water.

Experiment 7.3 The preparation of magnesium sulphate crystals

Measure 25 cm^3 of dilute sulphuric acid into a beaker. Add magnesium carbonate, a spatula-full at a time, until no more will react.

How can you tell that you have added enough magnesium carbonate to use up all the sulphuric acid? What is the gas produced in the reaction? If necessary carry out a suitable test to confirm the identity of the gas produced. Write a word equation for the reaction which has taken place.

Heat the mixture until it is nearly boiling (why?) and filter it into a crystallising dish. Put the dish aside for a day or two to allow the solution to crystallise.

Experiment 7.4 The preparation of copper(II) sulphate crystals

Measure 25 cm³ of dilute sulphuric acid into a beaker. Place the beaker on a gauze on a tripod and heat the acid until it is nearly boiling.

Stop heating and add copper(II) oxide, a spatula-full at a time, stirring after each addition. Continue adding copper(II) oxide until, after stirring, some remains undissolved.

Boil the solution for about half a minute, then filter it into an evaporating dish. Evaporate until the solution is reduced to about half of its original volume. Put the dish aside for a day or two to allow the solution to crystallise.

Why is the sulphuric acid heated before the copper oxide is added? Why is an excess of copper(II) oxide used? Write a word equation to represent the reaction which has taken place.

Experiment 7.5 The preparation of sodium chloride (salt) crystals

Measure 10 cm³ of sodium hydroxide solution into a conical flask. Fill a burette with dilute hydrochloric acid and clamp it above the flask as shown in Figure 7.5.

Add 1 cm³ of hydrochloric acid to the flask and swirl the solution around the flask to mix it. Use a glass rod to transfer one drop of the solution to a piece of universal indicator paper. Has sufficient acid been added to neutralise the alkali?

Add a further 3 cm³ of acid, 1 cm³ at a time, testing the pH after each addition. Then continue adding acid ONE DROP at a time, again testing the pH after each addition, until the solution in the flask is neutral.

Pour the contents of the flask into an evaporating basin and evaporate until only a few drops of liquid remain. Using tongs transfer the basin to a steam bath as shown in Figure 7.6 and continue heating until the solution has been evaporated to dryness.

Examine the crystals with a hand lens and make a drawing of their shape.

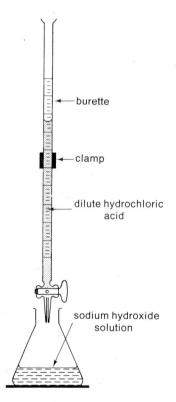

Figure 7.5

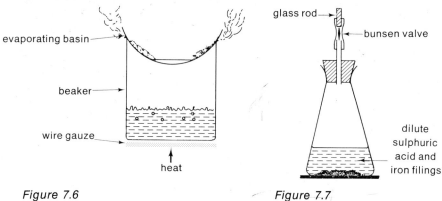

Figure 7.6   Figure 7.7

Why is a steam bath used in the last stages of the evaporation? Write a word equation for the reaction which has taken place.

Experiment 7.6 The preparation of iron(II) sulphate crystals

Measure 25 cm^3 of dilute sulphuric acid into a conical flask and add 5 spatulas-full of iron filings. Put a rubber bung with a bunsen valve into the flask (Figure 7.7).

A bunsen valve consists of a piece of rubber tubing with a small vertical slit, closed at the upper end with a piece of glass rod. It allows gas to escape from the flask but prevents air from getting in, which could oxidise iron(II) sulphate to iron(III) sulphate.

Stand the flask on a gauze, on a tripod; warm the flask until gas is given off steadily. Continue warming until the reaction stops, **but do not boil**.

Allow the solution to cool, then filter it into another conical flask. Refit the bunsen valve and boil the solution until about half of the water has evaporated away. Allow the solution to cool, pour it into a crystallising dish, and put the dish aside for a day or two to allow the solution to crystallise.

Why is more iron used than is needed? Why is the mixture heated while the reaction is taking place? What is the gas given off? (**Do not try to test it**.) Write a word equation for the reaction which has taken place.

The reaction between hydrochloric acid and sodium hydroxide

If dilute hydrochloric acid is added, a little at a time, to a solution of sodium hydroxide, which is a base, the acid and base neutralise each other. The progress of the neutralisation can be followed with the aid of universal indicator. Sodium hydroxide solution turns the indicator purple. As hydrochloric acid is added the colour of the indicator changes through blue to green, which is neutral.

The hydrochloric acid and sodium hydroxide have reacted together to produce a salt, sodium chloride (sometimes called *common salt*) and water:

sodium hydroxide + hydrochloric acid → sodium chloride + water

NaOH(aq) + HCl(aq) → NaCl(aq) + H_2O(l).

The sodium chloride can be recovered from the solution by evaporation. Other salts can be made in the same way. For example:

potassium hydroxide + sulphuric acid → potassium sulphate + water

calcium hydroxide + nitric acid → calcium nitrate + water.

Acid	Name of salts
Acetic acid	Acetate
Hydrobromic acid	Bromide
Hydrochloric acid	Chloride
Hydriodic acid	Iodide
Nitric acid	Nitrate
Phosphoric acid	Phosphate
Sulphuric acid	Sulphate

Salts are named according to the acid from which they came. Salts from hydrochloric acid are called CHLORIDES, salts from sulphuric acid are called SULPHATES, and so on. Figure 7.8 gives the names of salts formed from various acids.

Figure 7.8

The reaction between metals and acids

Metals which are fairly high in the activity series (from tin upwards in the table on page 141) *react with acids to form a solution of a salt and hydrogen gas.*

If iron is added to dilute sulphuric acid it dissolves and bubbles of gas can be seen. When sufficient iron has been added to react with all the acid the reaction stops and a solution of iron(II) sulphate remains, from which solid iron(II) sulphate can be recovered by evaporation:

iron + sulphuric acid → iron(II) sulphate + hydrogen

Fe(s) + H_2SO_4(aq) → $FeSO_4$(aq) + H_2(g).

Other metals and acids react similarly. For example:

zinc + hydrochloric acid → zinc chloride + hydrogen

magnesium + nitric acid → magnesium nitrate + hydrogen.

The reaction between metal oxides and acids

Metal oxides are bases, and usually react with acids to form salts, together with water.

If copper(II) oxide is added to dilute sulphuric acid it dissolves, forming a blue solution. Eventually no more copper(II) oxide will dissolve: the acid has been neutralised by the base. When any excess copper(II) oxide which has not reacted is filtered off, a clear blue solution remains, which can be evaporated down to leave crystals of copper(II) sulphate:

copper(II) oxide + sulphuric acid → copper(II) sulphate + water
$$CuO(s) + H_2SO_4(aq) \rightarrow CuSO_4(aq) + H_2O(l).$$

The reaction between metal carbonates and acids

Metal carbonates react with acids to form salts, together with carbon dioxide gas and water.

If zinc carbonate is added to dilute sulphuric acid there is a vigorous reaction and carbon dioxide is produced. After filtering off any unreacted zinc carbonate a clear solution of the salt, zinc sulphate, remains, from which solid zinc sulphate can be obtained by evaporation:

zinc carbonate + sulphuric acid →

zinc sulphate + carbon dioxide + water
$$ZnCO_3(s) + H_2SO_4(aq) \rightarrow ZnSO_4(aq) + CO_2(g) + H_2O(l).$$

Ways of making soluble salts
1 Metal hydroxide + acid   3 Metal oxide + acid
2 Metal + acid   4 Metal carbonate + acid

Making insoluble salts

All of these four methods can only be used to make salts which are soluble in water, because all of them involve making a solution of the salt, from which the solid salt can be obtained by evaporation. Salts which are INSOLUBLE in water have to be made by another method.

Silver chloride, for instance, can be made by adding sodium chloride solution to a solution of silver nitrate:

silver nitrate + sodium chloride → silver chloride + sodium nitrate.

This process is known as DOUBLE DECOMPOSITION or a *change of partners* reaction. Solid silver chloride is thrown out of solution (PRECIPITATED) and can be filtered off.

All salts are electrolytes. A solution of silver nitrate contains silver ions and nitrate ions:

$$AgNO_3(aq) = Ag^+(aq)\ NO_3^-(aq).$$

Sodium chloride solution contains sodium ions and chlorine ions:

$$NaCl(aq) = Na^+(aq)\ Cl^-(aq).$$

When the two solutions are mixed the ions change partners. The silver ions and chlorine ions join up to form silver chloride, which is insoluble in water and is precipitated:

$$Ag^+(aq) + Cl^-(aq) \rightarrow AgCl(s).$$

This is called an IONIC EQUATION. It shows us what is happening to the ions in the reaction. The other ions, the sodium ions and nitrate ions, are not involved in the reaction; they are left as a solution of sodium nitrate at the end.

Similarly the insoluble salt, lead chloride, can be made by mixing solutions of lead nitrate and copper chloride. Lead chloride is precipitated and can be filtered off:

lead nitrate + copper chloride → lead chloride + copper nitrate.

The lead ions from the lead nitrate and the chlorine ions from the copper chloride have joined up:

$$Pb^{2+}(aq) + 2Cl^-(aq) \rightarrow PbCl_2(s).$$

The copper ions and nitrate ions are left as a solution of copper nitrate.

Lead nitrate and copper chloride are not the only combination that will produce lead chloride. Lead acetate and sodium chloride would do just as well, in fact any soluble lead salt and any soluble chloride can be used.

Experiment 7.7   Preparing some insoluble salts

(a) To 1 cm depth of lead nitrate solution in a test tube add 1 cm depth of sodium chloride solution. Describe what you see happening.

Write a word equation for the change of partners reaction which has taken place. Which of the two products of this reaction has been precipitated? (If you are not sure refer to Figure 7.10, page 208.)

Write an ionic equation to show how the insoluble salt was formed. (If you do not know the formula of this salt work it out from Figure 5.11, page 122.)

(b) Put 1 cm depth of silver nitrate solution into each of three test tubes. To the first tube add 1 cm depth of sodium chloride solution, to the second the same amount of potassium chloride solution, and to the third the same amount of dilute hydrochloric acid. Describe what you see happening in the three tubes.

Write word equations for the reactions which have taken place in the three tubes. What is the insoluble salt formed in each case? Write an ionic equation to show how this salt was formed. Does the same ionic equation explain what has happened in all of the tubes?

(c) Mix equal volumes of the following pairs of solutions. Try to work out what the precipitate is in each case, and write an ionic equation to show how it is formed.
  (i) Lead nitrate solution and potassium iodide solution.
  (ii) Calcium chloride solution and copper sulphate solution.
  (iii) Sodium carbonate solution and copper sulphate solution.

Questions

1 Name *one* alkali and name the salts formed when it is neutralised by (i) nitric acid and (ii) hydrochloric acid. (*Wales*)

2 (a) Salts are prepared by various methods. From the following list select:
  (i) a salt which may be prepared by the action of an acid on a metal,
  (ii) one which may be prepared by the action of an acid on an oxide,
  (iii) one which may be prepared by precipitation (double decomposition).
In each case, name the salt and the substances you would use for the preparation.
  Zinc sulphate, copper(II) sulphate, lead chloride, iron(II) sulphide, lead nitrate, zinc chloride.
  (b) Write a step-by-step account of the way in which you would prepare a pure sample of zinc sulphate.
  (c) Solutions of salts contain ions. Name and give the formulae of the ions present in a solution of (i) zinc sulphate, (ii) potassium chloride, (iii) any other salt of your choice. (*South East*)

3 Describe, giving full details of the experiment, how you would prepare a crystalline specimen of sodium sulphate, starting from sodium hydroxide. (*Middlesex*)

4 Read the following instructions for the preparation of copper(II) sulphate crystals.
'Add copper(II) oxide a little at a time to warm dilute sulphuric acid until no more will dissolve. Filter off the excess copper(II) oxide and evaporate the solution until a small sample produces crystals when cooled. Allow the bulk of the solution to stand until crystals are formed. Remove the crystals from the remaining solution, wash them with a little distilled water and dry between sheets of filter paper.'
  (a) (i) Why is *warm* sulphuric acid used?
  (ii) Why is the copper(II) oxide added to excess?
  (iii) Why is the solution not evaporated to dryness to obtain the copper(II) sulphate?
  (iv) Why are the crystals washed before final drying?
  (b) The excess copper(II) oxide is removed by filtration. What is the principle on which filtration works?
  (c) In the above reaction the formula mass of copper(II) oxide reacts exactly with the formula mass of sulphuric acid. Calculate the weight of copper(II) oxide which will just neutralise one-tenth of a mole of sulphuric acid. (Atomic masses: Cu = 64, O = 16.)
  (*North West*)

5 Magnesium sulphate can be prepared in the laboratory by adding excess magnesium to dilute sulphuric acid.
   (a) What is the other product of this chemical change?
   (b) Why is it important to avoid excess acid?
   (c) Describe the necessary steps to obtain a pure sample of the crystals starting with magnesium sulphate solution.
   (d) What would be the colour of the crystals obtained?
   (e) For which of the following metals could the sulphate salt be prepared in the laboratory by the action of metal and dilute acid: aluminium, copper, iron, lead, zinc? *(East Anglia)*

6 When dilute solutions of calcium chloride and sodium carbonate are mixed which of the following is formed?
   A A white precipitate of sodium chloride
   B Carbon dioxide
   C A white precipitate of calcium carbonate
   D Precipitates of sodium chloride and calcium carbonate
   E Sodium and calcium metals *(North West)*

7 People who suffer from too much hydrochloric acid in the stomach take medicine containing magnesium hydroxide, $Mg(OH)_2$. Explain how this compound would counteract acidity. *(Wales)*

8 The following are full equations for double decompositions in which insoluble salts are produced:
   (a) $AgNO_3(aq) + NaCl(aq) \rightarrow AgCl(s) + NaNO_3(aq)$
   (b) $BaCl_2(aq) + CuSO_4(aq) \rightarrow BaSO_4(s) + CuCl_2(aq)$
   (c) $MgCl_2(aq) + Na_2CO_3(aq) \rightarrow MgCO_3(s) + 2NaCl(aq)$
   (d) $Pb(NO_3)_2(aq) + 2KI(aq) \rightarrow PbI_2(s) + 2KNO_3(aq)$.
In each case write an ionic equation to show the formation of the insoluble salt from its ions.

9 Consider the following equations which all involve the reaction of solids with acids.
   (i) $CaCO_3(s) + 2HCl(aq) \rightarrow CaCl_2(aq) + CO_2(g) + H_2O(l)$
   (ii) $Na_2SO_3(s) + 2HCl(aq) \rightarrow 2NaCl(aq) + SO_2(g) + H_2O(l)$
   (iii) $FeS(s) + H_2SO_4(aq) \rightarrow FeSO_4(aq) + H_2S(g)$
   (iv) $PbO(s) + 2HNO_3(aq) \rightarrow Pb(NO_3)_2(aq) + H_2O(l)$
These equations can be written more correctly by showing ions where they exist and may then be simplified by leaving out any ions which do not react.

e.g. (i) $CaCO_3(s) + 2HCl(aq) \rightarrow CaCl_2(aq) + CO_2(g) + H_2O(l)$
$Ca^{2+} . CO_3^{2-} + 2H^+ + 2Cl^- \rightarrow Ca^{2+} + 2Cl^- + CO_2 + H_2O$
$CO_3^{2-}(s) + 2H^+(aq) \rightarrow CO_2(g) + H_2O(l)$

Carry out the same process with the reactions (ii), (iii) and (iv) respectively. *(North West)*

3 Water and salts

Experiment 7.8 Investigating the action of heat on copper(II) sulphate crystals

About one-third-fill a test tube with copper(II) sulphate crystals and set up the apparatus shown in Figure 7.9.

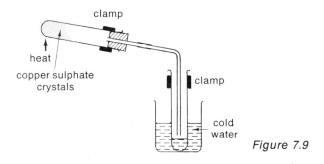

Figure 7.9

Heat the test tube containing the copper(II) sulphate crystals *gently* until the liquid driven off collects in the other test tube. Continue to heat gently until no more vapour is given off. **Make sure the delivery tube does not dip below the level of the liquid being collected.** Describe the changes you see in the appearance of the copper(II) sulphate.

Take the apparatus apart. Clamp vertically the test tube containing the liquid. Clamp a thermometer inside this test tube with the bulb about 5 cm above the level of the liquid. Heat the test tube with a small flame until the liquid is boiling. The thermometer will now be registering the boiling point of the liquid. What is it? What do you think the liquid is?

When the residue left from heating the copper(II) sulphate is cold put two spatulas-full of it on to a watch glass. Hold the watch glass in the palm of your hand. With a teat pipette, add 5 or 6 drops of the liquid you have been collecting to the solid in the watch glass. In what way does the appearance of the solid change? Do you notice anything else?

Experiment 7.9 Investigating the action of heat on cobalt chloride crystals

Put one spatula-full of cobalt chloride crystals into a small test tube. Describe the appearance of this salt.

Using a test tube holder, heat the tube *gently*. Watch carefully what happens, and note down everything you see.

Allow the tube to cool, then add two drops of water from a teat pipette. What happens now?

Water of crystallisation

When a solution of a salt is allowed to evaporate, water molecules often join on to ions of the salt, and the crystals produced contain water, called WATER OF CRYSTALLISATION. Any particular salt usually has a fixed amount of water of crystallisation. In copper(II) sulphate crystals, for instance, each mole of copper(II) sulphate crystallises with five moles of water, so the formula of the crystals is $CuSO_4.5H_2O$. Other examples are given in Figure 7.10.

Salt	Solubility in water	Number of moles of water of crystallisation per mole of salt	Formula of crystals of salt
Calcium chloride	Soluble	6	$CaCl_2.6H_2O$
Calcium sulphate	Insoluble	2	$CaSO_4.2H_2O$
Copper(II) sulphate	Soluble	5	$CuSO_4.5H_2O$
Iron(II) sulphate	Soluble	7	$FeSO_4.7H_2O$
Lead chloride	Insoluble	None	$PbCl_2$
Lead nitrate	Soluble	None	$Pb(NO_3)_2$
Magnesium sulphate	Soluble	7	$MgSO_4.7H_2O$
Potassium bromide	Soluble	None	KBr
Potassium chloride	Soluble	None	KCl
Potassium iodide	Soluble	None	KI
Potassium nitrate	Soluble	None	KNO_3
Sodium chloride	Soluble	None	$NaCl$
Sodium sulphate	Soluble	10	$Na_2SO_4.10H_2O$
Zinc sulphate	Soluble	7	$ZnSO_4.7H_2O$

Figure 7.10

Salts containing water of crystallisation are known as HYDRATES (from the Greek work *hydor*, meaning *water*). Salts which have no water of crystallisation are called ANHYDROUS (meaning *without water*). If a hydrated salt is heated the water of crystallisation can often be driven off, leaving the anhydrous salt. Addition of water to the anhydrous salt gives the hydrated salt back again.

The blue crystals of hydrated copper(II) sulphate lose their water of crystallisation when they are heated, leaving anhydrous copper sulphate, which is a white powder. Addition of water to the white powder gives blue hydrated copper(II) sulphate back again. This colour change is used as a test for water. *If an unknown liquid turns anhydrous copper(II) sulphate blue it must contain water.*

Cobalt chloride also changes colour when it loses its water of crystallisation. The hydrated salt is red. When it is heated it loses its water of crystallisation and turns blue, and the blue anhydrous salt turns red again when water is added. Cobalt chloride can therefore also be used as a test for water. However, these colour changes are unusual; most salts do not change colour when they lose their water of crystallisation.

Losing water of crystallisation to the air

Many hydrated salts, such as copper(II) sulphate, are quite stable. They only lose their water of crystallisation when they are heated fairly strongly. But some hydrated salts lose their water of crystallisation slowly without any heating. These are called EFFLORESCENT compounds. Sodium sulphate crystals ($Na_2SO_4.10H_2O$), for instance, which are colourless and transparent, lose all of their water of crystallisation when they are exposed to the air and become white and powdery. Sodium carbonate crystals ($Na_2CO_3.10H_2O$, known as *washing soda*) also effloresce when they are exposed to the air. From colourless transparent crystals they change to a white powder containing only one mole of water of crystallisation per mole of sodium carbonate ($Na_2CO_3.H_2O$).

Taking in water from the air

Some substances, instead of losing water to the air, take in water from the air and dissolve in it to form a solution. These are known as DELIQUESCENT substances. Examples are calcium chloride and sodium hydroxide. Deliquescent substances are useful as drying agents. They can be used to remove traces of water from gases, liquids, or solids. Solids can be dried by leaving them for a few hours in a DESSICATOR (Figure 7.11), gases by passing them through a tube containing anhydrous calcium chloride, and liquids by shaking them with lumps of anhydrous calcium chloride.

Figure 7.11

Some liquids also take in water from the air. They are called HYGROSCOPIC. Examples of hygroscopic substances are concentrated sulphuric acid and pure alcohol. They also find uses as drying agents.

Acids 209

Questions

1 Which of the following best describes an anhydrous substance?
A It never contains water.
B It is always a powder.
C It is not a salt.
D It always changes colour when water is put on it. *(Lancashire)*

2 Anhydrous copper sulphate turned blue when added to a colourless liquid. Which of the following best explains this?
A The liquid was water.
B The liquid was dilute sulphuric acid.
C The liquid was impure.
D The liquid was pure.
E The liquid contained water. *(North West)*

3 Sodium hydroxide, carbon dioxide, sodium carbonate (washing soda), calcium sulphate, calcium carbonate, calcium hydroxide.
   Choosing from the above list (and you may choose a given substance more than once) name
   (a) an efflorescent substance
   (b) a deliquescent substance
   (c) a hydrated salt. *(East Anglia)*

4 The following substances are allowed to remain in open dishes in the laboratory for two weeks:
   (a) concentrated sulphuric acid
   (b) iron filings
   (c) ether
   (d) sodium hydroxide pellets.
   State (i) whether or not each substance would increase or decrease in weight and (ii) what the residue, if any, would be in each dish.
   (West Yorkshire)

5 Washing-soda crystals (sodium carbonate decahydrate $Na_2CO_3.10H_2O$) effloresce in air to become the monohydrate ($Na_2CO_3.H_2O$). A packet of the crystals is marked *572 g (when packed)*.
   (i) Calculate the formula mass of the sodium carbonate decahydrate. (Atomic masses: $H=1$, $C=12$, $O=16$, $Na=23$)
   (ii) Explain why the words 'when packed' are included on the packet.
   (iii) Show all your working and calculate the weight of the packet if all the crystals had completely turned to the monohydrate.
   (Lancashire)

6 You will need to use the following atomic masses: $Ca=40$, $Mg=24$, $Cl=35.5$, $S=32$, $H=1$, $O=16$.

(a) The formula of anhydrous calcium chloride is $CaCl_2$. Write down its formula mass. When water vapour is passed over anhydrous calcium chloride its formula mass becomes 219. Calculate the number of moles of water of crystallisation per mole of calcium chloride.

(b) Magnesium sulphate crystallises out of solution as $MgSO_4.7H_2O$. Calculate the percentage loss in weight when one mole is completely dehydrated. (*East Anglia*)

4 Modern ideas about acids

Experiment 7.10   Are acids always acidic?

(a) Put about 1 cm depth of anhydrous acetic acid into a *dry* test tube and drop in a piece of universal indicator paper. (Do not touch the indicator paper: handle it with tweezers.)

Hold a crystal of tartaric acid with tweezers and rub it on a piece of universal indicator paper.

What happens to the universal indicator paper in each case? Do these acids have the effect on universal indicator paper that you expected?

(b) Put about 1 cm depth of anhydrous acetic acid into each of two *dry* test tubes. Drop a piece of magnesium ribbon into one tube and a marble chip (calcium carbonate) into the other.

What happens in each of the two tubes? What might you have expected would happen?

(c) In each of three test tubes put 1 cm depth of anhydrous acetic acid with 2 cm depth of water. Drop a piece of universal indicator paper into one tube, a piece of magnesium ribbon into the second, and a marble chip into the third.

Describe what happens in each of the three tubes. Are the results what you might expect?

(d) Dissolve a few crystals of tartaric acid in half a test tube of water. Divide the solution between three test tubes. Drop a piece of universal indicator paper into the first tube, a piece of magnesium ribbon into the second, and a marble chip into the third.

Describe what happens in each of the three tubes. Are the results what you might expect? What conclusions can you draw as to the conditions necessary for a substance called an acid to show acidic properties?

Experiment 7.11   Investigating the part that water plays in acidity

You will be provided with two solutions of the gas hydrogen chloride. One is a solution in water. The other is a solution in a liquid called toluene. This second solution contains no water at all.

(a) Carry out the following tests on each of the two solutions separately. Use CLEAN, ABSOLUTELY DRY test tubes for each test.
(i) Drop a piece of universal indicator paper into 1 cm depth of each solution in a test tube. (Handle the indicator paper with tweezers: do not touch it with your fingers.)
(ii) Drop a piece of magnesium ribbon into 1 cm depth of each solution in a test tube.
(iii) Drop a marble chip (calcium carbonate) into 1 cm depth of each solution in a test tube.
(iv) Use the apparatus of Figure 5.1, page 110, to test the electrical conductivity of the two solutions. Use 3 cm depth of each solution in a boiling tube. Test the toluene solution first.

Make a note of the results of each of these eight tests. Does the solution of hydrogen chloride in water show the properties of an acid? Does the solution of hydrogen chloride in toluene show the properties of an acid? What conclusion can you draw as to the part that water plays in the behaviour of acids?

(b) Put 3 cm depth of the solution of hydrogen chloride in toluene into a boiling tube and add an equal volume of water. Cork the tube and shake it. Pour away the upper toluene layer. Carry out the four tests in experiment (a) on the lower water layer.

Make a note of the results of these four tests. What are the properties of the water layer? Can you explain how the water layer has come to acquire these properties?

The properties of acids

These are the main properties of acids.

1 They have a sour taste.
2 They change the colour of indicators.
3 They are neutralised by bases.
4 They react with some metals (those which are high in the activity series), giving hydrogen gas.
5 They react with carbonates (e.g. calcium carbonate), giving carbon dioxide gas.
6 They conduct electricity and are decomposed by the current, giving hydrogen gas at the cathode.

Acids do not always show these acidic properties. Absolutely pure acetic acid, for instance, does not affect indicators, will not react with metals or metal carbonates, and does not conduct electricity. Only when acetic acid is dissolved in water does it show all the usual properties of acids. Similarly, hydrogen chloride gas shows none of the usual properties of acids in the absence of water, but a solution of hydrogen chloride in water (which we call hydrochloric acid) does all the things we normally expect of an acid.

It seems that *water must be present for acids to show acidic properties*. In 1923 a Norwegian chemist, Johannes Brønsted, put forward a theory to explain this. He suggested that when, for example, hydrogen chloride dissolves in water, it gives up a hydrogen ion to the water, forming a HYDRONIUM ION, together with a chloride ion:

$$HCl(g) + H_2O(l) \rightarrow \underset{\text{hydronium ion}}{H_3O^+(aq)} + Cl^-(aq).$$

Two main pieces of evidence support Brønsted's theory.
1 A solution of an acid in water conducts electricity, and this suggests that it contains ions. In the absence of water acids do not conduct, which suggests that pure acids do not contain ions.
2 When acids dissolve in water heat is produced, suggesting that a reaction takes place between the water and the acid.

A hydrogen atom consists of a single proton and an electron (page 84). In forming a hydrogen ion the atom loses its electron, so a hydrogen ion is really just a proton. Brønsted defined an acid as a substance which can give up a hydrogen ion, or proton, to another substance.

An acid is a proton donor.

He defined a base as a substance which can receive a hydrogen ion, or proton, from an acid.

A base is a proton acceptor.

In the example above, hydrogen chloride is an acid, since it gives up a proton. Water is a base, because it receives the proton from the acid.

Water is the only common solvent to which acids can donate protons. There are a few other solvents which can behave in this way, but they are much less important. One is liquid ammonia, but this only remains liquid below $-33°C$.

The six properties listed on page 212 are really the properties of the hydronium ion. Acids show these properties only in the presence of water, because only then are they capable of forming hydronium ions. It is the hydronium ion which our taste buds register as sour, and which changes the colour of indicators.

The reaction of an alkali, such as sodium hydroxide (NaOH), with an acid is really the reaction between the hydroxide ions of the alkali and the hydronium ions from the acid:

$$H_3O^+(aq) + OH^-(aq) \rightarrow 2H_2O(l).$$

The hydronium ion is acting as an acid, donating a proton to the hydroxide ion, which is a base.

The reactions of acids with metals, metal oxides, and metal carbonates are all reactions of the hydronium ion, for example:

$$Fe(s) + 2H_3O^+(aq) \rightarrow Fe^{2+}(aq) + H_2(g) + 2H_2O(l)$$
$$CuO(s) + 2H_3O^+(aq) \rightarrow Cu^{2+}(aq) + 3H_2O(l)$$
$$ZnCO_3(s) + 2H_3O^+(aq) \rightarrow Zn^{2+}(aq) + CO_2(g) + 3H_2O(l)$$

Questions

1 Which of the following is *not* a property of acids?
A They turn universal indicator paper red.
B They effervesce with magnesium ribbon.
C They conduct electricity.
D Their solutions have a pH greater than 7.
E They dissolve sodium carbonate.

2 Without using an indicator, describe *two* different experiments you would carry out to prove that the liquid from an accumulator is an acid. *(South East)*

3 Under which of the following conditions will acetic acid conduct electricity best?
A Below its freezing point.
B Above its boiling point.
C As a liquid at room temperature.
D In solution in water.
E In solution in toluene.

4 Complete Figure 7.12, describing the reactions of a solution of hydrogen iodide in water and a solution of hydrogen iodide in xylene.

	Solution in water	Solution in xylene
Action on dry litmus paper		
Action on zinc		
Action on sodium carbonate		
Electrical conductivity		

Figure 7.12

5 In 1777 Antoine Lavoisier wrote that *all acids contain oxygen*. In 1816 Humphry Davy said that *all acids contain hydrogen*. By the

middle of the nineteenth century an acid was defined as *a substance which contains hydrogen atoms that can be replaced by metal atoms.*

(a) What common acid does not fit Lavoisier's definition?

(b) What common substance which is not generally considered to be an acid would fit Davy's definition of an acid?

(c) Does the substance you gave in answer to (b) fit the third definition above, even though it is not an acid?

(d) What was Brønsted's definition of an acid? Explain exactly what this definition means.

6 Which of the following equations *best* describes the neutralisation of 1 mole of hydrochloric acid by 1 mole of sodium hydroxide, both substances being in aqueous solution?

A $Na^+(aq) + Cl^-(aq) \rightarrow NaCl(aq)$
B $NaOH(aq) + HCl(aq) \rightarrow NaCl(aq) + H_2O(l)$
C $Na^+OH^-(aq) + H^+Cl^-(aq) \rightarrow Na^+Cl^-(aq) + H_2O(l)$
D $H^+(aq) + OH^-(aq) \rightarrow H_2O(l)$
E $H_3O^+(aq) + OH^-(aq) \rightarrow 2H_2O(l)$

.5 The concentration of solutions

Sometimes, in the laboratory, we use pure acids, which we call CONCENTRATED acids. But more often, we use solutions of acids in water, called DILUTE acids. It is useful to have a method of describing how concentrated a solution is. We can, of course, describe the concentration of a solution of hydrochloric acid as, for example, 36 g per 1000 cm^3. But, as we have seen in chapters 2 and 3, it is more useful to know how many MOLES of a substance we have, rather than the number of grams. The formula of hydrochloric acid is HCl and its formula mass is 36 (atomic masses: H = 1, Cl = 35). A solution containing 1 mole (36 g) of hydrochloric acid in 1000 cm^3 is called a MOLAR SOLUTION.

Any solution containing 1 mole in 1000 cm^3 is called a molar solution.

Making solutions of known concentration

How much sulphuric acid is needed to make 1000 cm^3 of molar solution?

The formula of sulphuric acid is H$_2$SO$_4$.
(Atomic masses: H = 1, S = 32, O = 16.)
The formula mass of sulphuric acid is $(2 \times 1) + 32 + (4 \times 16) = 98$.
1 mole of sulphuric acid is 98 g.
98 g of sulphuric acid in 1000 cm^3 is a molar solution.

We do not always use solutions which are molar. Often we want to use solutions of other concentrations. The concentration of any solution can be expressed as a MOLARITY: the number of moles in 1000 cm^3 of the solution. The following are some commonly-used molarities.

5 M	(5 moles of the substance in 1000 cm^3)
2 M	(2 moles of the substance in 1000 cm^3)
M	(1 mole of the substance in 1000 cm^3)
0.5 M or $\frac{1}{2}$ M	($\frac{1}{2}$ mole of the substance in 1000 cm^3)
0.2 M or $\frac{1}{5}$ M	($\frac{1}{5}$ mole of the substance in 1000 cm^3)
0.1 M or $\frac{1}{10}$ M	($\frac{1}{10}$ mole of the substance in 1000 cm^3)
0.01 M or $\frac{1}{100}$ M	($\frac{1}{100}$ mole of the substance in 1000 cm^3)

How much sulphuric acid is needed to make 100 cm^3 of $\frac{1}{2}$ M solution?
1000 cm^3 of a $\frac{1}{2}$ M solution contains $\frac{1}{2}$ mole.
100 cm^3 of a $\frac{1}{2}$ M solution contains $\frac{100}{1000} \times \frac{1}{2} = \frac{1}{20}$ mole.
1 mole of sulphuric acid 98 g.
$\frac{1}{20}$ mole of sulphuric acid is $\frac{1}{20} \times 98 = 4.9$ g.
4.9 g of sulphuric acid is needed to make 100 cm^3 of $\frac{1}{2}$ M solution.

Using molarities

The advantage of expressing the concentration of a solution as a molarity is that we know how many moles of the dissolved substance we have.

Equal volumes of solutions of equal molarity contain the same number of moles.

Example 1 How much M potassium hydroxide solution will be needed to neutralise 25 cm^3 of M hydrochloric acid?

The equation is	KOH	+	HCl	→	KCl	+ H₂O.
This tells us	1 mole	+	1 mole	→		
So	25 cm^3 of M	+	25 cm^3 of M	→		

25 cm^3 of M potassium hydroxide will be needed to neutralise 25 cm^3 of M hydrochloric acid.

Example 2 How much M sodium chloride solution will be needed to react with 100 cm^3 of $\frac{1}{10}$ M silver nitrate solution?
1000 cm^3 of $\frac{1}{10}$ M silver nitrate solution contains $\frac{1}{10}$ mole.
100 cm^3 of $\frac{1}{10}$ M silver nitrate solution contains $\frac{1}{10} \times \frac{100}{1000} = \frac{1}{100}$ mole.

The equation is	$AgNO_3$	+	NaCl	→	$AgCl$ + $NaNO_3$.
This tells us	1 mole	+	1 mole	→	
So	$\frac{1}{100}$ mole	+	$\frac{1}{100}$ mole	→	

There is 1 mole of sodium chloride in 1000 cm^3 of M.
There is $\frac{1}{100}$ mole of sodium chloride in $\frac{1}{100} \times 1000 = 10 \text{ cm}^3$ of M.
10 cm^3 of M sodium chloride solution will be needed to react with 100 cm^3 of 0.1 M silver nitrate solution.

Example 3 How much copper(II) oxide is needed to react with 100 cm^3 of 2 M hydrochloric acid?

1000 cm^3 of 2 M hydrochloric acid contains 2 moles.
100 cm^3 of 2 M hydrochloric acid contains $\frac{100}{1000} \times 2 = \frac{1}{5}$ mole.

The equation is   CuO + 2HCl → CuCl$_2$ + H$_2$O.
This tells us   1 mole + 2 moles →
So $\frac{1}{10}$ mole + $\frac{1}{5}$ mole →

Atomic masses: Cu = 64, O = 16.
The formula mass of copper(II) oxide is (64 + 16) = 80.
1 mole of copper(II) oxide is 80 g.
$\frac{1}{10}$ mole of copper(II) oxide is $\frac{1}{10}$ × 80 = 8 g.

8 g of copper(II) oxide is needed to react with 100 cm^3 of 2 M hydrochloric acid.

Questions   (Where necessary, use the list of atomic masses, Figure 2.7, page 39)

1 A molar solution contains which of the following?
A   1 mole of a solute in 100 cm^3 of solution.
B   1g of a solute in 1000 cm^3 of solution.
C   1 mole of a solute in 1000 cm^3 of solution.
D   100g of a solute in 100 cm^3 of solution.
E   1 mole of a solute in 1000 cm^3 of solvent.   (North West)

2 Calculate the weight of substance required to make the following solutions:
  (a) 1000 cm^3 of M sodium chloride (NaCl),
  (b) 100 cm^3 of M hydrochloric acid (HCl),
  (c) 25 cm^3 of M potassium nitrate (KNO$_3$),
  (d) 100 cm^3 of 2M calcium chloride (CaCl$_2$),
  (e) 50 cm^3 of 5 M sulphuric acid (H$_2$SO$_4$),
  (f) 1000 cm^3 of 0.1 M alcohol (C$_2$H$_6$O),
  (g) 100 cm^3 of $\frac{1}{2}$ M sodium carbonate (Na$_2$CO$_3$),
  (h) 500 cm^3 of 0.2 M lead nitrate (Pb(NO$_3$)$_2$),
  (i) 10 cm^3 of $\frac{1}{100}$ M hydrated copper sulphate (CuSO$_4$.5H$_2$O).

3 What do we mean when we say that two solutions have the same molarity?
How many grams of solute per litre would there be in a half-molar solution of (a) sodium hydroxide, NaOH, (b) hydrochloric acid, HCl, (c) sodium sulphate, Na$_2$SO$_4$?   (Wales)

4 25 cm^3 0.1 M sodium hydroxide solution was added to 25 cm^3 0.1 M hydrochloric acid. Which of the following is the pH of the resultant liquid?

(i) 4   (ii) 5   (iii) 6   (iv) 7   (v) 8   (London)

5 Which of the following would need to be added to a mixture of

25 cm³ M hydrochloric acid and 26 cm³ M sodium hydroxide to make the solution neutral?
A 1 cm³ 2 M sodium hydroxide solution.
B 1 cm³ M hydrochloric acid.
C 1 cm³ M sodium hydroxide solution.
D 10 cm³ M hydrochloric acid.
E 1 cm³ 2 M hydrochloric acid.   (North West)

6 Using the following equations

$$NaOH(aq) + HCl(aq) \rightarrow NaCl(aq) + H_2O(l)$$
$$2NaOH(aq) + H_2SO_4(aq) \rightarrow Na_2SO_4(aq) + 2H_2O(l)$$
$$Ca(OH)_2(aq) + 2HNO_3(aq) \rightarrow Ca(NO_3)_2(aq) + 2H_2O(l),$$

calculate the volume of:
(a) M sodium hydroxide solution which will react with 100 cm³ of M hydrochloric acid;
(b) 0.5 M sodium hydroxide solution which will react with 100 cm³ of 0.5 M sulphuric acid;
(c) $\frac{1}{100}$ M calcium hydroxide solution which will react with 5 cm³ of $\frac{1}{100}$ M nitric acid;
(d) M sodium hydroxide solution which will react with 100 cm³ of 2 M hydrochloric acid;
(e) 0.1 M sulphuric acid which will react with 10 cm³ of M sodium hydroxide solution.

7 Aluminium sulphate solution and sodium hydroxide solution react according to the equation

$$Al_2(SO_4)_3(aq) + 6NaOH(aq) \rightarrow 2Al(OH)_3(s) + 3Na_2SO_4(aq).$$

(a) What volume of 0.2 M aluminium sulphate solution is required to react completely with 30 cm³ of 0.2 M sodium hydroxide solution?
(b) What weight of aluminium hydroxide is produced from this sodium hydroxide solution?

8 Excess dilute hydrochloric acid was added to 100 cm³ of $\frac{1}{10}$ M sodium carbonate. Carbon dioxide was produced according to the equation

$$Na_2CO_3(aq) + 2HCl(aq) \rightarrow 2NaCl(aq) + CO_2(g) + H_2O(l).$$

What volume of carbon dioxide was produced? (Volume of 1 mole of gas is 24000 cm³.)

9 Excess zinc oxide was dissolved in 100 cm³ of M sulphuric acid. After filtering off excess zinc oxide, hydrated zinc sulphate was allowed to crystallise out. The full reaction was:

$$ZnO(s) + H_2SO_4(aq) + 6H_2O(l) \rightarrow ZnSO_4 \cdot 7H_2O(s).$$

(a) What is the maximum weight of zinc sulphate crystals that might have been obtained?

(b) Why, in practice, would one be likely to get less than the maximum possible quantity?

10 A series of experiments were made with *sodium bisulphite* (NaHSO₃) and M/5 (or 0.2 M) *hydrochloric acid* (HCl). The sodium bisulphite dissolved and a gas was given off which was identified as sulphur dioxide. The measurements made at room temperature are recorded in Figure 7.13.

Weight of sodium bisulphite	Volume of acid used	Volume of gas given off
1.04 g	10 cm^3	48 cm^3
1.04 g	20 cm^3	96 cm^3
1.04 g	30 cm^3	144 cm^3
1.04 g	40 cm^3	192 cm^3
1.04 g	50 cm^3	240 cm^3
1.04 g	60 cm^3	240 cm^3
1.04 g	70 cm^3	240 cm^3

Figure 7.13

(a) Plot a graph of the volume of acid used against the volume of gas given off. Use your own scales and label them. Plot the volume of acid on the *horizontal* axis and the volume of gas on the *vertical* axis.

Using your graph, give the following:
(b) the volume of acid needed to make 120 cm^3 of gas;
(c) the maximum volume of gas produced;
(d) the smallest volume of acid needed to produce the maximum volume of gas.
(e) What do you think is the meaning of the fact that even though increasing volumes of acid are used, the volume of gas does not increase beyond a certain value?
(f) Give the formula mass of sodium bisulphite.
(g) What fraction of a mole of sodium bisulphite is 1.04 g?
(h) A molar solution is one which contains 1 mole of a substance per 1000 cm^3 of solution. How many moles does 1000 cm^3 of a $\frac{1}{5}$ molar solution contain?
(i) How many moles do 50 cm^3 of a $\frac{1}{5}$ molar solution contain?
(j) 1 mole of any gas occupies 24000 cm^3 at room temperature. How many moles are present in 240 cm^3?

(k) Complete the following equation by inserting the appropriate number in the space provided.

☐NaHSO$_3$(s) + HCl(aq) → ☐NaCl(aq) + ☐SO$_2$(g) + H$_2$O(l)

(*London*)

For the teacher

Experiment 7.1 Beetroot and red cabbage are two of the easiest plants to use as indicators. The extract is easy to make just by boiling with water and is reasonably stable. *Ribena* is quite a good indicator. Almost any coloured plant material can be used, except yellow flowers. In some cases more successful extraction is achieved by refluxing with 50:50 water and industrial methylated spirits. Most plant extracts 'go off' rather quickly and should be prepared not more than 24 hours before the lesson. For (b), 0.1 M hydrochloric acid and sodium hydroxide. Universal indicator paper is usually supplied with a small colour chart on each pack. It is worthwhile saving these and mounting them on cards for class use so that whole books or rolls need not be handed out. If required by the syllabus, testing with litmus, methyl orange, etc. can conveniently be added to this experiment.

A possible introduction to experiment 7.1 and to the whole topic is to give the class small pieces of acid drops to suck and then to provide them with a little sodium hydrogen carbonate (BP or AR grade) into which to dip their tongues. They can then 'taste the neutralisation process'.

Experiment 7.2 Either brown or white vinegar can be used, but it is worthwhile trying out the experiment in advance, as the acid content is somewhat variable.

Experiments 7.3 and 7.4 2 M sulphuric acid.

Experiment 7.5 M sodium hydroxide and hydrochloric acid give a reasonable amount of product with these quantities. 100 cm^3 conical flasks.

Experiment 7.6 2 M sulphuric acid. Iron reduced by hydrogen is preferable to commercial iron filings (see page 27). 100 cm^3 conical flasks.

Experiment 7.7 0.1 M solutions of calcium chloride, copper(II) sulphate, hydrochloric acid, potassium chloride, potassium iodide, silver nitrate, sodium carbonate, and sodium chloride and 0.05 M lead nitrate.

Experiment 7.8 Recrystallised copper(II) sulphate is preferable.

Deliquescence and efflorescence Deliquescence may be demonstrated by leaving samples of sodium hydroxide and anhydrous calcium chloride on watch glasses exposed to the atmosphere. Sodium sulphate decahydrate and sodium carbonate decahydrate can similarly be used to demonstrate efflorescence. The hygroscopic behaviour of sulphuric acid may be shown by half-filling a beaker with the concentrated acid, marking the acid level, and setting the beaker on one side for several days.

Experiment 7.10 An adequate supply of perfectly dry 100 × 16 mm tubes is necessary (at least 3 per working group). Universal indicator paper should be stored in a dessicator for at least 24 hours before use. To ensure that the acetic acid is dry 5 per cent of acetic anhydride should be added at least 24 hours before use.

Reaction of hydrogen chloride with water Evidence that water actually *reacts* with anhydrous acids rests partly on the heat change in the reaction. If a dry thermometer or a thermometer dipped in dry toluene is held in a gas jar of dry hydrogen chloride no temperature change will be registered. If a thermometer whose bulb has been dipped in water is held in a gas jar of dry hydrogen chloride the temperature will rise by about 10°C. The dilution of concentrated sulphuric acid is also a useful demonstration (**add acid to water**).

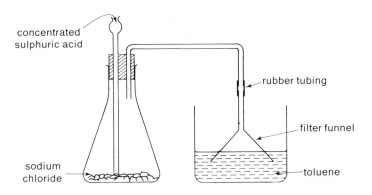

Figure 7.14

Experiment 7.11 For the solution labelled 'hydrogen chloride in water' use 2 M hydrochloric acid. The solution of hydrogen chloride in toluene can be prepared in a fume cupboard as shown in Figure 7.14. (Xylene is an acceptable alternative.) The solution must be dried over anhydrous calcium chloride for at least 48 hours before use and stored in well-stoppered bottles. Pupils should be warned not to leave the stoppers out any longer than is absolutely necessary during the experiment. An adequate supply of absolutely dry 100 × 16 mm tubes is needed (at least 8 per working group) together with one dry 150 × 25 mm tube. The universal indicator paper should be stored in a dessicator for at least 24 hours before use.

8 Sulphuric acid—the most important chemical in industry

In 1843 a German chemist, Justus von Liebig wrote: '*It is no exaggeration to say that we may judge the commercial prosperity of a country from the amount of sulphuric acid it consumes.*' This is still true today. Sulphuric acid is the most widely used of all manufactured chemicals. There are very few consumer goods that do not need sulphuric acid at some stage in their production.

Figure 8.1 shows the amount of sulphuric acid used each year in Britain since 1920. Each dip in the graph corresponds to a time of general industrial difficulty in the country. In 1926 it was the General Strike, in 1931 there was exceptionally high unemployment, in 1951 and 1952 it was because of a coal shortage, and so on.

Figure 8.2 shows the main uses of sulphuric acid. The largest single use is in the manufacture of fertilisers. It is involved in the manufacture of many plastics and man-made fibres. The *pickling* of metals, particularly iron and steel, is involved at some stage in the manufacture of a very wide range of articles. Pickling is treatment with sulphuric acid to remove any metal oxide from the surface before galvanising, plating, or enamelling. Unless the oxide film is removed the coatings do not stick to the metal properly. Sulphuric acid is needed in the preparation of many pigments for paints, such as barium sulphate and titanium dioxide. It is also involved in the manufacture of explosives, in the refining of petroleum, in the textile and leather trades, and for making all kinds of other chemicals, including drugs, dyestuffs and detergents. It is the electrolyte in car batteries, and it is a useful weedkiller.

8.1 The properties of sulphuric acid

To appreciate the tremendous part that sulphuric acid plays in the economy of a country it is necessary to understand something of the properties of this very important chemical. Sulphuric acid has two main properties. First it is the cheapest of all acids available to industry (about £20 per tonne in 1974). Secondly, it combines strongly with water, so it finds many industrial uses involving the removal of water from other substances.

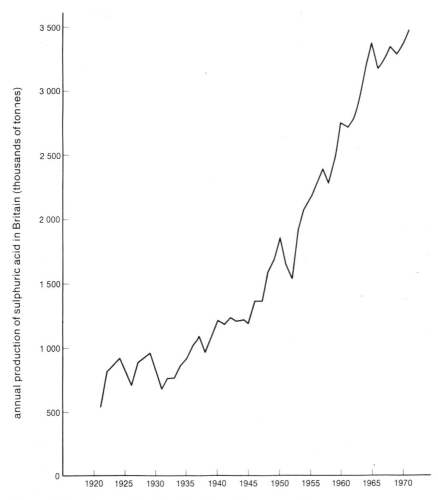

Figure 8.1 The growth of sulphuric acid consumption

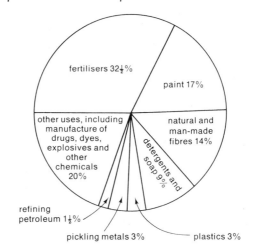

Figure 8.2 The uses of sulphuric acid

Sulphuric acid—the most important chemical in industry

Sulphuric acid as an acid

Concentrated sulphuric acid is a colourless oily liquid, almost twice as dense as water. Its formula is H_2SO_4. Since its molecule contains two hydrogen atoms it can donate two protons to a base. With water it forms two ions, the hydrogen sulphate ion, HSO_4^-, and the sulphate ion, SO_4^{2-}:

$$H_2SO_4(l) + H_2O(l) \rightarrow H_3O^+(aq) + \underset{\text{hydrogen sulphate ion}}{HSO_4^-(aq)};$$

or

$$H_2SO_4(l) + 2H_2O(l) \rightarrow 2H_3O^+(aq) + \underset{\text{sulphate ion}}{SO_4^{2-}(aq)}.$$

It can therefore produce two kinds of salt. With sodium hydroxide, for instance, it forms sodium hydrogen sulphate or sodium sulphate, depending on the proportions of acid and base used:

$$H_2SO_4(aq) + NaOH(aq) \rightarrow \underset{\text{sodium hydrogen sulphate}}{NaHSO_4(aq)} + H_2O(l);$$

or

$$H_2SO_4(aq) + 2NaOH(aq) \rightarrow \underset{\text{sodium sulphate}}{Na_2SO_4(aq)} + 2H_2O(l).$$

A salt such as sodium sulphate, in which all the hydrogen of the acid has been replaced by a metal, is called a NORMAL SALT. A salt like sodium hydrogen sulphate, in which not all of the hydrogen of the acid has been replaced, is called an ACID SALT.

Experiment 8.1 The preparation and properties of sodium sulphate and sodium hydrogen sulphate

(a) Sodium hydrogen sulphate can be prepared by the reaction:

$$NaOH(aq) + H_2SO_4(aq) \rightarrow NaHSO_4(aq) + H_2O(l).$$

Using a measuring cylinder put 10 cm³ of 2 M sodium hydroxide solution and 10 cm³ of 2 M sulphuric acid into an evaporating basin. Evaporate the solution down to about half its volume and put it aside to crystallise.

How many moles of sodium hydroxide and how many of sulphuric acid have you used? Are these the correct proportions as required by the equation?

How many moles of sodium hydrogen sulphate crystals will you obtain? What weight of sodium hydrogen sulphate is this? (It is an anhydrous salt.)

(b) Sodium sulphate can be prepared by the reaction:

$$2NaOH(aq) + H_2SO_4(aq) \rightarrow Na_2SO_4(aq) + 2H_2O(l).$$

Using a measuring cylinder put 20 cm³ of 2 M sodium hydroxide solution and 10 cm³ of 2 M sulphuric acid into an evaporating basin. Evaporate the solution down to about half its volume and put it aside to crystallise.

How many moles of sodium hydroxide and how many of sulphuric acid have you used? Are these the correct proportions as required by the equation?

How many moles of sodium sulphate crystals will you obtain? What weight of sodium sulphate is this? (It is a hydrated salt with the formula $Na_2SO_4.10H_2O$.)

(c) Examine your crystals of the two salts. If they are too small for the shapes to be seen clearly, use a microscope. Make drawings of the shapes of the crystals. Are they different?

Dissolve a few of the sodium hydrogen sulphate crystals in half a test tube of water. Find the pH of the solution with universal indicator paper. Divide the solution into two parts. Into one, drop a piece of magnesium ribbon and into the other a small piece of marble (calcium carbonate). Describe what happens. What can you say about the properties of sodium hydrogen sulphate?

Prepare a solution of your sodium sulphate crystals, find its pH, and test the action of the solution on magnesium and marble. What can you say about the properties of sodium sulphate?

Some useful salts of sulphuric acid

SODIUM HYDROGEN SULPHATE, $NaHSO_4$, is a strongly acid salt, used in many lavatory cleaners to dissolve away the calcium carbonate deposits produced by hard water.

SODIUM SULPHATE, $Na_2SO_4.10H_2O$, sometimes called *Glauber's salt*, and MAGNESIUM SULPHATE, $MgSO_4.7H_2O$, known as *Epsom salt*, are used in medicine as purgatives. ZINC SULPHATE, $ZnSO_4.7H_2O$, also finds a use in medicine as an emetic.

AMMONIUM SULPHATE, $(NH_4)_2SO_4$, is an important fertiliser.

BARIUM SULPHATE, $BaSO_4$, is used as a white pigment in paints. It is also used in taking X-ray photographs of the digestive system. Because barium is an element with very large atoms, its compounds are not transparent to X-rays. The patient is given a *barium meal*, consisting of a paste of barium sulphate. It sticks to the walls of the stomach and intestines, making their outline show up on the X-ray plate (Figure 8.3, overleaf), which would not otherwise happen.

CALCIUM SULPHATE is found in nature both as the anhydrous salt, $CaSO_4$, which is called *anhydrite*, and as a hydrated salt, $CaSO_4.2H_2O$, known as *gypsum*. When gypsum is carefully heated it loses part of its water of crystallisation, forming *Plaster of Paris*, $CaSO_4.\frac{1}{2}H_2O$. When water is added to Plaster of Paris the water of crystallisation is restored and the gypsum sets to a hard mass. Plaster of Paris is used to make plaster casts to hold broken bones in position while they set. It is also used for plastering walls and to make plasterboard for building, which is a mixture of wood pulp and calcium sulphate dried to a hard sheet.

COPPER(II) SULPHATE, $CuSO_4.5H_2O$, is a powerful fungicide, used in the form of *Bordeaux mixture* to kill the fungus responsible for potato blight.

Sulphuric acid—the most important chemical in industry

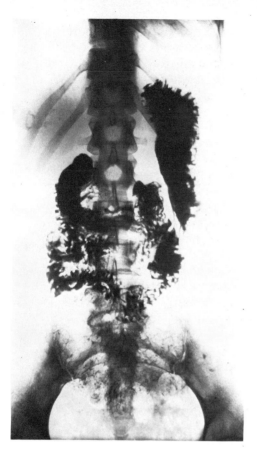

Figure 8.3 An X-ray picture of the human digestive system

Wood is treated with copper(II) sulphate to prevent the growth of the fungi which cause wet and dry rot.

IRON(II) SULPHATE, $FeSO_4.7H_2O$, is used in the manufacture of ink.

POTASSIUM ALUMINIUM SULPHATE, $K_2SO_4.Al_2(SO_4)_3.24H_2O$, known as *potash alum*, is used as a mordant in dyeing to help the dyestuffs to bite on to the fabric. This is a DOUBLE SALT of sulphuric acid. When a solution containing both potassium sulphate and aluminium sulphate is allowed to crystallise, the crystals contain a mixture of the two salts.

Sulphuric acid to remove water

The dilution of concentrated sulphuric acid with water produces a great deal of heat. **When diluting concentrated sulphuric acid the acid must always be added to water. Water must NEVER be added to the concentrated acid; the reaction is then so violent that it is extremely dangerous.**

Sulphuric acid combines so strongly with water that it is often used as a DEHYDRATING AGENT (drying agent), particularly to dry gases, both in the laboratory and in industry.

The concentrated acid will even remove water from compounds. Blue crystals of hydrated copper(II) sulphate slowly turn to the powdery white anhydrous salt when treated with concentrated sulphuric acid. The water of crystallisation is removed:

$$CuSO_4.5H_2O \xrightarrow{\text{concentrated sulphuric acid}} CuSO_4 + 5H_2O.$$

When sugar is treated with concentrated sulphuric acid water is removed, leaving a black mass of carbon:

$$C_{12}H_{22}O_{11} \xrightarrow{\text{concentrated sulphuric acid}} 12C + 11H_2O.$$

Paper is affected in the same way (and so is skin!).

Concentrated sulphuric acid is also used as a catalyst in many reactions in which water has to be removed, for example in the manufacture of explosives.

Questions

1 Read the following observations on reactions involving sulphuric acid and answer the questions asked.

(i) A gas was evolved when some pieces of magnesium ribbon were added to some dilute sulphuric acid.
Name the gas and the salt which would be left in the solution.

(ii) When some dilute sulphuric acid was warmed with a small amount of copper oxide powder, a blue solution appeared.
What was the salt present in the blue solution?
What type of reaction is involved? *(West Midlands)*

2 *Fill in the gaps.* Sulphuric acid forms two types of salt because it contains two _____ _____. 2 moles of sodium hydroxide (NaOH) react with 1 mole of sulphuric acid (H_2SO_4) to form the salt _____ _____, formula _____, which is called a _____ salt, because all of the _____ in the acid has been replaced.

1 mole of sodium hydroxide reacts with 1 mole of sulphuric acid to form the salt _____ _____ _____, formula _____, which is called an _____ salt because it contains a _____ atom. A solution of this salt has a pH of __ and has all the normal properties of an _____.

3 (a) When concentrated sulphuric acid is added to copper(II) sulphate crystals which of the following happens?
A   The crystals dissolve.   D   Copper(II) oxide is formed.
B   The mixture turns black.   E   The crystals slowly turn white.
C   Sulphur dioxide is evolved.

(b) The sulphuric acid is here acting as which of the following?
A   A dehydrating agent.   D   A solvent.
B   An oxidising agent.   E   An acid.
C   A catalyst.   *(North West)*

4 *Fill in the gaps.* When concentrated sulphuric acid is poured on to filter paper, the paper is turned black. The acid had removed _____ from the paper, leaving carbon. The process is called _____.
(West Yorkshire)

5 Some concentrated sulphuric acid was exposed to the air for four weeks in an open beaker and the level of the acid was marked as shown in Figure 8.4(a). Figure 8.4(b) shows the result after four weeks.

(a) Explain the increase in volume of liquid in the beaker. What is the liquid in the beaker after the four weeks?

(b) What is the word used to express the property of sulphuric acid illustrated by this experiment?

(c) Give a laboratory use for sulphuric acid based on this property. *(West Midlands)*

Sulphuric acid—the most important chemical in industry

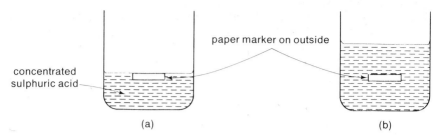

Figure 8.4

6 Sulphuric acid has been called 'the lifeblood of industry'. Write two or three paragraphs explaining what chemical properties of sulphuric acid are responsible for its importance.

8.2 The manufacture of sulphuric acid

There are three basic stages in the manufacture of sulphuric acid.
1 Burning sulphur in oxygen to make sulphur dioxide gas:

$$S(s) + O_2(g) \rightarrow SO_2(g).$$

2 Combining the sulphur dioxide with more oxygen to make sulphur trioxide:

$$2SO_2(g) + O_2(g) \rightleftharpoons 2SO_3(g).$$

3 Reacting the sulphur trioxide with water to produce sulphuric acid:

$$SO_3(g) + H_2O(l) \rightarrow H_2SO_4(l).$$

An outline of the whole process is shown in Figure 8.5.

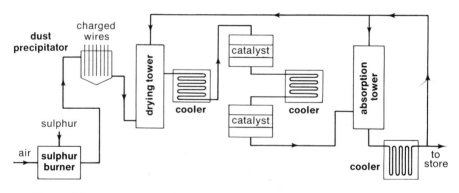

Figure 8.5

Making sulphur dioxide

Solid sulphur is fed into a large rotating drum, where it is burnt in a current of air. Before the second stage in the process the sulphur dioxide

228 Chemistry today

has to be purified. First dust is removed in an ELECTROSTATIC PRECIPITATOR. The gas is passed through a chamber in which are suspended a number of fine wires charged to 50 000 volts. The fine dust particles become electrically charged and are attracted to an oppositely charged wire. The charges on the dust particles are neutralised by the charge on the wire and they fall to the bottom of the chamber. Electrostatic precipitators find many uses in industry, for example in factory chimneys to remove dust from smoke.

After removal of dust the sulphur dioxide is passed up a tower down which concentrated sulphuric acid is trickling. This dries the gas and it is ready for the second stage of the process.

Conversion of sulphur dioxide to sulphur trioxide

The reaction:
$$2SO_2(g) + O_2(g) \rightleftharpoons 2SO_3(g)$$

reaches equilibrium. To make the process economic, conditions must be chosen so that this equilibrium is displaced as far as possible to the right.

The mixture of sulphur dioxide and air, which is still very hot from the sulphur burner, is cooled to 450°C and passed through the CONVERTER. This is a cylindrical steel vessel with perforated shelves carrying layers of the catalyst, vanadium pentoxide. Part of the sulphur dioxide is converted to sulphur trioxide. Heat is produced in the reaction so the mixture of gases is again cooled to 450°C before being passed through a second converter. Two or three converters are used in succession, and under these conditions 98 per cent of the sulphur dioxide is converted into sulphur trioxide and the second stage of the process is complete. It is this stage which gives this method of manufacturing sulphuric acid the name CONTACT PROCESS, from the contact between the gas mixture and the catalyst.

Sulphur trioxide to sulphuric acid

In the early contact process plants, the final stage was reaction of sulphur trioxide with water to give sulphuric acid. This reaction is extremely violent and difficult to control and modern plants use a slightly different method. The sulphur trioxide is passed up a tower down which concentrated sulphuric acid is trickling, and *oleum* is formed:

$$SO_3(g) + H_2SO_4(l) \rightarrow \underset{\text{oleum}}{H_2S_2O_7(l)}.$$

As the oleum leaves the tower it is mixed with just the right amount of water to produce sulphuric acid of the required strength:

$$H_2S_2O_7(l) + H_2O(l) \rightarrow 2H_2SO_4(l).$$

A great deal of heat is produced in these two reactions so the acid is finally cooled by passing it through banks of water-cooled pipes. Some of the acid is returned to the washing and absorption towers, the rest is drawn off into storage tanks.

Questions

1 (i) How is sulphur dioxide obtained in large quantities for industrial use?

(ii) When sulphur dioxide is mixed with air and passed over a suitable heated catalyst what is formed?

(iii) Write an equation for the reaction in (ii) above.

(iv) Write down the name of a catalyst used when the reaction in (ii) above is carried out industrially, and give the name of the process.

(v) Say what is formed when the gas formed in (ii) above is dissolved in water, and write an equation for the reaction.

(vi) What difficulties are encountered in carrying out the reaction in (v) above on a large scale and how are they overcome?

(Southern)

2 (a) The stages in the process for manufacturing sulphuric acid can be represented as in Figure 8.6.

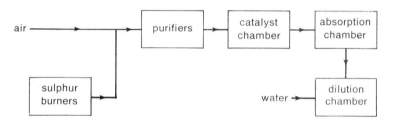

Figure 8.6 Stages in the manufacture of sulphuric acid

(i) Why is it necessary to include the purifying stage?
(ii) What kind of impurity will be removed?
(iii) Write a word equation for the reaction that takes place in the catalyst chamber.
(iv) Why is the absorption stage necessary?

(b) Name *two* purposes for which the sulphuric acid may be used on a large scale. *(East Anglia)*

3 What weight of pure sulphuric acid *could* be obtained from 32 tonnes of pure sulphur? (Atomic masses: H=1, O = 16, S = 32.)

(South East)

8.3 Sulphur for sulphuric acid

Sulphur makes up 0.05 per cent of the earth's crust. It is found *native* (as the free uncombined element). It also occurs combined with other elements. A number of metals are mined as sulphide ores (Figure 6.5, page 145), and in the United Kingdom the most important source of sulphur is calcium sulphate in the form of a mineral called *anhydrite*. Coal, petroleum, and natural gas all contain some sulphur and sulphur compounds.

Extracting native sulphur: the Frasch process

In 1868 huge underground deposits of native sulphur were discovered in Louisiana, in the United States, by prospectors looking for oil. The sulphur was found in limestone rock, into which it must have soaked, while the earth was hot, like water into a sponge. The sulphur beds lie about 150 metres below the surface, so mining companies set out to try to mine the sulphur rock in the ordinary way. Unfortunately the sulphur beds turned out to be covered with layers of quicksand so it was impossible to sink ordinary mineshafts.

The problem of getting the sulphur out was not solved until 1890. The method discovered then is still in use today, and is named after its inventor, Herman Frasch. A 15 cm hole is drilled down into the sulphur deposit, just as for an oil well. Three pipes are fitted into this hole, one inside the other (Figure 8.7). Superheated water, at a pressure 15 times as great as the normal pressure of the atmosphere, is forced down the outside pipe. At this pressure water remains liquid up to 170°C (page 50). The superheated water enters the porous sulphur-bearing rock and melts the sulphur. Molten sulphur rises partway up the middle tube and is forced to the surface by blasting compressed air down the innermost tube.

A froth of molten sulphur, water, and air flows out at the top of the well into enormous vats (Figure 8.8), where the sulphur solidifies and the water drains away. The huge blocks of sulphur, over 99.8 per cent pure, are broken up by explosives to be transported. A modern Frasch well brings up as much as 6000 tonnes of sulphur a day and will pump for three or four months before a new shaft has to be sunk.

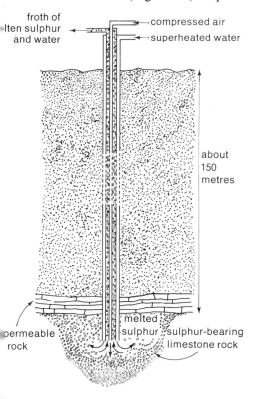

Figure 8.7 The Frasch sulphur pump

Figure 8.8 'Vats' of solid sulphur in Texas. Molten sulphur is pumped from underground into aluminium-sided tanks 350 metres long, 50 metres wide and 15 metres high

About 90 per cent of the world's sulphur is produced by the Frasch process. Most of it is used for making sulphuric acid, but some is required for vulcanising rubber, for the manufacture of matches, and in making carbon disulphide for use in the production of rayon.

Experiment 8.2   Making sulphur crystals

(a) Place a spatula-full of powdered sulphur in half a test tube of carbon disulphide. Shake the tube to dissolve as much of the sulphur as possible and then filter the solution into another test tube.

Use a teat pipette to put a drop of the filtered solution on to a microscope slide. Examine the drop at once under the microscope, using the lowest power lens, and watch the solution crystallise out.

Make a drawing of the best sulphur crystal you can see under the microscope.

(b) One-third-fill a test tube with powdered sulphur. Prepare a filter paper as though you were going to put it in a funnel, but instead fasten it with a paper clip and hold it in tongs.

Heat the sulphur *very slowly* until it has just melted. Quickly pour the molten sulphur into the folded filter paper.

Hold the filter paper still and blow gently on the surface of the molten sulphur to cool it. As soon as the upper surface has solidified hold the filter paper over the sink and open it out. **Take care not to burn your fingers. The sulphur will still be hot and some of it may even be molten.**

Examine the sulphur crystals that will have formed on the filter paper. Are they the same shape as those you obtained by crystallisation from carbon disulphide? Make a drawing of one of these sulphur crystals.

Experiment 8.3   Investigating molten sulphur

One-third-fill a test tube with powdered sulphur. Heat the sulphur *very slowly* until it has just melted. Notice the colour of sulphur at its melting point and tilt the tube slightly to see how runny the liquid is.

Continue heating the molten sulphur slowly until it is almost boiling. Notice the colour of the liquid as it gets hotter and tilt the tube every now and again to see how runny it is.

Pour the nearly boiling sulphur quickly into half a beaker of cold water. **(Take care! The temperature of the liquid sulphur will be about 400°C!)**

As soon as the sulphur is cold enough to touch, pick it out of the water and examine it. Leave it lying on the bench for 5 minutes and then examine its properties again.

Describe how the colour of molten sulphur changes as it gets hotter. Describe how the runniness of liquid sulphur changes as it gets hotter.

Describe the properties of the sulphur just after it has been picked out of the water. How do its properties change after it has been lying on the bench for 5 minutes?

Sulphur crystals

Like nearly all solids, sulphur is crystalline. The almost pure sulphur made by the Frasch process is made up of tiny RHOMBIC crystals (Figure 8.9). Larger rhombic crystals can be grown in much the same way as large crystals of, for instance, copper(II) sulphate or *alum*. To obtain large crystals of a salt a saturated solution in water is allowed to evaporate slowly. Sulphur is not soluble in water, but it does dissolve in another liquid, carbon disulphide. Large crystals of rhombic sulphur can be made by allowing a solution of sulphur in carbon disulphide to evaporate slowly.

When molten sulphur is slowly cooled and allowed to solidify, the crystals are quite different. They are long and needle-shaped (Figure 8.10). This is another form of sulphur, called MONOCLINIC sulphur. Different crystalline forms of the same element are called ALLOTROPES.

When sulphur is crystallised below 96.5°C rhombic crystals are formed. When it is crystallised above 96.5°C monoclinic crystals are formed. If the rhombic allotrope is heated above 96.5°C it changes into the monoclinic

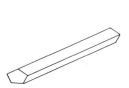

Figure 8.9 (above, left) A crystal of rhombic sulphur

Figure 8.10 (below, left) A crystal of monoclinic sulphur

Figure 8.11 Model of a sulphur molecule

Sulphuric acid—the most important chemical in industry 233

allotrope, and if the monoclinic form is cooled below this temperature it changes into the rhombic form. At room temperature, therefore, the stable allotrope is rhombic sulphur, but the change from one allotrope into the other is quite slow.

The sulphur molecule is made up of eight atoms, which we know from X-ray diffraction to be arranged in a puckered ring (Figure 8.11). Rhombic and monoclinic sulphur crystals are both made up of these 8-membered rings. The only difference between the two allotropes is the way in which the rings are packed together.

Molten sulphur

Sulphur melts at 113°C, and the liquid behaves in a very curious way. When first melted it is a pale yellow runny liquid. As the temperature increases the colour gets darker and darker, from orange to red and then almost black. When the temperature reaches 160°C the runny liquid suddenly becomes very thick and treacly. If it is being melted in a test tube the tube can be turned upside down without the molten sulphur running out. This is very unusual. Most liquids get runnier as the temperature rises, not thicker. As the temperature is raised even higher the liquid becomes slightly more runny again, but even at its boiling point it is not as runny as when it was just melted.

The explanation of this curious behaviour is as follows. When solid sulphur is heated, the 8-membered rings making up the crystals vibrate more and more vigorously. At the melting point the vibration of the molecules has become so great that the crystals are shaken to pieces. The pale yellow runny liquid is still made up of 8-membered ring molecules, which can move quite freely over each other. As the temperature rises the

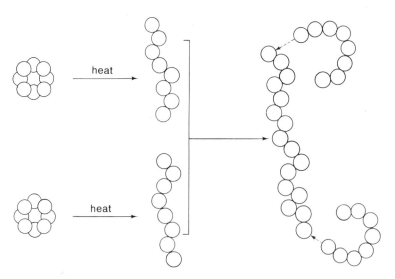

Figure 8.12

individual rings vibrate more and more until they are broken apart into chains of eight atoms. These short chains then link together to form much longer chains (Figure 8.12). The chains tangle up with one another, and this causes the molten sulphur to become thick and treacly. As the temperature rises even further the increased vibration shakes the chains apart to some extent and the liquid becomes slightly runnier, but there is still some tangling of the molecular chains right up to the boiling point.

Plastic sulphur

If molten sulphur near its boiling point is poured into cold water, so that it is cooled very rapidly, another allotrope is produced, called *plastic sulphur*. This is a flexible, rubbery material, rather like elastic. It keeps its elastic properties only for a few minutes, then slowly becomes hard and brittle.

If molten sulphur is allowed to cool slowly, as in the preparation of monoclinic sulphur, the chain molecules have time to reform into 8-membered rings, but if it is cooled rapidly there is not time for the rings to reform. Plastic sulphur is made up of long concertina-like chain molecules (Figure 8.12) and it is this molecular shape which gives it its elastic properties. Gradually, however, the chains break up and reform into 8-membered rings and the plastic sulphur turns into rhombic crystals, becoming hard and brittle.

Questions

1 The output of sulphur in the world is over 2 000 000 tons a year, about nine-tenths of it being produced in the USA. In Texas the sulphur is not mined but is extracted by the Frasch process.
 (a) Give the two most important uses of the sulphur produced.
 (b) Give two reasons for the difference in the methods of obtaining coal and sulphur from underground.
 (c) The Frasch sulphur pump consists of three concentric tubes. Explain how these three tubes are used to raise the sulphur from the sulphur beds. (*West Midlands*)

2 Sulphur is soluble in the two liquids carbon disulphide and xylene. Sulphur crystals can be obtained by allowing a saturated solution of sulphur in carbon disulphide to evaporate at room temperature. Another type of sulphur crystals can be obtained by evaporating a saturated solution of sulphur in xylene at 100°C.
 (a) Describe (or draw) the crystals obtained from each solution.
 (b) What would happen to the crystals from the xylene solution if they were allowed to cool to room temperature?
 (c) Describe briefly another method by which the same type of sulphur crystals as were obtained from the solution in xylene may be made.

Sulphuric acid—the most important chemical in industry

3 (a) What is meant by an *allotrope*? Name *two* allotropes of sulphur. Mention *two* distinct ways in which sulphur occurs naturally, naming any other substance involved.

(b) (i) Explain what happens when sulphur is heated in contact with air. What process has taken place?

(ii) Describe what happens when a lump of sulphur is heated slowly in a large test tube. (Draw diagrams if necessary.)

(c) Describe the preparation of plastic sulphur and state what happens when it stands for a prolonged period. Hence explain whether a chemical or a physical change has taken place.

(*East Anglia*)

4 The structure of a material determines its physical properties. Given below are some details of the structure of two different forms of sulphur.

Form 1 When just molten, the weight of a mole of molecules of Form I is 256 grams. The atomic mass of sulphur is 32.

(a) How many atoms are there in a molecule of Form I? These atoms are joined together in a smooth ring.

(b) If you attempted to pour this liquid, how would you expect it to flow?

(c) Explain your answer.

Form 2 At higher temperatures the ring breaks into zig-zag chains.

(d) How would you expect the liquid to flow?

(e) Explain your answer.

(f) At higher temperatures still, the flow characteristics of the liquid change again; but this is not entirely due to a further change in structure. Suggest how its flow characteristics will alter and why.

(*London*)

8.4 Sulphur dioxide for sulphuric acid

The sulphur beds in Louisiana and Texas are vast, and recently new deposits have been found in Mexico, but they are not inexhaustible and other sources of sulphur for making sulphuric acid are becoming increasingly important.

Sulphur dioxide from sulphide ores

The first stage in the manufacture of a metal from a sulphide ore is roasting (page 148), for example:

zinc sulphide + oxygen → zinc oxide + sulphur dioxide

$2ZnS(s) + 3O_2(g) \rightarrow 2ZnO(s) + 2SO_2(g)$.

The sulphur dioxide produced in this process is not wasted. Most plants for the extraction of zinc, lead, copper, etc. have a contact process plant on the site so that the sulphur dioxide produced as a by-product is used directly, after purification, to make sulphuric acid.

Sulphur dioxide from anhydrite

Anhydrite, a form of calcium sulphate, is the most important source of sulphur in the United Kingdom. Calcium sulphate is a very stable compound and cannot easily be broken down by heat alone. It can, however, be decomposed into calcium sulphide by heating with carbon in the form of *coke*:

calcium sulphate + carbon → calcium sulphide + carbon dioxide

$$CaSO_4(s) + 2C(s) \rightarrow CaS(s) + 2CO_2(g).$$

Calcium sulphide will react with more calcium sulphate to produce sulphur dioxide:

calcium sulphide + calcium sulphate → calcium oxide + sulphur dioxide

$$CaS(s) + 3CaSO_4(s) \rightarrow 4CaO(s) + 4SO_2(g).$$

Silicon dioxide (sand) and aluminium oxide (present in ashes) act as catalysts to this reaction.

Anhydrite is crushed and mixed with the correct proportions of coke, sand, and ashes. The mixture is heated at 1400°C in a ROTARY KILN (Figure 8.13), a huge cylinder $3\frac{1}{2}$ metres in diameter and 100 metres long. The mixture is fed in at the top and the kiln is heated by a burning mixture of powdered coal and air blown in at the bottom. The gases from the kiln contain about 10 per cent of sulphur dioxide, together with nitrogen (from the air) and carbon dioxide. The solid residue from the bottom of the kiln is just the right mixture for making cement.

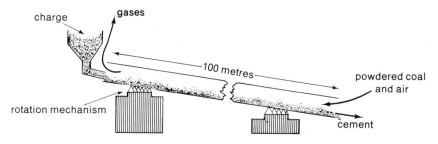

Figure 8.13 A rotary kiln

The properties and uses of sulphur dioxide

Most of the sulphur dioxide manufactured from sulphide ores and anhydrite is used to make sulphuric acid. It is a colourless gas, about twice as dense as air, with a choking smell. It is poisonous to both plants and animals.

Sulphur dioxide is a serious cause of atmospheric pollution. Most fuels, including coal and oil, contain small quantities of sulphur and when the fuel is burnt this is converted into sulphur dioxide which often escapes into the air. Where there are high concentrations of sulphur dioxide in the atmosphere plant life can be badly affected. The poisonous properties of sulphur dioxide can, however, be put to use. Gardeners sometimes burn a *sulphur candle* in greenhouses to kill insects. A small quantity of sulphur dioxide is dissolved in fruit juice to kill bacteria and act as a preservative.

Sulphur dioxide is an acidic gas. It is soluble in water, forming an acidic solution known as *sulphurous acid*, which contains hydronium and sulphite ions:

$$SO_2(g) + H_2O(l) \rightarrow H_2SO_3(aq)$$
$$H_2SO_3(aq) + 2H_2O(l) \rightleftharpoons 2H_3O^+(aq) + SO_3^{2-}(aq).$$

The acidic solution is neutralised by bases to form salts called *sulphites*. For example with sodium hydroxide solution:

$$2NaOH(aq) + H_2SO_3(aq) \rightarrow Na_2SO_3(aq) + 2H_2O(l).$$
<small>sodium sulphite</small>

The acidic properties of sulphur dioxide are another cause of pollution. In the atmosphere it corrodes both metal and stone (Figure 8.14).

Figure 8.14 The effect of corrosion by sulphur dioxide on stonework

Small quantities of sulphur dioxide can conveniently be prepared in the laboratory by reacting sodium sulphite with an acid, such as hydrochloric acid:

$Na_2SO_3(s) + 2HCl(aq) \rightarrow 2NaCl(aq) + SO_2(g) + H_2O(l).$

Sulphur dioxide is also a bleach. It removes oxygen from some coloured substances, converting them into colourless substances:

$H_2O(l) + SO_2(g) + O(\text{from dye}) \rightarrow H_2SO_4(aq).$

Large quantities are used to bleach wood pulp for the manufacture of paper, as well as to bleach delicate materials such as wool, silk, and straw which would be damaged by the use of stronger bleaches. Unfortunately the bleaching action is only temporary. Gradually the oxygen removed by the sulphur dioxide is replaced by oxygen from the air, hence the slow yellowing of old paper.

Experiment 8.4 Some properties of sulphur dioxide

Put a spatula-full of sodium sulphite in a test tube. Using a teat pipette add 3 or 4 drops of concentrated hydrochloric acid. Warm the tube **very gently**.

(a) Smell the gas **very cautiously**. Describe its smell.

(b) Hold a piece of damp universal indicator paper near the mouth of the tube. What pH is registered? What does this tell you about the gas?

(c) Mix equal volumes of potassium permanganate solution and dilute sulphuric acid in a test tube. Dip a piece of filter paper into this solution and hold it at the mouth of the tube in which sulphur dioxide is being produced. Describe what happens. (This is a useful test for sulphur dioxide.)

(d) Put 1 cm depth of distilled water into a test tube. Hold this tube together with the one containing sodium sulphite and hydrochloric acid as shown in Figure 8.15 and warm gently.

After about half a minute put your thumb over the end of the tube

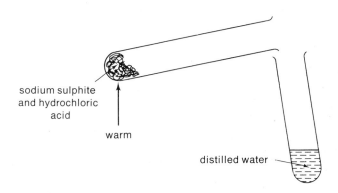

Figure 8.15

containing distilled water and shake it. Drop a piece of universal indicator paper into the water. What happens? What two pieces of information about sulphur dioxide can you deduce from this experiment?

Questions

1 State two methods by which sulphur dioxide is produced for use in the contact process. *(West Midlands)*

2 In an experiment 1.6 g of sulphur were burned completely in a free supply of air.
   Write the equation of the reaction.
   What weight of oxygen would combine with the sulphur? (Atomic masses: S = 32, O = 16.) *(London)*

3 Both the oxides of sulphur are acidic, reacting with water to form sulphurous and sulphuric acids respectively.
   (a) Write an equation for the reaction of sulphur dioxide with water.
   (b) Write an equation for the reaction of sulphur trioxide with water.
   (c) How would you show that the solutions formed in (a) and (b) were acidic? *(North West)*

4 A solution of sulphur dioxide in water acts as an acid and is called sulphurous acid. Describe how this solution reacts when added to each of the following:
   (a) magnesium
   (b) sodium carbonate crystals
   (c) a solution of sodium hydroxide coloured blue with litmus solution.
In each case, state how you would know a reaction had occurred and name *one* of the products of the reaction.
   Name *two* large-scale uses of sulphur dioxide. *(Wales)*

5 *Fill in the gaps.* Rusting and corrosion of metal occurs more quickly in industrial cities than in country areas because of the presence of ——————— gas. This is the product of burning ——————— and to a less extent ———————, both of which are common fuels. *(West Yorkshire)*

8.5 Sulphuric acid from fuels

The coal used in Britain in a year contains about half a million tonnes of sulphur. Crude oil and natural gas also contain a significant amount of sulphur. The amount varies from one oilfield to another, but the average is about 2 per cent.

Sulphur dioxide from coal

In the manufacture of coke for the extraction of metals such as iron most of the sulphur in the coal is converted into a gas called HYDROGEN SULPHIDE (H_2S). This gas is absorbed in iron(III) oxide, with which it reacts to form a mixture of iron(II) sulphide and sulphur called *spent oxide*:

$$Fe_2O_3(s) + 3H_2S(g) \rightarrow \underbrace{2FeS(s) + S(s)}_{\text{spent oxide}} + 3H_2O(g).$$

The spent oxide is roasted in the same way as a metal sulphide ore (p. 148) to obtain sulphur dioxide. The iron(III) oxide is recovered and can be used again.

Sulphur from natural gas

In natural gas most of the sulphur is in the form of hydrogen sulphide. The gas is passed through a liquid called ethanolamine, in which hydrogen sulphide is soluble. The other gases in natural gas do not dissolve. The hydrogen sulphide is recovered from the ethanolamine and mixed with air, with which it reacts to form sulphur dioxide:

$$2H_2S(g) + 3O_2(g) \rightarrow 2SO_2(g) + 2H_2O(g).$$

This sulphur dioxide is mixed with more hydrogen sulphide. The mixture is passed over a hot catalyst which helps the gases to react together to form sulphur:

$$2H_2S(g) + SO_2(g) \rightarrow 3S(l) + 2H_2O(g).$$

Sulphur from crude oil

In crude oil the sulphur is present both as the free uncombined element and in various compounds. The oil is vaporised, mixed with hydrogen, and passed over a catalyst which converts all the sulphur to hydrogen sulphide. Sulphur is obtained from this hydrogen sulphide in exactly the same way as from the hydrogen sulphide in natural gas.

The properties of hydrogen sulphide

Hydrogen sulphide is a colourless gas, a little more dense than air. It smells like bad eggs. (When eggs and other protein foods decay, hydrogen sulphide is one of the products.) It is extremely poisonous, and is particularly dangerous because it quickly destroys the sense of smell, in spite of the fact that its smell is at first very noticeable.

In the laboratory a small amount of hydrogen sulphide can

conveniently be prepared by the reaction of iron(II) sulphide with an acid. For example with hydrochloric acid:

$$FeS(s) + 2HCl(aq) \rightarrow FeCl_2(aq) + H_2S(g).$$

Many metal sulphides are insoluble in water, and are precipitated when hydrogen sulphide is passed into a solution of one of their salts. With a solution of lead nitrate:

$$Pb(NO_3)_2(aq) + H_2S(g) \rightarrow PbS(s) + 2HNO_3(aq).$$

This is another example of a double decomposition or change of partners reaction.

Though most of the sulphur in fuels is converted into sulphur dioxide when the fuel is burnt, some escapes as hydrogen sulphide. The hydrogen sulphide in the atmosphere reacts with many metals and their compounds to form metal sulphides. This is particularly noticeable with silver. The tarnishing of silver objects is due to reaction with hydrogen sulphide to form silver sulphide, which is black. Paints containing lead compounds are also discoloured, due to the formation of black lead sulphide. (Modern *super-white* paints contain titanium dioxide instead, which does not react with hydrogen sulphide.)

Questions

1 Sulphur can be obtained from the mixture of gases associated with petroleum deposits, generally known as *natural gas.*
 (a) What is the gas present in the largest proportion in natural gas?
 (b) Name the gas present in the petroleum gases which is used as a source of sulphur.
 (c) How is sulphur obtained from this gas?

2 A mixture of iron filings and sulphur was heated. A red glow spread through the mixture and a dark grey solid was left when the reaction was over. The residue was allowed to cool and dilute hydrochloric acid was added, evolving a gas which produced a black precipitate when passed into a solution containing lead ions.
 (a) What did the red glow indicate?
 (b) What is the name of the dark grey solid?
 (c) Write an equation for the reaction between iron and sulphur.
 (d) What gas was evolved when the residue reacted with dilute hydrochloric acid?
 (e) Write an equation for the reaction between the residue and the dilute hydrochloric acid.
 (f) What was the name of the black precipitate formed?
 (g) Write an equation to represent the formation of the black precipitate. *(North West)*

3 Explain the chemical change involved and name the product when paint which contains lead compounds darkens when exposed for some time to the atmosphere of towns. (*South East*)

For the teacher

The hydrates of calcium sulphate Pupils can mix Plaster of Paris with water and allow it to set ($CaSO_4.\frac{1}{2}H_2O \rightarrow CaSO_4.2H_2O$). The block can then be powdered and heated strongly in an evaporating basin ($CaSO_4.2H_2O \rightarrow CaSO_4$ or $CaSO_4.\frac{1}{2}H_2O$). After cooling, the residue can be mixed with water again and left to set. Results usually vary from almost complete success to complete failure, emphasising the criticality of the degree of hydration. Ordinary Plaster of Paris must be used, not a proprietary substitute such as *Polyfilla*.

Concentrated sulphuric acid as a dehydrating agent (a) Sugar: put 1 cm depth of sucrose or glucose into a 100 cm³ beaker. Add 5 cm³ of concentrated sulphuric acid and stir with a glass rod. After a minute or two the reaction commences, becoming quite violent. (b) Paper: put a spot of concentrated sulphuric acid on a filter paper. If there is no effect within a minute or so warm the paper gently. (c) Copper(II) sulphate: put 1 cm depth of small crystals in a 150 × 25 mm tube, add 5 cm³ of concentrated sulphuric acid, and stir. Within 5 minutes the copper(II) sulphate will be almost completely white. These tests are *not* suitable for class experiment.

Contact process A laboratory imitation of the process may be helpful, but it is the industrial process that should be emphasised: too many pupils when asked to describe the contact process in an examination have drawn the laboratory apparatus.

The apparatus in Figure 8.16 should be as dry as possible before assembly. The platinised asbestos should be heated strongly in a bunsen flame for a few seconds before use. Further drying of the apparatus can be achieved by allowing a slow stream of oxygen to pass through and heating the tube containing the platinised asbestos strongly for 5 minutes. Oxygen from a cylinder and sulphur dioxide from a canister should both be bubbled through concentrated sulphuric acid. This ensures that they are perfectly dry and also assists control of the rate of flow. There should be a slight excess of oxygen. Heat the tube containing the platinised asbestos, using a moderate bunsen flame. Under these conditions it is possible to obtain a deposit of solid sulphur trioxide on the sides of the cooled tube. At the very least a white smoke will be seen when this tube is disconnected, due to reaction of sulphur trioxide with moisture in the air. If a little water is added to the tube a solution of sulphuric acid will be obtained which can be tested with indicator paper and barium chloride solution.

Sulphuric acid—the most important chemical in industry

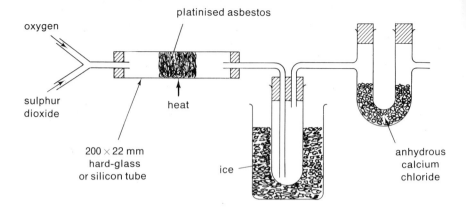

Figure 8.16

Experiments 8.2 and 8.3 Powdered roll sulphur must be used, not flowers of sulphur.

Molecular models Several ball-and-spoke models of the S_8 molecule are useful to demonstrate the breaking of the rings and linking into chains when molten sulphur is heated. The concertina effect of the zig-zag chains also helps to explain the properties of plastic sulphur. Linnell-type sulphur atoms (yellow, 2-hole, tetrahedral) and 38-mm springs are available from Gallenkamp. A tangential model of the S_8 molecule can be constructed from a convenient size of expanded polystyrene spheres. Using a protractor template as shown in Figure 8.17 the spheres are marked at 105° round the 'equator'. They can be joined by sharpened matchsticks or cocktail sticks dipped in adhesive and pushed half-way into each sphere. The most satisfactory adhesive is the polystyrene cement sold for use with ceiling tiles. if a less permanent model is required the spheres can be joined by lengths of pipe-cleaner, which need not be glued in. Griffin and George's *Grifzote* spheres are supplied are supplied with an instruction leaflet with full details for making scale space-filling models.

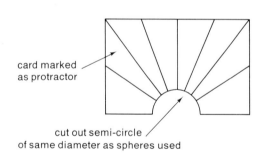

Figure 8.17

Preparation of sulphur dioxide If required, the full laboratory preparation of dry sulphur dioxide can be demonstrated as shown in Figure 8.18.

Alternatively the gas may be dried through concentrated sulphuric acid. Note colour, odour, and density. Test gas jars with damp indicator paper, a lighted splint, and filter paper dipped in acidified

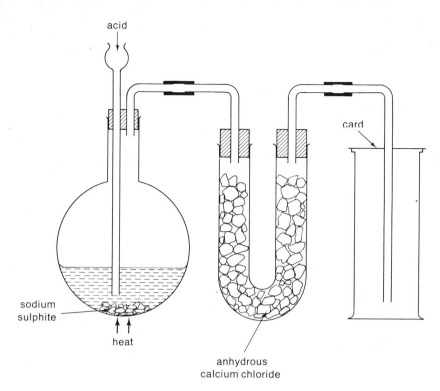

Figure 8.18

potassium manganate(VII) (equal volumes of 0.01 M KMnO₄ and M H₂SO₄). Solubility may be demonstrated by bubbling the gas through distilled water and testing the solution with indicator paper. Magnesium will continue to burn in the gas, leaving a deposit of sulphur on the sides of the jar.

Experiment 8.4 For (c) 0.01 M potassium manganate(VII) and M sulphuric acid. (d) demonstrates the density and solubility in water of sulphur dioxide.

Properties of hydrogen sulphide Pupils should **not** be allowed to prepare hydrogen sulphide. It is too poisonous. **All demonstrations should be performed in a fume cupboard**. Add 4 M hydrochloric acid to iron(II) sulphide in a test tube. Note colour and smell of gas. Test with damp indicator paper and filter paper dipped in 0.5 M lead nitrate or lead acetate solution. Bubble through 0.5 M lead nitrate (or acetate) and silver nitrate solutions. Invert test tube of hydrogen sulphide over test tube of sulphur dioxide; sulphur will be deposited on the sides of the tubes, demonstrating the reaction by which sulphur is obtained from oil and natural gas.

Sulphuric acid—the most important chemical in industry

9 Carbon

Carbon is not a very common element in the earth's crust, forming less than one part per thousand (0.08 per cent) by weight. In spite of this it is the most interesting of all elements. All living things are made up of carbon compounds. Over 90 per cent of the world's power is obtained from carbon and its compounds in the form of coal, petroleum, and natural gas.

Because it is so important, carbon has been studied more fully than any other element, and over one million carbon compounds are known – more than the number of compounds of all the other elements put together. Many of these carbon compounds have proved useful to man in a variety of ways: as drugs, plastics, dyes, detergents, etc.

Figure 9.1 Some of the many compounds of carbon in everyday use

Carbon compounds are called *organic compounds*, and their study is called *organic chemistry*. The study of the rest of the elements is called *inorganic chemistry*.

9.1 The element carbon

Even as a pure element, carbon is unusual. It exists in several different forms. Although much of the carbon in the earth's crust is combined in compounds, the free element is found, mainly as GRAPHITE and DIAMOND. Graphite is a soft black mineral with a rather greasy texture, whereas diamond is the hardest of all natural substances. Some of the properties of graphite and diamond are shown in Figure 9.2.

	Graphite	Diamond
Density	2.2 g cm^{-3}	3.5 g cm^{-3}
Conduction of electricity	good	bad
Conduction of heat	good	bad
Ignition temperature in air	700°C	900°C
Ignition temperature in fluorine	500°C	700°C

Figure 9.2

In spite of these differences in properties they are chemically both pure carbon. For example, they both burn in oxygen (or air) to produce carbon dioxide gas only, and in fluorine gas to produce carbon tetrafluoride. The differences between the properties of graphite and diamond are due to the different ways in which their carbon atoms are packed together. Substances with the same chemical composition but with the atoms packed in different structures are called ALLOTROPES. The structures of the allotropes of carbon have been discovered by X-ray diffraction (page 169).

Graphite

The structure of graphite is shown in Figure 9.3. The carbon atoms are arranged in layers, with the atoms in each layer forming a hexagonal pattern. Strong forces hold the atoms in a layer together, but the forces between the layers are very weak. They can therefore slip past one another very easily and this makes graphite very useful as a lubricant, either by itself or mixed with oil or water. It is also used in making the 'leads' of pencils (which, in spite of the name, have nothing at all to do with lead). As the pencil is drawn across the paper, layers of carbon atoms are wiped off on to it. To make pencil leads, graphite is mixed with clay; the greater the proportion of clay the 'harder' the pencil.

This allotrope of carbon has become such an important material that supplies of natural graphite are no longer sufficient. It is made artificially by

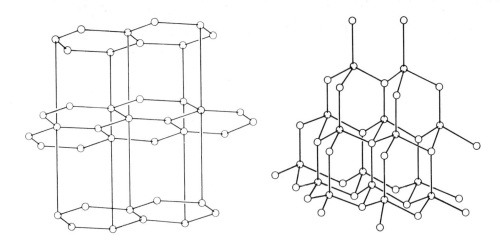

Figure 9.3 The arrangement of carbon atoms in graphite

Figure 9.4 The arrangement of carbon atoms in diamond

the Acheson process, in which coke (a fairly pure form of carbon) is heated electrically to about 2500°C for several hours.

CARBON FIBRE consists of long single crystals of graphite. These were first made by striking an electric arc between carbon rods under certain special conditions, when they form as whiskers up to 3 cm in length on the carbon rods. Carbon fibre is tremendously strong, and unlike most materials becomes stronger as the temperature rises. At 1500°C it is one of the strongest substances known. Since the whiskers are very thin and easily damaged they are of little use by themselves, but it has recently been found possible to embed them in metal. This produces a new engineering material with immense strength.

Diamond

The structure of diamond is shown in Figure 9.4. It is a regular three-dimensional structure in which the forces holding the carbon atoms together are equally strong in all directions. It is this regular structure which makes diamonds so hard that they can be used for cutting and drilling materials such as rock and glass (Figure 9.5). Very few diamonds are the clear transparent kind seen in a jeweller's window. Most are dark-coloured and opaque and of no ornamental value, but they are all equally hard and valuable for industrial purposes.

Natural diamonds are found in the cores of extinct volcanoes, which suggests that great pressure and heat are needed to produce this allotrope of carbon. In the nineteenth century there were many unsuccessful attempts to make artificial diamonds, some of which led to a good deal of fraud and trickery. Success has now been achieved and their manufacture has become an important industry. Carbon is placed in a small capsule together with some metal, usually nickel. The capsule is subjected to a

Figure 9.5 Cutting a block of slate with a diamond frame saw

pressure of 50 000 atmospheres and a temperature of over 1 500°C for several minutes in a huge press. It is not certain why the metal is necessary. It may be that, since the metal is melted in the press, it acts as a solvent in which the carbon dissolves and then crystallises out as diamond. Most artificial diamonds are very small. Sizes up to about half a centimetre across have been produced, but not of a high enough quality to be used as gemstones.

Other forms of carbon

Carbon occurs in a number of other forms which have no obvious crystalline structure.

When wood is heated out of contact with air CHARCOAL remains. This is fairly pure carbon. Its most useful property is its ability to absorb smoke and poisonous gases. For this reason it is used in gas masks. Sticks of charcoal are used for drawing. Before the discovery of coal it was an important fuel and today it is still used in barbecues.

When crushed bones are heated in the absence of air, ANIMAL CHARCOAL (also known as *activated charcoal*) remains; this is a mixture of carbon and calcium phosphate. It has the property of absorbing colouring matter and is used, for example, to decolourise crude sugar.

COKE is fairly pure carbon produced by heating coal in the absence of air. It is an important fuel and is used as a reducing agent in the manufacture of many metals (page 150).

Experiment 9.1 Are graphite and charcoal made of the same substance?

(a) Mix a spatula-full of powdered graphite with two spatulas-full of copper(II) oxide in a hard glass test tube.

Carbon   249

Have ready a teat pipette and a small test tube one-third filled with lime water.

Heat the mixture of graphite and copper(II) oxide with a fierce bunsen flame for two minutes, holding the tube almost vertical.

Remove the tube from the flame, keeping it vertical. Squeeze in the bulb of the teat pipette, then put the pipette into the test tube with the nozzle near to, but not quite touching, the mixture. Release the bulb so that the gas in the tube is sucked into the teat pipette. Squirt this gas into the lime water.

What happens to the lime water? What gas has been produced by heating graphite with copper(II) oxide? Did you see anything else being produced? Write a word equation for the reaction which has taken place between the graphite and the copper(II) oxide.

(b) Repeat the experiment using powdered charcoal instead of graphite.

What happens to the lime water this time? What gas has been produced by heating charcoal with copper(II) oxide? Did you see anything else being produced?

What do these two experiments tell you about graphite and charcoal?

Questions

1 (a) A carbon atom has six protons, six neutrons, and six electrons. Draw a sketch to show how these particles are arranged within an atom of carbon.

(b) How is it possible for carbon to exist as graphite and diamond, which differ so much in terms of appearance and hardness?

(East Anglia)

2 The structure of a material determines its physical properties. Given below are some details of the structures of different forms of carbon.

Form 1 is made up of flat planes of atoms. The planes are separated by quite large distances compared with the distance between atoms within a plane, so that the forces holding the planes together are small. Forces are applied to this material as shown in Figure 9.6.

(a) What would you expect to happen?

(b) Suggest a use to which the material may be put.

Form 2 is made up of a very compact three-dimensional network of atoms. It has a very even and close-knit structure (Figure 9.7).

(c) The relationship between the densities of Form 1 and Form 2 is best described by which of the following?

A   Form 1 is denser than Form 2.
B   They have the same density.
C   Form 2 is denser than Form 1.

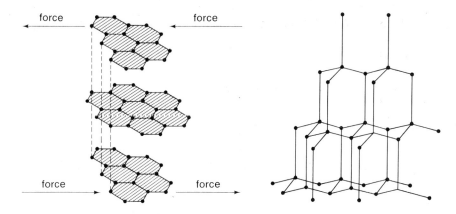

Figure 9.6   Figure 9.7

D The ratio of the density of Form 1 to Form 2 is 8:1.
E Form 2 is 9 times heavier than Form 1.

  (d) What can you say about the hardness of these two materials?
  (e) Suggest a use for Form 2.   (*London*)

3 (a) The element carbon exists in allotropic forms. Explain what this statement means.
  (b) Describe an experiment you have done or seen which demonstrated an important property of activated charcoal.
                                                    (*West Midlands*)

4 Which of the following statements are true and which are false?
A Graphite will burn to form carbon dioxide.
B Graphite is a carbon compound.
C Graphite is used as a lubricant.
D Graphite is used in pencils.   (*Lancashire*)

5 Some black copper(II) oxide was mixed with powdered carbon in a test tube and the mixture was heated strongly. The solid remaining after the experiment contained a reddish-brown powder.
  (a) Name the reddish-brown powder.
  (b) An invisible gas was produced by the reaction. Name this gas.
  (c) Write an equation for the reaction.   (*West Midlands*)

9.2 Carbon compounds in living things

Experiment 9.2 What elements are present in various foodstuffs?

  (a) You will be provided with a number of different foodstuffs. Put a spatula-full of one of the substances on a crucible lid, supported on a pipeclay triangle on a tripod.

Heat, at first gently, and then more strongly, until no further change seems to take place.

Repeat with two or three other substances. Do the residues left behind after heating the various substances resemble one another in any way? What do you think the black stuff left behind is?

(b) Mix a spatula-full of one of the substances with a spatula-full of copper(II) oxide. Put the mixture in a hard-glass test tube and add another spatula-full of copper(II) oxide on top of the mixture.

Before you start heating the mixture be prepared to test for carbon dioxide as in experiment 9.1, page 249. Heat the tube gently at first, then more strongly. Continue heating until there appears to be no further change. Test the gas in the tube to see if it contains carbon dioxide.

Does the lime water turn cloudy? What element is therefore present in the substance?

What do you notice in the upper part of the test tube in which the mixture was heated? What do you think this is? What experiment could you do to see if you are right? Carry out the experiment you suggest. Were you right? What element does this suggest must have been present in the original substance?

Repeat the experiment with two or three of the other substances.

The carbon cycle

You may have carried out experiments in which you burned sugar, starch, fat, and so on. One of the products of combustion of these substances is always carbon dioxide, which indicates that living things, from which these substances are derived, contain compounds of carbon.

Plants obtain their carbon in the form of carbon dioxide, which they absorb from the air through their leaves. They use the carbon dioxide, along with water taken in through their roots, to make SUGAR:

carbon dioxide + water + energy → sugar + oxygen.

The oxygen produced is released to the atmosphere through the leaves of the plants. The energy is provided by sunlight. CHLOROPHYLL, the green pigment in the leaves of plants, acts as a catalyst to help to bring about the reaction. The formation of sugar by plants is not fully understood, and chemists have not been able to imitate it in the laboratory. It is, however, one of the most important of all chemical reactions, since all life depends on it. It has a special name: PHOTOSYNTHESIS (*synthesis* = building up, *photo* = with light).

Some plants store sugar, but more often they convert it into STARCH, which they store in their seeds for use by the new generation of seedlings. Plants also use sugar to make CELLULOSE, which is the main material from which their stalks and leaves are made. Sugar, starch, and cellulose are all members of a group of chemical substances called CARBOHYDRATES, which are compounds of carbon, hydrogen, and oxygen.

Animals cannot make sugar from carbon dioxide; they are not capable of photosynthesis. They obtain carbohydrates by eating plants. They get sugar from fruit, sugar cane, etc. and starch from a wide variety of plants such as wheat, corn, rice, and potatoes. All animals can convert starch into sugar. Try chewing a piece of bread (made from wheat, a plant seed containing starch) for a long time without swallowing it. You will soon find it becoming sweet. Enzymes (page 64) in your saliva are decomposing the starch to sugar. Some animals can also break down cellulose into sugar. Cows, sheep, horses, and other animals that eat mainly grass get most of their sugar from the cellulose in grass.

Sugar is the basic 'fuel' which keeps plants and animals going. An animal cannot 'go' without sugar any more than a car can go without petrol. The energy of sunshine used in photosynthesis is stored within the sugar molecules as chemical energy. Plants and animals break the sugar down to release the energy for use in moving, keeping warm, and growing. This process of breaking down sugar to release energy is called RESPIRATION:

sugar + oxygen → carbon dioxide + water + energy.

Respiration is exactly the reverse of photosynthesis. Plants absorb oxygen through their leaves, fish through their gills, and animals through their lungs and get rid of the carbon dioxide and water vapour by the same route.

The carbohydrates which animals do not immediately require are stored as FAT. Some fat is needed as a food reserve and a layer of fat underneath the skin is necessary for some animals to insulate them against the cold. However, if an animal eats more sugar and starch than it needs, the extra fat produced just becomes useless extra weight to carry around.

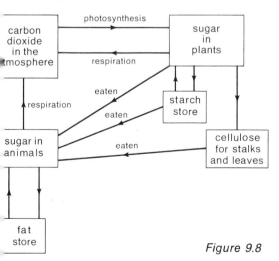

Figure 9.8

The sequence of reactions by which plants and animals take carbon dioxide from the air, use it, and return it to the air is called the CARBON CYCLE (Figure 9.8). The same carbon dioxide (0.03 per cent of the atmosphere by volume) is used and re-used over and over again. When plants and animals die, all their carbon compounds are sooner or later returned to the air as carbon dioxide.

Carbon 253

Experiment 9.3 Investigating breathing

(a) One-third-fill a test tube with lime water and blow slowly and steadily through the lime water with a straw. Note the time it takes for the lime water to turn milky.

Your teacher will demonstrate the time it takes for lime water to turn milky when ordinary air is drawn through it. Why is this time different from the time you have found for exhaled air?

(b) Stand a clean, dry test tube in a beaker full of iced water. Blow into the test tube through a straw. Note the time it takes for the appearance of condensation on the inside of the tube.

Test the condensation with a piece of cobalt chloride paper. What happens to the cobalt chloride paper? What does this tell you about the condensation?

Your teacher will demonstrate the time it takes for condensation to appear when ordinary air is drawn through a cold test tube. Why is this time different from the time you have found for exhaled air?

(c) Measure the temperature of the air in the room. Breathe out gently on to the thermometer bulb. What temperature is the air you breathe out? What do these temperatures tell you about the process of breathing?

The manufacture of sugar

Figure 9.9 Sugar cane

Some sugar is present in all plants, but it is obtained on a large scale mainly from two. SUGAR CANE (Figure 9.9), which grows only in tropical climates, provides 60 per cent of the world's sugar. The rest comes from SUGAR BEET, which grows in cooler regions.

Sugar cane is a gigantic grass, resembling bamboo. The sugar is in the soft fibre inside the stalk. The canes are cut up and crushed with water. The sugar, which is very soluble in water, dissolves, and the juice from the crushers contains up to 20 per cent of sugar, together with some impurities.

The juice is heated in large tanks and *lime* is added. This helps to collect the insoluble impurities together as a sludge at the bottom of the tank. The clear juice is then evaporated down until the sugar crystallises out. Evaporation is carried out under low pressure so that the liquid boils at a temperature well below 100°C (page 50). This is necessary because sugar begins to decompose into carbon and water at 100°C. It also economises on heat. The evaporation produces a mixture of sugar crystals

254 Chemistry today

and a thick, syrupy liquid called *molasses*. The impure crystals are separated from the molasses in a centrifuge and transported to a refinery. The molasses is used to make rum and cattle food.

Sugar is extracted from sugar beet in the same way as from cane. Crude sugar from beet or cane (*Demerara sugar*) is a moist brown substance consisting of fairly pure sugar crystals coated with molasses. In the refinery it is mixed with syrup to soften the molasses and centrifuged again. This removes most of the molasses. The crystals are dissolved in water and the solution is filtered to remove any insoluble impurities. The filtrate is a clear brownish-yellow colour. The colour is removed by allowing the solution to trickle slowly over animal charcoal (page 249) which absorbs the coloured impurities. The clear colourless solution is evaporated at low pressure. Sugar crystallises out and the crystals are separated by centrifuging.

Paper making

The basic material of paper is CELLULOSE, obtained from a wide variety of plants including trees, grasses, cotton, and flax. By far the most important source is wood, which provides over 95 per cent of the world's paper. Wood is made up of long thin cells; their appearance under the microscope is shown in Figure 9.10. The walls are composed largely of cellulose. These are the wood *fibres* which are important in paper making.

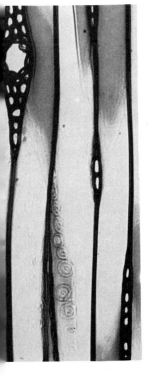

For poor-quality paper (newsprint, etc.) the wood is pulped mechanically by disintegrating the logs with a grindstone. For better quality paper chemical pulping is used to remove the other substances in the cells, so that only the cellulose is left. In chemical pulping, wood chips are 'cooked' under pressure in a solution of sodium sulphite at a temperature well above the boiling point of water for several hours. All the constituents of the wood fibres except the cellulose dissolve. The cellulose *pulp* is filtered from the solution, washed, and bleached. It is then beaten to break up the fibres (Figure 9.11). The amount of beating controls the properties of the finished paper: blotting paper results from very little beating, greaseproof paper from heavy beating.

After mixing various additives with the pulp, such as fillers (e.g. calcium sulphate) to make the paper less transparent and dyes to colour it, it is mixed into a thin paste with water. This paste, which contains

Figure 9.10 Photograph (magnification × 130) of wood cells

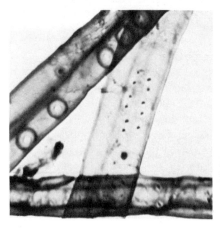

Figure 9.11(a) Photograph (magnification × 250) of unbeaten wood pulp

(b) Photograph (magnification × 250) of wood pulp which has been well beaten

between 1 per cent and 2 per cent of pulp, is fed onto a fast-moving wire gauze. Water begins to drain out, helped by suction applied from underneath. The *web* then passes through presses which squeeze out most of the remaining water, and finally through ovens to dry it.

Manufacturing rayon

Cellulose fibres from certain plants have been used for centuries to make clothing. In India records of the manufacture of cotton date back to before 500 BC and the Egyptians wrapped their mummies in linen cloth, made from the cellulose fibres of the flax plant.

Wood is also composed of cellulose fibres and is cheaper, more abundant, and easier to cultivate than cotton and flax. Why, then, not make thread from wood? There are two problems. The fibres in the wood are matted and tangled together and difficult to separate. Also they are too short to be spun into thread. Figure 9.12 shows the lengths of the cellulose fibres from various plants.

	Length of fibre
Wood	1 mm–2 mm
Cotton	6 mm–8 mm
Flax	10 mm–15 mm

Figure 9.12

Towards the end of the nineteenth century these problems were overcome by making a solution of wood cellulose and then reprecipitating the cellulose in the form of a continuous thread. The product is called RAYON. The first solvent to be used was copper(II) hydroxide dissolved in a solution of ammonia. Highly purified cellulose had to be used, which made the product rather expensive.

The modern process, now carried out on a large scale, was discovered in 1892. Wood pulp is prepared as in the manufacture of paper (page 255). The pulp is steeped in a solution of sodium hydroxide and then in carbon disulphide. This produces a crumbly yellow solid called *cellulose xanthate*, which is soluble in sodium hydroxide solution. The xanthate solution, called *viscose*, is left to 'ripen' for several days at a carefully controlled temperature. It is then squeezed through a spinneret into a bath of dilute sulphuric acid. The acid neutralises the alkali and pure cellulose is precipitated in the form of continuous threads, ready for weaving (Figure 9.13).

Figure 9.13 Rayon filaments being wound out of the bath of dilute sulphuric acid

Experiment 9.4   Imitating rayon production

(a) To 1 cm depth of copper(II) sulphate solution in a test tube add two drops of ammonia solution, and shake it. What do you observe?

To the mixture in the tube add ammonia solution to almost fill the tube, and shake it again. What do you observe now?

This solution is called ammoniacal copper(II) sulphate.

(b) One-quarter-fill a beaker with ammoniacal copper(II) sulphate solution. Tear up filter papers into small pieces and put the pieces into the beaker until the solution will take no more. Stir the mixture up and put it on one side for at least 24 hours.

At the end of this time pour the liquid off the remains of the filter paper into another beaker. Have ready a beaker half-filled with dilute sulphuric acid. Suck some of the blue solution into a syringe. Dip the tip of the syringe into the acid and **very slowly** squirt the solution into the acid.

What happens to the filter paper when it is left in the ammoniacal copper(II) sulphate solution? What happens when the blue solution is squirted into the acid? Why?

Questions

1 (a) Why are carbon compounds so important to living organisms?

Two chemicals can be joined together by plants to start a food cycle.

(b) What are these substances?
(c) Write an equation for the reaction.
(d) What provides the energy to allow this chemical reaction to take place? (London)

2 (a) State one way in which carbon dioxide is constantly being taken from the atmosphere.
(b) State two ways in which carbon dioxide is constantly being put into the atmosphere. (West Yorkshire)

3 Explain why green plants increase the amount of oxygen present in the atmosphere during the day, but not during the night.
(North West)

4 Starches and sugars belong to which of the following classes of compounds?
A   allotropes
B   catalysts
C   carbohydrates
D   hydrocarbons
E   indicators (London)

5 In the process of breathing, a comparison of exhaled air with inhaled air shows which of the following?
A   Increase in oxygen content and decrease in carbon dioxide content.
B   Decrease in oxygen content and decrease in carbon dioxide content.
C   Increase in oxygen content and increase in carbon dioxide content.
D   No change in oxygen content but decrease in carbon dioxide content.
E   Decrease in oxygen content and increase in carbon dioxide content. (West Midlands)

6 A sample of a foodstuff was heated alone. It was seen to blacken. When heated with black copper oxide, it gave off a gas which turned lime water milky. A colourless liquid was seen to condense in the cooler part of the tube. This liquid turned anhydrous copper sulphate blue. The copper oxide turned to a bright pink colour.
(a) What does the blackening suggest?
(b) What gas was given off?
(c) What do you think the liquid was that condensed in the cooler part of the tube?
(d) What elements does this information suggest are present in the foodstuff? (London)

3 Alcohol

Experiment 9.5 Fermentation

Half-fill a 100 cm³ conical flask with water and dissolve 6 spatulas-full of glucose (a type of sugar) in the water. Add one spatula-full of yeast to the solution.

Set up the apparatus shown in Figure 9.14 with the delivery tube dipping below the surface of the lime water.

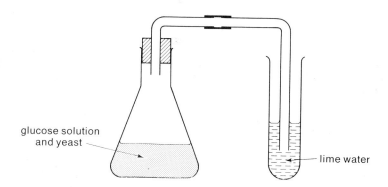

Figure 9.14

Leave the apparatus in a warm place, preferably near a radiator. Examine the test tube after about an hour. What can you deduce from the appearance of the contents of the test tube?

Fermentation will be complete in two or three days.

Alcohol from carbohydrates

The group of chemical compounds known as CARBOHYDRATES (sugars and starches) can be broken down into alcohol by FERMENTATION:

carbohydrate → alcohol + carbon dioxide + energy.

Some simple forms of life such as bacteria, fungi, and tapeworms are not capable of respiration because they often live in places where they cannot get oxygen. Instead, they obtain their energy by decomposing carbohydrates into alcohol, using enzymes as catalysts.

Since primitive times man has been using fungi to produce alcohol. You may have made alcohol by fermenting glucose (a sugar) with yeast, which is a type of fungus. The carbon dioxide produced can be detected by its effect on lime water. After a day or two, when fermentation stops, a dilute solution of alcohol in water remains. The alcohol can be separated from the water by FRACTIONATION (fractional distillation) a method of distillation which enables liquids with different boiling points to be separated. Figure 9.15 shows typical laboratory apparatus for fractionation.

Water boils at 100°C and alcohol at 78°C. Both liquids are vaporised in the flask. As the vapours pass up the column they are cooled; the small glass rings or beads with which the column is filled slow down the progress of the vapour. The liquid with the higher boiling point (water) tends to condense first, and runs back into the flask. By the time the vapour reaches the top of the column it is almost pure alcohol.

Commercial forms of alcohol

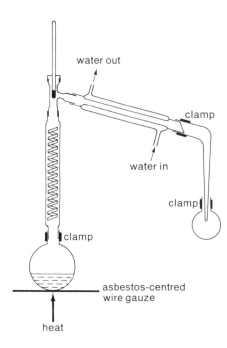

Figure 9.15

The proper chemical name for alcohol is ETHANOL. Absolutely pure ethanol cannot be obtained by fractionation; the most concentrated product contains 95.6 per cent ethanol with 4.4 per cent water by weight. This is known as *rectified spirit*. It is difficult and expensive to remove the last traces of water, but for special purposes *absolute alcohol*, which is 99.5 per cent pure, is produced.

Rectified spirit and absolute alcohol, like alcoholic drinks, are taxed at a very high rate. For many industrial purposes pure alcohol is not needed and to avoid tax it is made undrinkable by adding various other chemicals. *Industrial methylated spirit* is a mixture of 95 per cent of rectified spirit with 5 per cent methanol, which is extremely poisonous. *Mineralised methylated spirit*, or *meths*, is a mixture of 90 per cent rectified spirit with 10 per cent methanol, together with a violet dye and small amounts of other chemicals to give it an unpleasant taste. This is the familiar form of alcohol which can be purchased at ironmongers.

The main use of alcohol is as a solvent: for paints, varnishes, perfumes, and so on. It is a useful fuel, since it burns to form carbon dioxide and water with the production of a considerable amount of heat. It is added in small quantities to some types of petrol. It is also used in the manufacture of other chemicals.

For taxation purposes the ethanol content of industrial alcohols and alcoholic drinks is assessed in DEGREES PROOF. 100° proof spirit used to be defined as the weakest alcohol-water mixture which would not prevent the ignition of gunpowder moistened with it. Nowadays it is defined as the alcohol-water mixture which, at 51°F, weighs exactly $\frac{12}{13}$ of the weight of the same volume of distilled water. It contains 49 per cent ethanol with 51 per cent water by weight. The average strengths of some alcoholic drinks are given in Figure 9.16.

	°proof	% ethanol (by weight)
Whisky, gin, rum, brandy	70°	34
Sherry, port	35°	17
Wine	20°–25°	10–12
Beer, cider	5°–9°	2–4

Figure 9.16

The effects of alcohol on the body

The effects of alcohol on the body are very complex. In small quantities it is a stimulant and can be beneficial. We enjoy alcoholic drinks before and with meals because alcohol increases the flow of the digestive juices and stimulates our appetites. The Saint Bernard dog carries his flask of brandy for the use of people suffering from exposure to the cold because alcohol sets the heart beating more strongly and rapidly.

In larger quantities alcohol makes the whole nervous system less sensitive. Before the discovery of anaesthetics a patient was made drunk before an operation. Alcohol slows down reactions too: *Don't Drink and Drive.* An excessive intake of alcohol over a long period can cause addiction (*alcoholism*), a deterioration of the liver and kidneys, and eventually damage to the brain (*delirium tremens*, or DTs).

Fermentation in baking

A cooked mixture of flour and water (*dough*) is a basic food in most parts of the world. In some Mediterranean countries, especially Italy, a dried dough in various shapes called *pasta* (macaroni, spaghetti, etc.), cooked by boiling in water, is popular. Bread is also made from dough, but when cooked it is very much less dense than pasta. The lightness of bread is achieved by producing carbon dioxide gas inside the dough by fermentation. The dough becomes filled with tiny bubbles so that when it is cooked the bread is light in texture.

The dough is mixed with yeast and a little sugar and kept warm for an hour or two. The yeast grows, obtaining its energy by decomposing the sugar to alcohol and carbon dioxide. When the dough has risen to about twice its original size it is ready for cooking. At the high temperature of the oven the yeast is killed so that fermentation stops and the bread ceases to rise.

Brewing

The carbohydrate for fermentation to alcohol in beer comes from barley. Grains of barley (the plant seeds) are provided with water and warmth so that they start to grow. The seed cases split and the seeds begin to grow a stem and roots. This is called *malting* in the brewing industry. After a week or ten days the small plants are killed by heating in an oven, or *kiln*.

The partly-grown barley, called *malt*, is crushed and mashed with water. The starch dissolves and the husks of the seeds are filtered off. The malt extract, called *wort*, is run into huge copper boilers. Sugar and roasted hops are added and the mixture is boiled to extract the flavour from the hops.

The hot wort is cooled and run into fermentation vessels (Figure 9.17) where yeast is added. After about a week fermentation is complete and the beer is pumped into conditioning tanks and allowed to mature before being put into barrels or bottles.

There are a number of important by-products from the brewing industry. The husks of the barley are dried and sold to farmers as cattle food. Yeast multiplies about six times during fermentation; the surplus is

Figure 9.17 A fermenting room in a brewery, showing the yeast head on a fermenting vessel

Figure 9.18 Traditional pot stills of the type in which malt whisky has been made for centuries

sold, mainly to the baking industry. The carbon dioxide produced during fermentation is also a marketable by-product.

The manufacture of whisky

The first stages in the manufacture of whisky are the same as in the brewing of beer. Barley is malted and then kilned. The kiln is heated with burning peat; the peat smoke is partly responsible for the final smell and flavour of the whisky. The malt is crushed and mashed with water to produce wort, which is pumped into fermenting vessels. Yeast is added and fermentation is allowed to proceed for two or three days. The fermented liquor, or *wash*, now contains about 15 per cent of ethanol by weight and must be concentrated by distillation.

The wash is run into a copper *pot still* (Figure 9.18). It is boiled, and just as in the laboratory fractionating column (page 260) the vapour condensed at the top of the still is richer in alcohol. After the first fractionation the whisky is neither strong enough or pure enough, so the fractionation process is repeated in another pot still. The middle part of the distillate from the second still is considered to be the best whisky. It is run into barrels where it is stored for several years to mature before being bottled for sale.

Wine production

The grapes from which wine is made, like other fruits, contain sugar. After the grapes have been picked they are crushed to squeeze out the juice, which is known as *must*. This is still sometimes done by treading the grapes in a large tank, though mechanical presses are often used nowadays. A fungus, rather like yeast, grows naturally on the skins of the grapes; its enzymes start the fermentation of the sugar in the must as soon as the grapes have been pressed.

After several days fermentation is complete and all the sugar in the must has been converted into alcohol. If sweet wine is required, extra sugar is added. The wine is stored in barrels for several months, or even years, to mature before being bottled for sale.

Red wine is produced from black grapes by mixing some of the skins with the must during fermentation so that the colouring matter from the skins passes into the wine. Sparkling wines, like champagne, are made by bottling the wine before fermentation is complete. Fermentation continues inside the bottle and the carbon dioxide produced dissolves in the wine. When the bottle is opened the pressure is released and the carbon dioxide bubbles slowly out. Wine contains between 10 per cent and 12 per cent of alcohol by weight. To make stronger wines like sherry and port, brandy is added to increase the ethanol content to about 17 per cent by weight.

Questions

1 In the fermentation of sugar, frothing or effervescence takes place for which of the following reasons?
A   The process is very hard to control.
B   The sugar is initially very impure.
C   The temperature of the reaction rises very rapidly.
D   Large quantities of carbon dioxide are produced.
E   The process takes place at reduced pressure.   (*East Anglia*)

2 Substance P is a white solid made up of small crystals. On heating, it turns BLACK and gives off vapours which condense out to give a colourless liquid which turns anhydrous copper sulphate BLUE. When yeast is added to a solution of P in water and the mixture is placed in a warm cupboard bubbles of GAS appear, which when tested, turn lime water MILKY. If after a few days the liquid mixture is filtered and distilled a colourless sweetish-smelling liquid is produced. Identify P by the steps below.
The gas is _____ .
The liquid produced by distilling the yeast mixture is _____ .
The formation of a black substance _____ together with a liquid which is _____ should give the conclusion that P is a _____ , and being crystalline it is most likely to be _____ .
(*West Yorkshire*)

3 Ethanol and water mixed in equal volumes can be largely separated by the method called _____ .   (*London*)

9.4 Coal

The origin of coal

About 250 million years ago the Northern hemisphere is believed to have had a tropical climate. The land was covered with dense forests of primitive plants, mainly mosses and huge ferns. These plants grew, reproduced, and died to form thick layers of decaying vegetation.

Sand and mud, washed down by rivers, covered these layers. More vegetation grew from the sand and mud, died and was covered in the same way. Over a period of millions of years this process repeated itself many times. Then, as a result of earthquakes, which in those times were far more frequent and violent than they are now, the alternating layers of mud, sand, and decaying vegetation were covered with rock, a few metres or thousands of metres thick.

The weight of the rock compressed the sand into *sandstone* and the mud into *shale*. The heat at the great depth at which the decaying vegetation was buried and the pressure of the rock above, turned it into

Figure 9.19 Fossilised ferns found in coal seams

coal. Evidence of this origin of coal is found in the form of fossilised ferns and other plants embedded in the coal itself (Figure 9.19).

There are two main types of coal. *Bituminous coal* is hard and black, and is believed to have been formed earlier than *lignite*, which is soft and brown. *Peat* is another form of decayed vegetation, less old and less compressed, which is only part-way towards being converted into coal. Figure 9.20 shows the average composition by weight of wood, peat, lignite, and bituminous coal. The main difference between them lies in the oxygen content.

	Per cent carbon	Per cent oxygen	Per cent hydrogen
Wood	53	42	5
Peat	60	34	6
Lignite	67	28	5
Bituminous coal	88	6	6

Figure 9.20

Mining for coal

Coal has been used as a fuel for thousands of years. Some lies close to the surface; some is mined up to depths of one kilometre. The layers, or SEAMS, may be only a few centimetres thick or as much as 25 metres thick. Coal was first gathered where seams outcropped as a result of earthquakes or erosion.

Mining began about a thousand years ago, but only for coal lying near the surface. In the nineteenth century, with the invention and growing use of the steam engine, coal became extremely important as a fuel and mining was carried out on a large scale. It was a combination of the abundant supply of coal and mechanical know-how that then made Britain a great industrial power.

In its early years large-scale coal-mining was a primitive and dangerous business. Miners, including women and young children, worked in conditions of near-slavery. Poor ventilation in the mines led to suffocation. Explosions and fires below ground were frequent due to the presence of *firedamp* (methane), an inflammable gas produced during the formation of coal from decaying vegetation. Dust in the mines caused a lung disease called silicosis. Rock falls were common. Even today mining has its dangers. 80 miners were killed in Britain in 1969 (though this compares with over 600 killed in 1947, the year in which the coal industry was nationalised).

Nowadays most of the coal is cut from the coal face by machines, instead of by men with picks. The coal is loaded automatically on to conveyor belts which carry it back to the mine shaft. In some British mines the whole process is fully automated so that it can be controlled by a single man on the surface.

Uses of coal

Coal is used for three main purposes: to burn, as a source of energy, to make coke for the extraction of metals from their ores (page 150), and as a raw material for the production of other chemicals. Figure 9.21 shows the proportions used for various purposes in the United Kingdom in 1969.

Power stations	76 million tonnes
Coke production	29
Industrial fuel	23
Domestic fuel	21
Gasworks	6
Other uses	4
Export	$3\frac{1}{2}$
Total	$162\frac{1}{2}$ million tonnes

Figure 9.21

Three-quarters of Britain's electricity is still produced from coal, though oil-fired, gas-fired, and nuclear power stations are gradually taking a larger share. As can be seen from Figure 9.21, nearly half the coal mined in the United Kingdom is burnt directly in power stations to produce electricity. Heat from the burning coal is used to convert water into steam. The steam is made to drive turbines which in turn drive the dynamos.

If coal is burnt completely, as it can be in power stations and other industrial furnaces, there need be no smoke, but it is difficult to design domestic fires and boilers to burn coal completely. The smoke from burning coal contains sulphur dioxide and other poisonous and corrosive gases, and in large quantities is damaging to health, corrodes brickwork and metal, and produces smog. In 1956 the *Clean Air Act* made it illegal to release certain types of smoke, including the smoke from burning coal, into the atmosphere.

This led to the development of various kinds of smokeless fuel made from coal. *Anthracite* and *steam coal* are two smokeless forms of coal that are obtained directly from the ground but they are not very abundant and are therefore expensive. Anthracite, for example, is only found in a few seams within 30 kilometres of Swansea. The first man-made smokeless fuel was coke, produced by heating coal in the absence of air. Over the last few years many more smokeless fuels have appeared on the market, under names like *Phurnacite, Homefire* and *Coalite.* Some of these are designed for closed furnaces and some for open fires. They are made from powdered coal, which is heated to about 400°C and compressed into briquettes bound together with pitch. The heating drives off the impurities in the coal which cause it to smoke when burnt. The Clean Air Act and the increased use of smokeless fuel have had dramatic effects. Smog has almost been eliminated. The death rate from diseases like bronchitis has fallen. Corrosion has been reduced and buildings remain clean.

You may have carried out an experiment to heat coal in the absence of air. When coal is heated in this way (DESTRUCTIVELY DISTILLED) there are four main products: COAL GAS; AMMONIACAL LIQUOR; COAL TAR; and COKE. Coal gas was discovered at the end of the seventeenth century and was first developed for street lighting by James Watt and William Murdock in 1802. Its composition varies but on average it contains about 50 per cent of hydrogen, 35 per cent of methane, and 8 per cent of carbon monoxide, together with smaller amounts of other gases, including nitrogen and carbon dioxide. Until recently all gas for household and industrial purposes was produced from coal, but by 1970 the gas industry had converted almost completely to natural gas and gas manufactured from oil.

About 20 per cent of the coal produced in Britain is converted into coke for use in the extraction of metals, especially iron, from their ores (page 150). The coal is heated in rows of coke ovens, each holding up to 20 tonnes of coal which is loaded in from the top. Burning gas is blown between the ovens to heat the coal. When the gas has been driven off, the coke is emptied into a coke-car, cooled by a spray of water, and transported to the blast furnaces. Some of the gas is used to heat the ovens, the remainder is purified and sold.

Coal tar and ammoniacal liquor, produced as by-products in the manufacture of coal gas and coke, contain many different chemicals which are used in the manufacture of an enormous variety of products. Figure 9.22 gives some indication of the vast scope of the coal by-

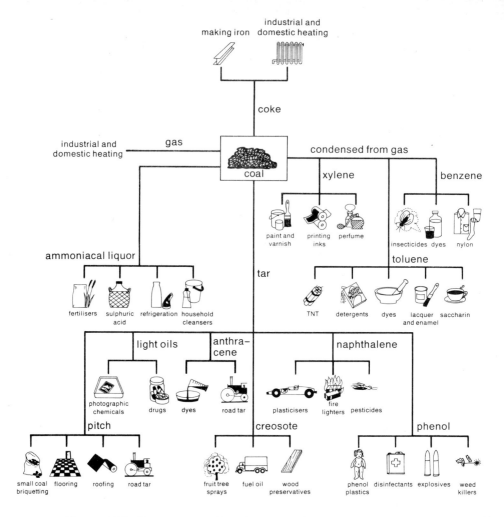

Figure 9.22 By-products of coal

products industry. The tar and liquor go through a long and complicated series of separation and purification processes, including fractionation, for these chemicals to be obtained. At present only about 16 per cent of the coal mined in the United Kingdom is treated in this way; 84 per cent is burnt completely. This is rather a waste of our natural resources, but as time goes by more and more coal is being fully utilised.

Experiment 9.6 The destructive distillation of coal

Set up the apparatus shown in Figure 9.23.

The test tube should be about one-third full of pieces of coal. The side-arm tube should contain enough water to cover the bottom of the delivery tube.

Heat the coal, gently at first, then more and more strongly. What do you see (a) coming out of the side-arm, (b) at the bottom of the side-arm tube, (c) on the sides of the tube containing the coal?

Light the gas coming out of the side-arm. Describe the way in which it burns.

When there is no more gas remove the side-arm tube, then stop heating. (Why must the apparatus be disconnected before you stop heating?)

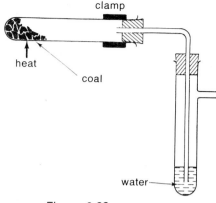

Figure 9.23

Add an equal volume of sodium hydroxide solution to the liquid in the side-arm tube. Warm the mixture, smell **cautiously**, and test the gas given off, using a piece of moist indicator paper. What gas is it?

Questions

1 (a) What is meant by photosynthesis?
  (b) Explain, in detail, how photosynthesis led to the formation of coal. (*South East*)

2 (a) What evidence is there that coal was formed in prehistoric times?
  (b) The chemical energy released from fossil fuel is obtained from the sun's energy. Why is it important that alternative sources of fuels are developed? (*London*)

3 When coal is heated in the absence of air the process is called which of the following?
A   fractional distillation
B   combustion
C   reduction
D   destructive distillation
E   purification. (*North West*)

4 (*Use Figure 9.22 to help you to answer this question.*)
When coal is heated strongly in the absence of air it forms solid, liquid, and gaseous products.
   (a) What is the solid product?
   (b) Give the two different uses of the solid product.
   (c) What are the two main liquid products?

Carbon   269

(d) Give the main use for the 'thinner' liquid product.
(e) What is the main gaseous product?
(f) Give one use for this product.
(g) Name an important gaseous by-product.
(h) Give the main use of this by-product.

The 'thicker' of the two liquid products of the destructive distillation of coal provides a rich source of chemicals.

(i) How is this 'thicker' liquid treated to yield up its components?
(j) Give the chemical names of four chemicals obtained in this process.
(k) What is the residue from this process?
(l) What is its main use?   (London)

5 It is said that the use of North Sea Gas by the gas boards is resulting in a shortage of smokeless fuel such as coke. Explain why the increasing use of natural gas should cause a shortage of coke.
(Wales)

9.5 Petroleum

Petroleum, or crude oil, is one of the most useful materials found in the earth's crust. It can be burnt as a fuel, and, indeed, as a fuel it produces about one-third of all the world's power. It can also be converted by the chemist into an enormous range of products: fertilisers, plastics, artificial rubber, washing powder, paint, and many others.

Petroleum is found in many parts of the world. Sometimes it is visible, seeping through the surface; sometimes it is buried at a depth of several kilometres. Usually it is hard to find and even when found it has to be refined and transported to where it will be used. Finding, refining, and transporting are all expensive operations, but the end-products are well worth it. Petroleum has been nicknamed *black gold*.

The origin of petroleum

No one is certain how petroleum came to be formed in the earth. Petroleum-bearing rock contains fossils of sea-creatures (Figure 9.24), just as coal contains fossils of plants.

This suggests that petroleum may have been produced by the decay of sea-creatures in the same way that coal was formed by the decay of plants. The most likely theory is that when the tiny creatures died and fell to the bottom of the sea they became covered with mud and decomposed into oil. Over millions of years the mud was compressed into a type of rock, called *shale*. Earthquake activity heaped more rock on top of the shale. The pressure of the overlying rock sometimes forced the oil up to the surface. But if, on its way up, the oil met a layer of non-porous rock it became trapped, soaked into the porous rock beneath like water in a sponge (Figure 9.25).

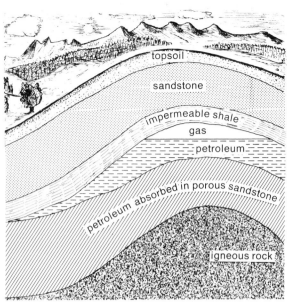

Figure 9.24 (left)   Oil-bearing rock seen through a microscope
Figure 9.25 (right)   A typical oil-bearing rock structure

Drilling for oil

Oil from natural seepages has been used by man for thousands of years. It was traces of oil in spring water in Texas which led the early prospectors to the oil there, and it was in Texas in the middle of the nineteenth century that drilling for oil trapped underground was first attempted.

Often petroleum is trapped far underground and there is no seepage to indicate where to drill. The geologist has to make an intelligent guess. He learns a good deal by studying the shape of the earth's surface and by examining the fossils in the surface rock. He can get some idea of what lies beneath the surface in various other ways. He studies slight variations in the earth's gravity and magnetism which result from different rock formations below the surface. He analyses the way in which the sound from an explosion on the surface is reflected by the rock layers beneath. But however much evidence he collects the geologist can never be certain of the presence of oil. The only way to be certain is to drill.

The first well to be drilled, in 1859, struck oil at a depth of 12 metres. Modern *rigs* can drill up to depths of ten kilometres and for oil under the sea bed they can operate through 200 metres depth of water or more.

The oil underground may be under pressure, in which case it will force its way to the surface as soon as the drill reaches it. This type of well is called a *gusher*. If the oil is not under pressure it must be pumped to the surface.

Once at the surface the oil has to be transported to where it is required. It usually leaves the oilfield by pipeline. Sometimes the pipeline takes it all the way to the refinery but often oil is discovered in out-of-the-way places and has to be transported for part of its journey by sea in a tanker. Oil tankers make up more than a third of all the world's shipping tonnage. Some of them are among the largest ships in the world, half a million tonnes or more.

Experiment 9.7 Fractionation of crude oil

Figure 9.26

Put mineral wool to a depth of 3 cm in a side-arm tube, and add 2 cm^3 of crude oil. Set up the apparatus as shown in Figure 9.26.

Label four small test tubes 1, 2, 3, and 4.

Warm the bottom of the tube gently until liquid starts to distil over and collects in the small test tube labelled 1. When the temperature reaches 70°C remove this tube and replace it with the tube labelled 2. At 120°C change to tube 3, and at 170°C change to tube 4. Stop heating at 220°C.

You will now have collected four fractions with the boiling ranges:

1 Room temperature to 70°C
2 70°C to 120°C
3 120°C to 170°C
4 170°C to 220°C.

Describe the appearance of the four fractions. Test their runniness (viscosity) by tilting the tubes. Arrange the four fractions in order of viscosity, from the most runny to the thickest. Why do you think the viscosity of the fractions is different?

Pour each fraction in turn on to a watch glass and light it with a splint. Describe the appearance of the flames and note if they are easy or difficult to light.

Refining petroleum

Petroleum is a mixture of thousands of different compounds. Most of them are HYDROCARBONS, compounds of carbon and hydrogen which differ from each other only in the number of carbon and hydrogen atoms in their molecules. The molecules are of different sizes and have different boiling points; use is made of this in separating them.

It is not profitable, or necessary, to separate out all the individual hydrocarbons in crude oil. It is separated into a few FRACTIONS which boil over different ranges of temperature. Each of the fractions is still a mixture of many different compounds.

The industrial process is carried out on a huge scale. Figure 9.27 shows an industrial FRACTIONATING COLUMN. The petroleum is first heated to about 300°C. At this temperature a considerable proportion of the oil boils. The hot mixture of liquid and vapour is led into the column. The liquid drops down to the bottom of the tower where it is tapped off. The vapour rises up the tower where it meets a series of trays containing BUBBLE CAPS.

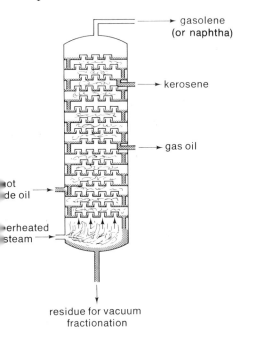
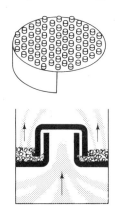

Figure 9.27 (left) An industrial fractionating column

Figure 9.28 (above) A tray of bubble caps

Figure 9.29 (below) A single bubble cap

A tray (Figure 9.28) may consist of hundreds of bubble caps (Figure 9.29). As vapour passes up through the bubble cap it is forced to pass through liquid in the tray before it can go further up the tower. The liquid in the tray has formed by condensation of previous supplies of vapour. As the vapour passes through the bubble cap it cools and the compounds in the vapour with the highest boiling points condense. The compounds with lower boiling points remain as vapour and carry on up the tower to the next tray. This process continues up through the tower. Hydrocarbons with the highest boiling points condense in the lower trays; this is the fraction called *gas oil*. These with medium boiling points condense in the upper trays; this is the fraction called *kerosene*. Those with lower boiling points still do not condense, and are led off from the top of the tower as vapour; this is the fraction called *gasolene*, or *naphtha*.

The gasolene, kerosene, and gas oil fractions are purified to remove unwanted materials, mainly sulphur compounds. If these were not removed the petroleum products would have a foul smell and on combustion would produce the very corrosive gas, sulphur dioxide (page 238).

Carbon   273

The liquid drawn off from the bottom of the first fractionating column, the PRIMARY FRACTIONATING COLUMN, is fractionated again in another column under low pressure. This is known as VACUUM FRACTIONATION. If this liquid were heated to a high enough temperature to boil it under normal atmospheric pressure many of the hydrocarbons in it would decompose. Lowering the pressure has the effect of lowering the boiling point of all the compounds in the liquid (page 50) so that they can be boiled without decomposition. Vacuum fractionation of the liquid from the base of the primary tower produces three more fractions: *lubricating oil*, *fuel oil*, and *paraffin wax*, and leaves a residue of *bitumen*, which does not boil even under the reduced pressure.

Figure 9.30 summarises the uses of the principal fractions obtained in the primary and vacuum fractionations of petroleum.

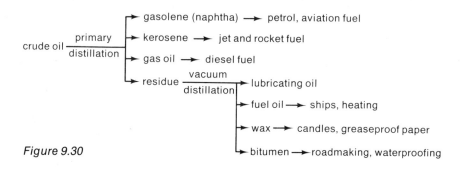

Figure 9.30

The chemistry of petroleum

The simplest hydrocarbon is the gas methane, CH_4. Some hydrocarbons, with more complicated molecules, are liquids; octane, C_8H_{18}, for example, which is a component of petrol. Those with still larger molecules, with 20 or more carbon atoms per molecule, are solids; vaseline and paraffin wax are mixtures of these complex hydrocarbons.

The carbon atom has 6 electrons. Two form a complete shell. When carbon forms a compound the other 4 electrons are *shared* with electrons from four other atoms to complete the second shell of 8 electrons (page 132); carbon compounds are therefore COVALENT. In methane the four hydrogen atoms are arranged regularly around the central carbon atom (Figure 9.31). The methane molecule is described as TETRAHEDRAL in shape. Models of some other hydrocarbons are shown in Figures 9.32 to 9.36.

Carbon is the only element whose atoms can form long chains of up to 60 atoms (as in butane, Figure 9.34 and octane, Figure 9.32, or branched chains (as in *iso*-butane, Figure 9.33, and *iso*-octane, Figure 9.35), or rings in which the ends of the chain are linked up (as in cyclohexane, Figure 9.36). You can begin to appreciate why there are so many different hydrocarbons.

 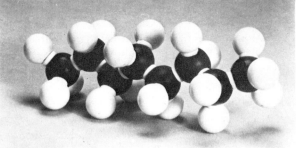

Figure 9.31 Model of a methane molecule

Figure 9.32 Model of an octane molecule

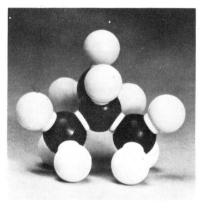 

Figure 9.33 Model of a molecule of methylpropane, also known as iso-butane

Figure 9.34 Model of a butane molecule

 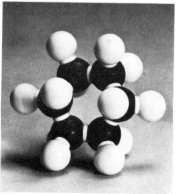

Figure 9.35 Model of a molecule of 2,2,4-trimethylpentane, also known as iso-octane

Figure 9.36 Model of a cyclohexane molecule

Carbon 275

The models of butane and *iso*-butane molecules (Figures 9.34 and 9.33) illustrate another aspect of the variety of hydrocarbons. Both molecules are composed of four carbon and ten hydrogen atoms; the formula of both is C_4H_{10}. The only difference between them is the way in which their atoms are joined together. Compounds with the same formula but different molecular structures are called ISOMERS. Isomers have different physical properties; they are different substances. Figure 9.37 compares the properties of butane and *iso*-butane. The bigger the molecule, the more isomers are possible. Octane, C_8H_{18}, has 18 isomers (two of which are shown in Figures 9.32 and 9.35), and the hydrocarbon with the formula $C_{30}H_{62}$ has over four thousand million isomers. It should now be clear why there is such a large number of different hydrocarbons!

	Butane	*Iso*-butane
Density	0.579 g cm^{-3}	0.557 g cm^{-3}
Melting point	$-138°C$	$-159°C$
Boiling point	$0°C$	$-12°C$

Figure 9.37

If we want to represent one mole of methane, it is quite sufficient to write CH_4. But C_4H_{10} is an ambiguous formula, as it represents one mole of two different compounds. It would be too time-consuming to draw the molecules as in Figures 9.33 and 9.34 when we want to write down the formulas, so we use a shorthand version:

```
    H H H H                    H H H
    | | | |                    | | |
  H-C-C-C-C-H              H-C-C-C-H
    | | | |                    |   |
    H H H H                    H   H
     butane                      |
                               H-C-H
                                 |
                                 H
                              iso-butane
```

These are called STRUCTURAL FORMULAS. The atomic symbols are being used to represent single atoms of carbon and hydrogen, not, as usual, 1 mole of carbon atoms and 1 mole of hydrogen atoms. It must also be remembered that structural formulas are only two-dimensional representations of three-dimensional molecules. The four atoms which surround each carbon atom are really arranged tetrahedrally and not, as the structural formulas suggest, in the same plane.

Hydrocarbons all have different melting points and boiling points; some are gases at room temperature, some liquids, and some solids. Their densities differ and so do some of their other physical properties.

Chemically, however, the hydrocarbons in petroleum are all remarkably similar to one another. They all belong to the same chemical family called the ALKANES, or PARAFFINS. They all burn in air or oxygen, and the products of combustion are always the same: water and carbon dioxide. For example:

$$CH_4(g) + 2O_2(g) \rightarrow CO_2(g) + 2H_2O(g),$$

$$2C_8H_{18}(l) + 25O_2(g) \rightarrow 16CO_2(g) + 18H_2O(g).$$

When they burn, the chemical energy which was locked up in their molecules is released as heat, hence the extensive use of hydrocarbons as fuels.

METHANE is the main constituent of *natural gas* which was formed, it is believed, in the same way as petroleum, by the decay of marine animals. It also forms about 35 per cent of coal gas (page 267). It is produced by the decay of organic material from plants or animals: it bubbles out of stagnant ponds (*marsh gas*), it occurs in coal mines (*firedamp*), and it is a useful by-product from modern sewage farms. Methane, often mixed with other gaseous hydrocarbons and with hydrogen, is extensively used as a fuel both in industry and in the home.

You may have seen cylinders of PROPANE gas (C_3H_8) on building sites or in scrap metal yards. It is burnt with oxygen in a torch for cutting and welding metal.

Calor gas and other brands of bottled gas are a mixture of BUTANE (C_4H_{10}) with some propane. Cylinders of this mixture are used for cooking in caravans, when camping, or in the country where there is no piped gas supply. The gas used for cigarette lighters is similar.

The sizes of the hydrocarbon molecules in the main petroleum fractions are shown in Figure 9.38.

	Number of carbon atoms per molecule
Petroleum gases	1 to 4
Gasolene (or naphtha)	5 to 10
Kerosene	10 to 14
Gas oil	14 to 19
Lubricating oil and fuel oil	19 to 35
Wax and bitumen	more than 35

Figure 9.38

Experiment 9.8 Investigating the effect of heat on large hydrocarbon molecules

Put mineral wool to a depth of 3 cm in a hard-glass test tube. Soak the mineral wool with 2 cm³ of medicinal paraffin (a hydrocarbon with large molecules). Fill up the tube, above the mineral wool, with pieces of pumice stone and set up the apparatus shown in Figure 9.39. Label four test tubes 1, 2, 3, and 4, and fill them with water; you will collect the gas in these tubes.

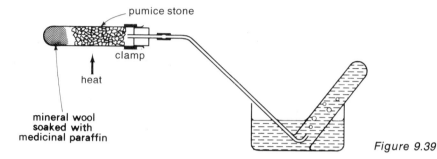

Figure 9.39

Heat the centre of the tube as strongly as possible. When it is beginning to glow red-hot move the bunsen flame down to the mineral wool for two or three seconds, then back to the centre of the tube. Repeat this at intervals.

When tube 1 is full of gas, put your thumb over the end, take it out of the water, and cork it. Put tube 2 in position as quickly as possible. Continue until you have filled all four tubes with gas. Remove the delivery tube from the water before you stop heating (why?).

Describe the appearance of the gas you have collected. What can you say about the size of the molecules of this gas as compared with the size of the molecules in medicinal paraffin?

Carry out the following tests on the gas you have collected. Do not use the first tube of gas you collected (tube 1). Why?

Tube 2 Smell the gas. Describe its smell.

Tube 3 Take out the cork and quickly put a lighted splint to the mouth of the tube. Does the gas burn? If so, describe the appearance of the flame.

Tube 4 Take out the cork, quickly pour about 1 cm depth of bromine solution into the tube, and replace the cork as quickly as possible. Shake the tube. What happens to the bromine solution?

Carry out this last test also on the medicinal paraffin. Place 1 cm depth of bromine solution in a clean test tube, add three drops of medicinal paraffin, and shake. What happens to the bromine solution this time?

Cracking

There is a larger market for some oil fractions than for others. The biggest demand is for gasolene, kerosene, and gas oil as fuel for cars, jet aeroplanes, and diesel engines respectively. These fractions are all made up of hydrocarbons with less than 20 carbon atoms per molecule (Figure

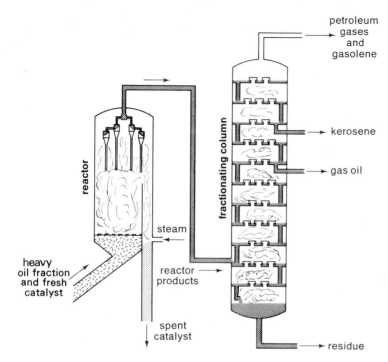

Figure 9.40 The cat-cracking process

9.38, page 277). The fractions made up of larger molecules (the HEAVY FRACTIONS) do, of course, have their uses (Figure 9.30, page 274), but are not required on such a large scale. Less than 5 per cent of the total tonnage of petroleum products sold is heavy fractions; crude oil, however, may contain 50 per cent or more of these fractions.

In the early days of the petroleum industry a large proportion of the heavy fractions was wasted. Methods have now been found to break up the less useful large molecules into the smaller, more useful ones. This process is known as CRACKING. It was first achieved simply by heating the heavy fractions to a very high temperature (THERMAL CRACKING). More recently it has been found that by using a catalyst, a mixture of aluminium oxide and silicon oxide, cracking can be achieved at a temperature of only 500°C. This is known as CATALYTIC CRACKING, or *cat-cracking* (Figure 9.40). The oil is pumped into the base of the reaction vessel where it meets a stream of hot catalyst and is vaporised. The hot, cracked vapours pass into a fractionating column for separation. The catalyst is blown into another vessel where it is cleaned by heating with steam to remove the oil which has stuck to it.

Petrol

Petrol and aviation spirit are carefully blended mixtures of hydrocarbons. When mixed with air and vaporised in the engine cylinders petrol should explode at just the right moment, when under full compression and ignited by the spark. An unsuitable blend of hydrocarbons will be ignited by the heat of the cylinders before it has been sparked, when the piston is in the wrong position. This is known as *knocking* or *pinking* and results in a loss of efficiency.

Straight-chain hydrocarbons are more prone to knocking than hydrocarbons with branched-chain or ring-shaped molecules. Unfortunately the majority of the hydrocarbon molecules in gasolene obtained by the fractionation of oil are straight-chain molecules. Petrol is produced from the gasolene fraction by a process known as REFORMING. The gasolene is heated under pressure in the presence of a platinum catalyst. Under these conditions the straight-chain molecules rearrange themselves into branched-chain and ring-shaped molecules, producing a more suitable mixture for use as petrol.

The anti-knock efficiency of petrol is measured by its OCTANE NUMBER. Pure *iso*-octane, which has a very low knocking tendency, is described as octane number 100. Other fuels are compared with *iso*-octane; the lower the octane number the greater the tendency to knocking. For motor spirit the octane number is usually between 90 and 100. For aviation spirit it may be as high as 130. To achieve high octane numbers other substances are added to the basic hydrocarbon mixture. The most important of these is tetraethyl lead, which has a dramatic effect in cutting down knocking (pages 21–2).

The manufacture of gas from oil

In 1960 90 per cent of all town gas was made from coal (page 267); by 1970 less than 10 per cent was being made in this way. Natural gas and processes for the manufacture of gas from petroleum have taken over. The raw material is the lowest boiling fraction from the primary fractionation of crude oil, called *gasolene*, or *naphtha*. This consists mainly of hydrocarbons with between 5 and 10 carbon atoms per molecule.

There are two main methods. To make *lean gas* gasolene is heated with steam in the presence of a catalyst which converts it mainly to hydrogen and carbon monoxide. For example, a hydrocarbon molecule with 7 carbon atoms will react as:

$$C_7H_{16}(g) + 7H_2O(g) \rightarrow 15H_2(g) + 7CO(g).$$

In the second stage of the process the mixture of carbon monoxide and hydrogen is passed, together with more steam, over another catalyst, which converts the carbon monoxide to carbon dioxide, producing more hydrogen:

$$CO(g) + H_2O(g) \rightarrow CO_2(g) + H_2(g).$$

The carbon dioxide is removed so that the final lean gas is almost pure hydrogen.

The other method produces *rich gas*, so called because it produces more heat when burnt. Gasolene is heated with hydrogen (obtained from the lean gas process). No catalyst is required, and the gasolene is converted mainly to methane. A hydrocarbon with 7 carbon atoms per molecule will react as:

$$C_7H_{16}(g) + 6H_2(g) \rightarrow 7CH_4(g).$$

Town gas produced from coal by the old method consisted mainly of a mixture of hydrogen and methane. The rich and lean gases from petroleum are mixed in suitable proportions to give a gas with the same heat value and burning characteristics as coal gas.

Questions

1 Which of the following best describes petroleum?
A It is a hydrocarbon.
B It is a mixture of alcohols.
C It is mainly a mixture of hydrocarbons.
D It is a mixture of liquids.
E It is a mixture of fats.   (*North West*)

2 The primary distillation in an oil refinery is carried out in tall towers with pipes leading away from different heights.
   (a) Describe what happens when a liquid is distilled.
   (b) Why are the pipes at different heights?
   (c) Why are the liquids leaving the tower called fractions?
   (d) State any two differences in the fractions obtained.
   (e) Name two of the fractions obtained and state a use for each fraction named.   (*Lancashire*)

3 Crude petroleum is a naturally occurring mixture of hydrocarbons. For laboratory demonstration a synthetic mixture of the following substances was used:

25 cm³ of petroleum ether   boiling range 60°C to 100°C
25 cm³ of petrol   boiling range 110°C to 160°C
25 cm³ of kerosene   boiling range 170°C to 250°C
25 g of vaseline.

The graph (Figure 9.41) shows the volume of distillate collected at different stillhead temperatures when the mixture was fractionated.
   (a) Why is there a steep rise at A?
   (b) What is the fraction being collected along B?
   (c) Why is there a break in the curve at C?
   (d) Why is the curve sloping upwards along D?
   (e) Why does the graph stop at E?   (*London*)

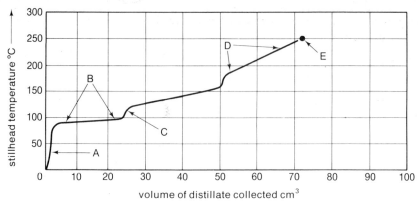

Figure 9.41

4 Carbon forms a large number of compounds for which of the following reasons?
A   It is a non-metal.
B   It exists in two allotropic forms.
C   It is present in all organic compounds.
D   It is a reducing agent.
E   It forms bonds with other carbon atoms.   *(North West)*

5 Organic chemistry is the chemistry of the element carbon and its compounds with very few other elements.
   (a) The compounds with hydrogen only are called _____.
   (b) Name one carbon/hydrogen compound.
   (c) Give its formula.
   (d) Carbohydrates are compounds of carbon with _____ and _____.
   (e) Name one such compound.
   (f) Draw the *two* structural formulae of the compounds with the formula C_4H_{10}.
   (g) What is the name given to the relationship between these two compounds?   *(London)*

6 Give a scientific reason for the fact that when a car is started up in cold weather drops of liquid come out of the exhaust.
          (South-East)

7 (a) Write an equation in words and symbols for the combustion of butane (C_4H_{10}).
   (b) If you are given the atomic masses $C = 12$, $H = 1$, $O = 16$, calculate the weight of oxygen required for the complete combustion of 5.8 g of butane.
   (c) Explain why a gardener might expect a better crop of tomatoes if he uses a butane burner rather than an electrical heater in his glasshouse.   *(Wales)*

8 Read the following passage about petrol production and then answer the questions asked:

'Petrol and light fuel oils are two of the fractions obtained by the fractional distillation of crude oil. The yield of petrol from this process is not sufficient to meet demand and so more petrol is obtained by the catalytic cracking of fuel oil.'

(a) Explain the meaning of fractional distillation.
(b) What is meant by cracking?
(c) What is the difference between thermal cracking and catalytic cracking? Briefly outline the process of catalytic cracking.

(*West Midlands*)

9.6 Chemicals from petroleum

Ethylene (or Ethene)

The effect of CRACKING (page 278) is often to produce hydrocarbons with less than the normal proportion of hydrogen in their molecules. These are called UNSATURATED HYDROCARBONS. You may have carried out an experiment to crack medicinal paraffin. The gas produced is more reactive than the original hydrocarbon; for example, it decolourises a solution of bromine in water whereas medicinal paraffin does not. This gas is an unsaturated hydrocarbon.

The simplest unsaturated hydrocarbon is ethylene, C_2H_4, also called ethene. A model of its molecule is shown in Figure 9.42 and its structural formula is:

$$\begin{array}{c} H \\ \diagdown   \diagup H \\ C=C \\ \diagup   \diagdown H \\ H \end{array}$$

Figure 9.42 Model of a molecule of ethene, also known as ethylene

The DOUBLE BOND between the carbon atoms represents 4 shared electrons in this region of the molecule. These electrons tend to repel each other with the result that the molecule is rather unstable and reactive. Ethylene and other UNSATURATED hydrocarbons are much more reactive than the SATURATED hydrocarbons in petroleum (paraffins, or alkanes) which do not contain a double bond.

Unsaturated hydrocarbons readily undergo ADDITION REACTIONS, in which an extra atom is added on either side of the double bond to saturate it.

Carbon 283

1 They react with hydrogen. For example, with ethylene:

$$\begin{array}{c}H\\H\end{array}\!\!>\!\!C\!=\!C\!<\!\!\begin{array}{c}H\\H\end{array} + H_2 \longrightarrow H-\underset{\underset{H}{|}}{\overset{\overset{H}{|}}{C}}-\underset{\underset{H}{|}}{\overset{\overset{H}{|}}{C}}-H$$
ethane

The reaction of unsaturated hydrocarbons with hydrogen is important in the manufacture of margarine.

2 They react with halogens (fluorine, chlorine, bromine, and iodine). For example, with ethylene and bromine:

$$\begin{array}{c}H\\H\end{array}\!\!>\!\!C\!=\!C\!<\!\!\begin{array}{c}H\\H\end{array} + Br_2 \longrightarrow H-\underset{\underset{Br}{|}}{\overset{\overset{H}{|}}{C}}-\underset{\underset{Br}{|}}{\overset{\overset{H}{|}}{C}}-H$$

The halogen compounds produced are colourless, and the decolourisation of bromine water (a solution of bromine in water) is a convenient test for unsaturated hydrocarbons. The compound produced from ethylene and bromine, which is called ethylene dibromide, is an important petrol additive. It combines with the lead from the tetraethyl lead (page 280) to form a gas, lead tetrabromide, which passes out of the engine with the other exhaust gases. If ethylene dibromide was not added to petrol, metallic lead would be deposited in the cylinders and would damage them.

3 Unsaturated hydrocarbons will also *react with themselves*. When two molecules of ethylene combine:

$$\begin{array}{c}H\\H\end{array}\!\!>\!\!C\!=\!C\!<\!\!\begin{array}{c}H\\H\end{array} + \begin{array}{c}H\\H\end{array}\!\!>\!\!C\!=\!C\!<\!\!\begin{array}{c}H\\H\end{array} \longrightarrow -\underset{\underset{H}{|}}{\overset{\overset{H}{|}}{C}}-\underset{\underset{H}{|}}{\overset{\overset{H}{|}}{C}}-\underset{\underset{H}{|}}{\overset{\overset{H}{|}}{C}}-\underset{\underset{H}{|}}{\overset{\overset{H}{|}}{C}}-$$

the carbon atom at each end of the chain is capable of adding on more ethylene molecules. The process can therefore continue until thousands of ethylene molecules have formed into a long chain. The product is polyethylene, commonly called *polythene*, a plastic:

$$\text{etc.} -\overset{H}{\underset{H}{|}}\overset{|}{C}-\overset{H}{\underset{H}{|}}\overset{|}{C}-\overset{H}{\underset{H}{|}}\overset{|}{C}-\overset{H}{\underset{H}{|}}\overset{|}{C}-\overset{H}{\underset{H}{|}}\overset{|}{C}-\overset{H}{\underset{H}{|}}\overset{|}{C}-\overset{H}{\underset{H}{|}}\overset{|}{C}-\overset{H}{\underset{H}{|}}\overset{|}{C}-\overset{H}{\underset{H}{|}}\overset{|}{C}-\overset{H}{\underset{H}{|}}\overset{|}{C}-\overset{H}{\underset{H}{|}}\overset{|}{C}-\overset{H}{\underset{H}{|}}\overset{|}{C}- \text{etc.}$$

These three reactions of unsaturated hydrocarbons are just some examples of the many different addition reactions that are possible. The variety of possible reactions makes unsaturated hydrocarbons very useful basic materials in the petroleum chemicals industry, since so many other substances can be made from them.

By far the most useful of these unsaturated hydrocarbons is ethylene itself. It is manufactured mainly by cracking the petroleum gases ethane and propane by heating them in the presence of a suitable catalyst. The principal reactions are:

$$C_2H_6(g) \rightarrow C_2H_4(g) + H_2(g)$$

and

$$C_3H_8(g) \rightarrow C_2H_4(g) + CH_4(g).$$

Small proportions of other hydrocarbons are also produced during the cracking process and the cracked gases require fractionation to produce pure ethylene. Hydrogen is a useful by-product of cracking. It is used in the manufacture of fertilisers and margarine.

Margarine

Margarine was named after its French inventor, Mège-Mouriès, who was awarded a prize in 1870 by Napoleon III for the production of a substitute for butter, then in short supply. Margarine was originally made by blending animal fats such as lard with soured milk and salt.

Later, it was discovered that vegetable oils like cottonseed oil and soya-bean oil, which are unsuitable for eating, could be converted into edible fats for use in margarine manufacture. The molecules of fats and oils both contain long hydrocarbon chains. The only difference is that in fats most of the hydrocarbon chains are saturated whereas in oils a large proportion of them are unsaturated; they contain double bonds between some carbon atoms in the chain. Treatment of liquid vegetable oils with hydrogen in the presence of a nickel catalyst at 150°C saturates the molecules to produce solid fats which can be used to make margarine.

Acetylene

Acetylene (C_2H_2) is an even more unsaturated hydrocarbon than ethylene. A model of its molecule is shown in Figure 9.43 and it is represented by the structural formula:

$$H-C\equiv C-H$$

with a TRIPLE BOND (6 shared electrons) between the carbon atoms. It is manufactured by cracking methane, obtained from natural gas:

Figure 9.43 Model of a molecule of ethyne, also known as acetylene

$$2CH_4(g) \rightarrow C_2H_2(g) + 3H_2(g).$$

Carbon 285

Like other unsaturated hydrocarbons, acetylene may be detected by its effect on a solution of bromine in water, which it decolourises. As in the case of ethylene, the unsaturation in the acetylene molecule makes it very reactive, and it is a valuable basic material in the petroleum chemicals industry. It is most familiar, however, as a welding gas. When burned with oxygen in the oxy-acetylene torch it releases so much chemical energy as heat that temperatures up to 3000°C are produced.

Experiment 9.9 Investigating the action of heat on ethanol

Repeat experiment 9.8, using 2 cm³ of ethanol on the mineral wool in place of medicinal paraffin.

Collect 4 tubes of gas. Reject the first one and use the other tubes to smell the gas, test it with a lighted splint, and examine its reaction with a solution of bromine in water, just as in experiment 9.8. Also test the effect of a solution of bromine on ethanol.

Describe the appearance and smell of the gas you have collected.
Does the gas burn? If it does, describe the appearance of the flame.
What happens to a solution of bromine which is shaken with the gas? What does this tell you about the molecules of the gas?
Does ethanol react with a solution of bromine?

Experiment 9.10 Oxidising ethanol

(a) Potassium permanganate is a powerful oxidising agent. It contains oxygen which it readily gives up to other compounds.

To 1 cm depth of ethanol in a test tube, add 1 cm depth of dilute sulphuric acid. Then add 3 or 4 drops of potassium permanganate solution from a teat pipette.

What do you observe? Has a reaction taken place? How do you know?

(b) Potassium dichromate is another oxidising agent. Dissolve two spatulas-full of potassium dichromate in 5 cm³ of dilute sulphuric acid in a large test tube. Add about 10 drops of ethanol from a teat pipette and warm *gently*.

Smell the tube **cautiously**. Do you recognise the smell? What other evidence is there that a reaction has taken place?

Experiment 9.11 Investigating the reaction of ethanol with acetic acid

Put 1 cm depth of ethanol in a test tube and add an equal quantity of acetic acid. **Very carefully** add 1 drop of concentrated sulphuric acid with a teat pipette.

Warm the mixture *gently* for 2 minutes (**do not boil**). Put 1 cm depth of water in a small beaker and pour the mixture from the tube into the water.

Can you identify the smell? What does it remind you of? What chemical compound is responsible for this smell?

Find out the formula of this substance, then use models to work out what happened when ethanol reacted with acetic acid.

What do you think was the function of the sulphuric acid?

Ethanol and other oxygen compounds made from petroleum

Ethanol is the proper chemical name for alcohol (page 259). You may have carried out an experiment to decompose ethanol by heat; it breaks down into ethylene and water:

$$\begin{array}{c}H\;H\\|\;\;|\\H-C-C-O-H\\|\;\;|\\H\;H\end{array} \longrightarrow \begin{array}{c}H\\ \diagdown\\ H\end{array}C=C\begin{array}{c}H\\ \diagup\\ H\end{array} + H_2O$$

Ethanol is manufactured on a large scale by reversing this process: adding water to ethylene obtained from petroleum. It is not an easy reaction to bring about, and must be aided by high pressure and a catalyst. This process is now replacing, to a large extent, the manufacture of ethanol for industrial purposes by fermentation (page 259).

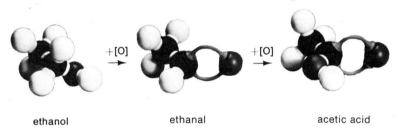

    ethanol   ethanal   acetic acid

Figure 9.44

If ethanol is treated with an oxidising agent, two reactions take place (Figure 9.44). First, an atom of oxygen from the oxidising agent removes two hydrogen atoms from the ethanol molecule, forming water and a substance called ETHANAL, or ACETALDEHYDE:

$$\begin{array}{c}H\;H\\|\;\;|\\H-C-C-O-H\\|\;\;|\\H\;H\end{array} + \underset{\substack{\text{(from}\\\text{oxidising}\\\text{agent)}}}{O} \longrightarrow \begin{array}{c}H\\|\\H-C-C\\|\\H\end{array}\!\!\begin{array}{c}\diagup O\\ \diagdown H\end{array} + H_2O$$

Then ethanal combines with another atom of oxygen from the oxidising agent to produce acetic acid:

Carbon 287

$$\text{H-}\underset{\underset{\text{H}}{|}}{\overset{\overset{\text{H}}{|}}{\text{C}}}\text{-C}\overset{\nearrow \text{O}}{\underset{\searrow \text{H}}{}} + \text{O} \longrightarrow \text{H-}\underset{\underset{\text{H}}{|}}{\overset{\overset{\text{H}}{|}}{\text{C}}}\text{-O-C}\overset{\nearrow \text{O}}{\underset{\searrow \text{O-H}}{}}$$

(from oxidising agent)

In industry the cheapest oxidising agent of all, oxygen from the air, is used to bring about these reactions. Acetic acid is manufactured by reacting ethanol with air in the presence of a catalyst.

Acetic acid is used mainly in the production of a group of chemicals called ESTERS. One of the simplest esters is ETHYL ACETATE, a solvent used in the preparation of paints and varnishes (including nail varnish). It is formed by reacting ethanol with acetic acid (Figure 9.45):

$$\text{H-}\underset{\underset{\text{H}}{|}}{\overset{\overset{\text{H}}{|}}{\text{C}}}\text{-C}\overset{\nearrow \text{O}}{\underset{\searrow \text{O-H}}{}} + \text{H-O-}\underset{\underset{\text{H}}{|}}{\overset{\overset{\text{H}}{|}}{\text{C}}}\text{-}\underset{\underset{\text{H}}{|}}{\overset{\overset{\text{H}}{|}}{\text{C}}}\text{-H} \longrightarrow \text{H-}\underset{\underset{\text{H}}{|}}{\overset{\overset{\text{H}}{|}}{\text{C}}}\text{-C}\overset{\nearrow \text{O}}{\underset{\searrow \text{O-}\underset{\underset{\text{H}}{|}}{\overset{\overset{\text{H}}{|}}{\text{C}}}\text{-}\underset{\underset{\text{H}}{|}}{\overset{\overset{\text{H}}{|}}{\text{C}}}\text{-H}}{}} + \text{H}_2\text{O}$$

Ethyl acetate has a pleasant, sweetish smell. Different esters have different smells and they have been found to be responsible for the smell and taste of many flowers and fruits. Artificial flavourings are often esters manufactured from petroleum.

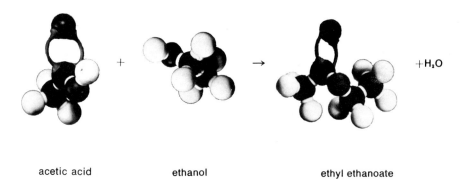

acetic acid   ethanol   ethyl ethanoate

Figure 9.45

Cellulose acetate

The molecule of cellulose (page 252) consists of several hundred rings of atoms linked together in a long chain. Figure 9.46 shows a model of part of it. Attached to each ring are three HYDROXYL (—OH) groups. Just as the hydroxyl group in ethanol will react with acetic acid to form the ester, ethyl acetate, so the hydroxyl groups in cellulose will react with acetic acid to form cellulose acetate.

This is a very useful material which goes by several names. Under the name of *celluloid* it is essential to the whole of the photographic industry; it is the material of which film is made. In thin sheets it is known as *cellophane*, a wrapping material for foodstuffs. More recently it has been produced in the form of a thread under the brand name *Tricel* (and in another, modified form *Dicel*), which can be woven or knitted into fabrics.

The raw materials for the production of cellulose acetate are wood pulp (page 255) and acetic acid made from petroleum (page 288). They react together to produce a thick syrupy liquid from which the cellulose acetate is precipitated by the addition of water. After drying, the white granular

Figure 9.46 Model of part of a cellulose molecule

flakes are dissolved in a liquid called methylene chloride, and to make *Tricel* this concentrated solution is forced through a spinneret into a warm atmosphere. The solvent evaporates, leaving threads.

Halogen compounds

A great variety of compounds, useful in all sorts of different ways, are made by reacting hydrocarbons with halogens, particularly chlorine and bromine. One example, ETHYLENE DIBROMIDE, a petrol additive, was mentioned on page 284.

Replacement of the hydrogen atoms in methane by chlorine atoms, a reaction which takes place quite easily, gives:
CH_3Cl, METHYL CHLORIDE, used in the manufacture of silicones;
CH_2Cl_2, METHYLENE CHLORIDE, used as a paint remover and grease remover in industry, and in the manufacture of cellulose acetate;
$CHCl_3$, CHLOROFORM, an early anaesthetic; and
CCl_4, CARBON TETRACHLORIDE, one of the earliest dry-cleaning fluids (a very good solvent for grease and oil).

The use of carbon tetrachloride as a grease remover always suffered from the disadvantage that its fumes are rather poisonous. Nowadays the two most important cleaning solvents are:

$$\begin{array}{c}Cl\\ \end{array}C=C\begin{array}{c}H\\ Cl\end{array}$$ TRICHLOROETHYLENE, used in industry for degreasing metals; and

$$\begin{array}{c}Cl\\ Cl\end{array}C=C\begin{array}{c}Cl\\ Cl\end{array}$$ PERCHLOROETHYLENE, the most widely used dry-cleaning fluid.

Both are manufactured from acetylene.

Insecticides

The halogen–hydrocarbon compounds from petroleum which have done most for man are the insecticides. DDT (short for **dichlorodiphenyltrichloroethane**) was discovered in the 1930s and was the first really effective insecticide. DDT, and similar chemicals discovered since, have been responsible for controlling the spread of some of the terrible diseases transmitted by insects to human beings, particularly malaria, yellow fever, and sleeping sickness.

Control of the anopheles mosquito, which carries malaria, reduced the death toll from six million people per year in 1939 to less than one and a half million in 1965, despite the increase in the world's population in that time. Nevertheless there is still a great deal to be done.

DDT and other more advanced insecticides have helped in another way too. It has been calculated that 20 per cent of all crops planted are eaten by insects. On a world-wide scale enough food to feed two hundred million people is lost every year. Spraying food crops with insecticides is improving this situation.

Questions

1 Most organic compounds are which of the following?
A black   D explosive
B electrovalent   E covalent
C unaffected by heat

(*North West*)

2 Which of the following describes ethene (ethylene), carbon dioxide and acetic acid?
A They are all gases at room temperature.
B They all contain oxygen.
C They all contain carbon.
D They all contain hydrogen.
E They are all liquids at room temperature.   (North West)

3 (a) Ethene (ethylene) is an unsaturated compound.
  (i) What does the term 'unsaturated' mean?
  (ii) Draw a simple diagram to show the structure of ethene.
  (b) When ethene reacts with hydrogen, a saturated compound ethane is obtained.
  (i) What does the term 'saturated' mean?
  (ii) Write an equation for the reaction between ethene and hydrogen to yield ethane.
  (iii) Draw a simple diagram to show the structure of ethane.
  (c) Ethene can be polymerised.
  (i) What is a polymer?
  (ii) Name the polymer formed from ethene.   (North West)

4 Two samples, one of ethane, the other of ethylene, were shaken separately with bromine water.
  (a) What would you see with the ethane sample?
  (b) What would you see with the ethylene sample?
  (c) Explain your answers to (a) and (b).   (London)

5 Oils are converted into fats for use in confectionery by hydrogenation.
  (a) What does *hydrogenation* mean?
  (b) Name a catalyst which would be appropriate for the process of hydrogenation.   (East Anglia)

6

(i) $H-\underset{H}{\overset{H}{\underset{|}{\overset{|}{C}}}}-H$

(ii) $H-\underset{H}{\overset{H}{\underset{|}{\overset{|}{C}}}}-\underset{H}{\overset{H}{\underset{|}{\overset{|}{C}}}}-H$

(iii) $\underset{H}{\overset{H}{\diagdown}}C=C\underset{H}{\overset{H}{\diagup}}$

(iv) $Cl-\underset{Cl}{\overset{H}{\underset{|}{\overset{|}{C}}}}-Cl$

(v) $H-C\equiv C-H$

(vi) $H-\underset{H}{\overset{H}{\underset{|}{\overset{|}{C}}}}-\underset{H}{\overset{H}{\underset{|}{\overset{|}{C}}}}-O-H$

(a) Give the names of chemical compounds (i) to (vi).
(b) Which of these compounds are members of the paraffin series?

(c) Give one important difference in chemical behaviour between compound (i) and compound (v).

(d) What important substance in everyday use is manufactured from compound (iii)?

(e) An enzyme is defined as a substance which has been derived from living matter and which, when present in only a minute proportion, can start or accelerate a given chemical change without undergoing a change itself.

(1) Which of the compounds (i) to (vi) could be obtained by the action of an enzyme?

(2) Describe briefly how the compound you have chosen would be prepared by enzyme action. *(East Anglia)*

7 Draw the *three* structural formulae of the compounds with the formula C_3H_8O. *(London)*

8 Acetylene is a hydrocarbon possessing a carbon-to-carbon triple bond. It is an unsaturated compound which has recently become a source of important chemicals on a large scale.

(i) What is meant by a 'hydrocarbon'? Write the molecular and structural formula of acetylene.

(ii) Describe and explain any test which would show acetylene to be unsaturated.

(iii) Name and give one use of an important chemical which requires acetylene in its preparation. *(North West)*

9 By which of the following reactions are esters obtained?
A   an alcohol with an oxidising agent
B   an alcohol with an organic acid
C   an alcohol with a dehydrating agent
D   yeast with sugars
E   acids with alkalis. *(West Midlands)*

9.7 Plastics and artificial fibres

The twentieth century may come to be known as the *plastic age*, just as earlier periods are now called the stone age, the iron age, and so on. Unlike wood, stone, and metals, plastics are man-made materials. To a greater and greater extent they are taking the place of these natural materials, and in many respects are better than the substances they replace.

Plastics are chemical compounds with very large molecules. These large molecules are called POLYMERS (meaning *many parts*) because they are made by joining together, or POLYMERISING, hundreds, or even thousands, of similar small molecules. The small molecules are often referred to as MONOMERS.

Polymers do occur in nature. On page 252 mention was made of the fact that plants store their surplus sugar in the form of starch. Starch (Figure 9.47b) is a polymer built up from hundreds of molecules of the monomer glucose, a sugar (Figure 9.47a). Cellulose (Figure 9.46, page 289) is another glucose polymer, and proteins are another type of natural polymer.

Figure 9.47(a) (above)   Model of a glucose molecule
(b) (below)   Model of part of a starch molecule

Artificial polymers can be made in two ways. On method is to join a lot of the same monomer molecules together. This process may be represented:

A + A + A + A + A + A + A + A → –A–A–A–A–A–A–A–A–.

The other method is to join together two different monomer molecules. This may be represented:

A + B + A + B + A + B + A + B → –A–B–A–B–A–B–A–B–.

Method 1: –A–A–A–A–A–A–A–A–

This method was mentioned on page 284. The unsaturated ethylene molecule can, with the aid of a suitable catalyst, add on to itself to produce a long chain-like molecule consisting of thousands of ethylene units linked together:

Carbon   293

$$\underset{H}{\overset{H}{C}}=\underset{H}{\overset{H}{C}} + \underset{H}{\overset{H}{C}}=\underset{H}{\overset{H}{C}} + \underset{H}{\overset{H}{C}}=\underset{H}{\overset{H}{C}} + \underset{H}{\overset{H}{C}}=\underset{H}{\overset{H}{C}} + \underset{H}{\overset{H}{C}}=\underset{H}{\overset{H}{C}} \quad \text{(ethylene molecules)}$$

$$\downarrow \text{catalyst}$$

$$\text{etc.} -\underset{H}{\overset{H}{C}}-\underset{H}{\overset{H}{C}}-\underset{H}{\overset{H}{C}}-\underset{H}{\overset{H}{C}}-\underset{H}{\overset{H}{C}}-\underset{H}{\overset{H}{C}}-\underset{H}{\overset{H}{C}}-\underset{H}{\overset{H}{C}}-\underset{H}{\overset{H}{C}}-\underset{H}{\overset{H}{C}}- \text{etc.} \quad \text{(polythene)}$$

This is an example of what is called ADDITION POLYMERISATION. Ethylene is the monomer; polythene the polymer.

Addition polymerisation always involves a monomer with an unsaturated molecule (a molecule containing a double bond). If one of the hydrogen atoms in ethylene is replaced by a chlorine atom, vinyl chloride is obtained. This can be polymerised, in the same way as ethylene itself, to produce polyvinyl chloride, or *PVC*:

$$\underset{H}{\overset{H}{C}}=\underset{Cl}{\overset{H}{C}} + \underset{H}{\overset{H}{C}}=\underset{Cl}{\overset{H}{C}} + \underset{H}{\overset{H}{C}}=\underset{Cl}{\overset{H}{C}} + \underset{H}{\overset{H}{C}}=\underset{Cl}{\overset{H}{C}} + \underset{H}{\overset{H}{C}}=\underset{Cl}{\overset{H}{C}} \quad \text{(vinyl chloride molecules)}$$

$$\downarrow \text{catalyst}$$

$$\text{etc.} -\underset{H}{\overset{H}{C}}-\underset{Cl}{\overset{H}{C}}-\underset{H}{\overset{H}{C}}-\underset{Cl}{\overset{H}{C}}-\underset{H}{\overset{H}{C}}-\underset{Cl}{\overset{H}{C}}-\underset{H}{\overset{H}{C}}-\underset{Cl}{\overset{H}{C}}-\underset{H}{\overset{H}{C}}-\underset{Cl}{\overset{H}{C}}- \text{etc.} \quad \text{(polyvinyl chloride, } PVC\text{)}$$

If all four of the hydrogen atoms in the ethylene molecule are replaced by fluorine atoms, the product is tetrafluoroethylene, which can be polymerised to **polytetrafluoroethylene**, or **PTFE**:

$$\underset{F}{\overset{F}{C}}=\underset{F}{\overset{F}{C}} + \underset{F}{\overset{F}{C}}=\underset{F}{\overset{F}{C}} + \underset{F}{\overset{F}{C}}=\underset{F}{\overset{F}{C}} + \underset{F}{\overset{F}{C}}=\underset{F}{\overset{F}{C}} + \underset{F}{\overset{F}{C}}=\underset{F}{\overset{F}{C}} \quad \text{(tetrafluoroethylene molecules)}$$

$$\downarrow \text{catalyst}$$

$$\text{etc.} -\underset{F}{\overset{F}{C}}-\underset{F}{\overset{F}{C}}-\underset{F}{\overset{F}{C}}-\underset{F}{\overset{F}{C}}-\underset{F}{\overset{F}{C}}-\underset{F}{\overset{F}{C}}-\underset{F}{\overset{F}{C}}-\underset{F}{\overset{F}{C}}-\underset{F}{\overset{F}{C}}-\underset{F}{\overset{F}{C}}- \text{etc.} \quad \text{(polytetrafluoroethylene, PTFE)}$$

Under the brand names *Teflon* and *Fluon* this plastic is used to line saucepans and frying pans to make them non-stick.

By replacing one, or more, of the hydrogen atoms in ethylene with another atom or group of atoms a variety of different addition polymers can be produced. Some turn out to be of little commercial value, others prove sufficiently useful to be manufactured on a large scale. Figure 9.48 shows some of the plastics made by addition polymerisation. In every case the monomer is manufactured from a petroleum hydrocarbon, often ethylene or acetylene.

Monomer	Polymer	Trade names
$CH_2{=}CH_2$ ethylene	Polyethylene	Polythene, Alkathene
$CH_2{=}CHCl$ vinyl chloride	Polyvinyl chloride	PVC, Geon, Contact, Fablon
$CF_2{=}CF_2$ tetrafluoroethylene	Polytetrafluoro-ethylene	PTFE, Fluon, Teflon
$CH_2{=}CHCN$ acrylonitrile	Polyacrylonitrile	Orlon, Acrilan, Courtelle
$CH_2{=}CHCH_3$ propylene	Polypropylene	Propathene
$CH_2{=}CHC_6H_5$ styrene	Polystyrene	Kotina, Lustrex
$CH_2{=}C{-}COOCH_3$ $\;\;\;\;\;\;\;\vert$ $\;\;\;\;\;\;CH_3$ methylmethacrylate	Polymethylmeth-acrylate	Perspex, Plexiglass

Figure 9.48

Method 2: –A–B–A–B–A–B–A–B–

The process in which two different monomer molecules are linked together to produce a polymer is called CONDENSATION POLYMERISATION. One of the commonest types of condensation polymer is a POLYESTER. On page 288 the formation of a simple ester, ethyl acetate, was mentioned. The hydroxyl (—OH) group of the alcohol and the hydroxyl group of the acid join up together, eliminating water:

$$H-\underset{H}{\overset{H}{C}}-\underset{}{\overset{O}{C}}-O{\vdots}H \;+\; H{-}O{\vdots}\underset{H}{\overset{H}{C}}-\underset{H}{\overset{H}{C}}-H \longrightarrow H-\underset{H}{\overset{H}{C}}-\underset{}{\overset{O}{C}}-O-\underset{H}{\overset{H}{C}}-\underset{H}{\overset{H}{C}}-H \;+\; H_2O.$$

acid　　　　　　alcohol　　　　　　ester

Carbon　295

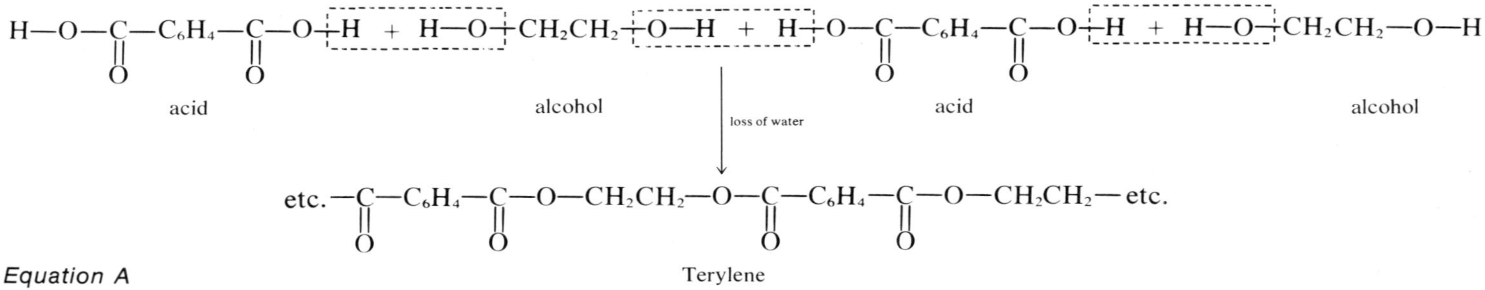

$$H-O-C-C_6H_4-C-O\underbrace{-H\ +\ H-O}_{}-CH_2CH_2-O\underbrace{-H\ +\ H-O}_{}-C-C_6H_4-C-O\underbrace{-H\ +\ H-O}_{}-CH_2CH_2-O-H$$

$$\underset{\text{acid}}{} \qquad \underset{\text{alcohol}}{} \qquad \underset{\text{acid}}{} \qquad \underset{\text{alcohol}}{}$$

$$\xrightarrow{\text{loss of water}}$$

$$\text{etc.}-C-C_6H_4-C-O-CH_2CH_2-O-C-C_6H_4-C-O-CH_2CH_2-\text{etc.}$$

Terylene

Equation A

$$H-O-C-C_4H_8-C-O\underbrace{-H\ +\ H}_{}-N-C_6H_{12}-N\underbrace{-H\ +\ H-O}_{}-C-C_4H_8-C-O\underbrace{-H\ +\ H}_{}-N-C_6H_{12}-N-H$$

$$\underset{\text{acid}}{} \qquad \underset{\text{amine}}{} \qquad \underset{\text{acid}}{} \qquad \underset{\text{amine}}{}$$

$$\xrightarrow{\text{loss of water}}$$

$$\text{etc.}-C-C_4H_8-C-N-C_6H_{12}-N-C-C_4H_8-C-N-C_6H_{12}-N-\text{etc.}$$

nylon 66

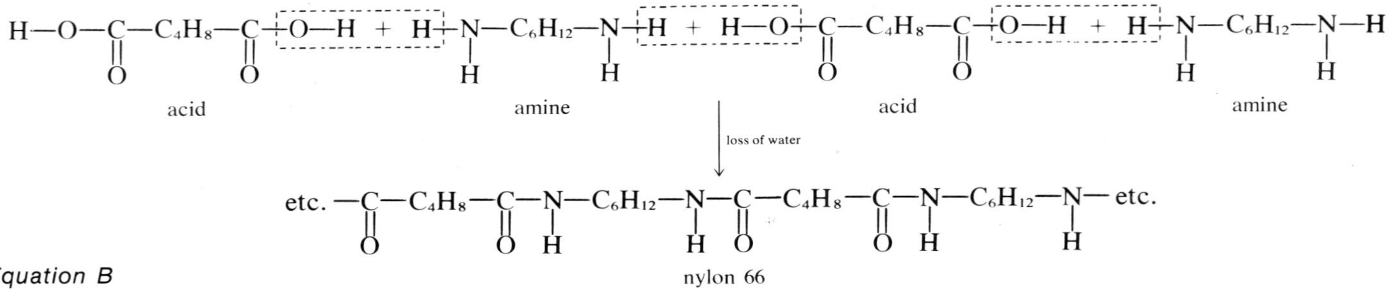

Equation B

In 1928 a brilliant American chemist, Wallace Carothers, suggested that if an alcohol with a hydroxyl group on both ends of its molecule was reacted with an acid, also with a hydroxyl group on both ends of its molecule, a polymer could be made. An example is shown opposite (equation A).

Although Carothers developed the idea of producing polyesters, it was not until 1940 that two British chemists, J. R. Whinfield and J. T. Dickson, succeeded in making the method work. The product was *Terylene*, also known as *Dacron* and *Crimplene*.

Nylon, also known under the trade names *Bri-Nylon*, *Celon*, and *Enkalon*, is another type of condensation polymer discovered by Carothers in 1935. One monomer is an acid with a hydroxyl group at both ends of its molecule. The other monomer is another hydrocarbon molecule with an AMINE group ($-NH_2$) at both ends. As with *Terylene* the two monomers polymerise into a long chain, eliminating water (see equation B, opposite).

Since each of the monomers has six carbon atoms in its molecule the product is known as nylon 66. Using monomers with the same reactive groups of atoms at the ends of their molecules, but different lengths of carbon chain in between, different types of nylon can be produced.

Nylon and Terylene are both polymers with long chain-like molecules. It is this molecular shape which makes them particularly suitable materials for use as artificial fibres, though they are used for making other things too.

A rather different type of condensation polymer can be produced by using a monomer with more than two reactive groups of atoms in its molecule. The monomers can then link into a three-dimensional network instead of into a chain. These plastics are sometimes called SYNTHETIC RESINS. A polyester resin can be produced, in a similar way to Terylene, by using glycerol, which has three hydroxyl groups:

$$\begin{array}{c} H \\ | \\ H \quad O \quad H \\ | \quad | \quad | \\ H-O-C-C-C-O-H \\ | \quad | \quad | \\ H \quad H \quad H \end{array}$$

Another type of resin, chemically similar to nylon, results from using melamine, which has three amine groups. *Melaware* and *Formica* are condensation polymers made from melamine. *Bakelite* and *Polyurethane* are other condensation polymers of this three-dimensional type.

The properties of plastics

Plastics may be divided into two main types, those that are made up of long chain-like molecules, such as polythene, PVC, Terylene, and nylon, and those that are made up of three-dimensional molecules, such as Bakelite and melamine resins.

If we could look at a piece of polythene under a microscope powerful enough to see the individual molecules it would probably look like a lot of pieces of string in a tangled mass (Figure 9.49). When polythene is heated the long string-like molecules move about more vigorously, so that the polythene becomes softer and more flexible, until it eventually melts. When it is cooled again the molecules slow down and the material solidifies and becomes rigid again. With polymers like polythene the processes of heating and cooling, softening and hardening, can be repeated almost indefinitely (unless the temperature becomes high enough to decompose the molecules). Polymers of this type are called THERMOPLASTICS. Thermoplastic materials are easy to work with. They are supplied to manufacturers as powders or granules which can be melted down and moulded into the final product or forced through holes (extruded) to produce a thread or a pipe. Alternatively, they are supplied in sheet or rod form and softened by heating so that they can be moulded.

The structure of a polymer such as Bakelite is rather different (Figure 9.50). The chains are linked together into what is, in effect, a single three-dimensional giant molecule. The structure of this type of polymer makes it very much stronger than the thermoplastic type, but it cannot be softened or melted by heat. There are no separate chains, so heating has almost no effect until a high enough temperature is reached to break some of the strong chemical bonds holding the molecule together. At this point the plastic is decomposed. Polymers of this type are called THERMOSETTING PLASTICS. They cannot be worked in the same way as thermoplastic materials. Usually the chemical reaction which produces the polymer is carried out inside a mould so that the finished product is produced directly. Alternatively this type of plastic is cut or machined to the required shape, rather like wood.

A product intermediate between a thermoplastic material and a thermosetting material can be made by introducing some chemical cross-linking between the molecular chains. The product will be stronger and more rigid than a thermoplastic polymer, but easier to work with than a thermosetting plastic.

The final properties of a plastic can be modified in other ways also. A PLASTICISER can be added to make it more flexible and rubbery. Plasticisers are usually high-boiling-point liquids with molecules that are small by comparison with the polymer molecules. The small molecules fit in between the polymer chains, pushing them apart so that the material becomes softer. Air or carbon dioxide can be forced into the monomer during the polymerisation process so that the polymer is filled with tiny bubbles. This produces light materials with good insulating properties such as expanded polystyrene (e.g. for ceiling tiles) and polyurethane foam (e.g. to fill the hulls of unsinkable boats). Dyes can be mixed with the monomers before polymerisation so that the plastic is produced ready coloured.

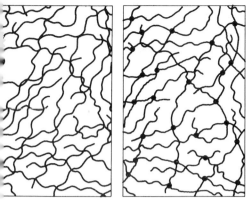 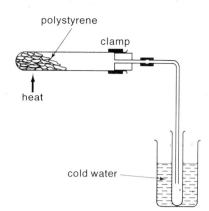

Figure 9.49 (left) Separate molecular chains in a thermoplastic polymer
Figure 9.50 (right) Extensive cross-linking of molecular chains in a thermosetting polymer

Figure 9.51

At present there are about 30 different types of basic polymer manufactured on a large scale, but the properties of these can be modified in so many ways that hundreds of plastics and artificial fibres can be produced from them, all with properties tailor-made to do a specific job.

Experiment 9.12 Investigating the action of heat on polystyrene

Set up the apparatus shown in Figure 9.51, with the test tube half-full of pieces of polystyrene.
Heat the tube containing the polystyrene, at first gently, then more strongly, until a liquid condenses in the other tube.
Describe what you see happening to the polystyrene as it is heated. Do you think polystyrene is a thermoplastic or thermosetting material? What is the liquid which condenses in the cooled test tube? What evidence is there that this is made up of smaller molecules than the polystyrene?

Experiment 9.13 Making nylon

Put 2 cm^3 of a solution of sebacoyl chloride in carbon tetrachloride in a 5 cm^3 beaker. Add 2 cm^3 of a solution of diaminohexane in water, using a teat pipette to run it very carefully down the side of the beaker, so that it forms a separate layer on top of the first solution.
Use a pair of forceps to pull out the whitish substance which forms between the two liquid layers in the beaker. With care you can pull this out as a filament and wind it round a glass rod.
The sebacoyl chloride molecule is:

Carbon 299

$$Cl-\underset{\underset{}{}}{\overset{\overset{O}{\|}}{C}}-\underset{\underset{H}{|}}{\overset{\overset{H}{|}}{C}}-\underset{\underset{H}{|}}{\overset{\overset{H}{|}}{C}}-\underset{\underset{H}{|}}{\overset{\overset{H}{|}}{C}}-\underset{\underset{H}{|}}{\overset{\overset{H}{|}}{C}}-\underset{\underset{H}{|}}{\overset{\overset{H}{|}}{C}}-\underset{\underset{H}{|}}{\overset{\overset{H}{|}}{C}}-\underset{\underset{H}{|}}{\overset{\overset{H}{|}}{C}}-\underset{\underset{}{}}{\overset{\overset{O}{\|}}{C}}-Cl,$$

and the diaminohexane molecule is:

$$H-N-\underset{\underset{H}{|}}{\overset{\overset{H}{|}}{C}}-\underset{\underset{H}{|}}{\overset{\overset{H}{|}}{C}}-\underset{\underset{H}{|}}{\overset{\overset{H}{|}}{C}}-\underset{\underset{H}{|}}{\overset{\overset{H}{|}}{C}}-\underset{\underset{H}{|}}{\overset{\overset{H}{|}}{C}}-\underset{\underset{H}{|}}{\overset{\overset{H}{|}}{C}}-N-H.$$

This is an example of condensation polymerisation. What small molecule is eliminated when these two monomers condense together? Draw a diagram of part of the nylon 610 molecule which you have made. Why is this type of nylon called nylon 610?

Experiment 9.14   Making synthetic resins

(a) Stand a small tin on an asbestos square. Put into the tin 5 cm³ of formaldehyde and 5 spatulas-full of urea and mix them together with a glass rod.

Very carefully add 10 drops of concentrated sulphuric acid from a teat pipette, stirring all the time.

Describe what happens. This is a condensation polymerisation. Why do you think it is necessary to add concentrated sulphuric acid before polymerisation will take place? Do you think your product is a thermoplastic or thermosetting material? How could you check this?

(b) Repeat experiment (a) with resorcinol in place of urea.

Questions

1 Quite a number of plastics have names that commence with the prefix *poly-*, e.g. polystyrene, polythene (short for polyethylene), PVC (short for poly-vinyl-chloride).

Explain in a few sentences why the structure of the molecules of these plastics justifies the use of the prefix *poly-* in their names.

(*Wales*)

2 Polyethylene (ICI trade name polythene) is made by polymerising the gas ethylene.

 (a) Under what conditions is polymerisation carried out?
 (b) What is the commercial source of ethylene?
 (c) Draw the equation for the reaction, using structural formulae.
 (d) About how many small repeating units are joined together in the polymer polythene?
 (e) What is the general name for the small units of which the large molecules of polymers are made?

(f) What kind of bond joins the small units together?

(g) Artificial fibres can be either man-made or naturally-occurring polymers reconstituted. Give one example of each. (Trade names are acceptable.) *(London)*

3 (a) Two methods of making artificial fibres and plastics are *addition polymerisation* and *condensation polymerisation.*

(i) What chemical feature must the monomer show for addition polymerisation to take place?

(ii) Give one example of a polymer made by this process.

(iii) What is the essential chemical feature of condensation polymerisation?

(iv) Give one example of a polymer made by this process.

(b) Polymers may be *thermoplastic* (thermosoftening), or *thermosetting*.

(i) What does thermoplastic mean?

(ii) Give an example of a thermoplastic polymer.

(iii) What does thermosetting mean?

(iv) Give an example of a thermosetting polymer.

(v) Using lines like this for the polymer chains ∼∼∼∼ draw representations of thermoplastic and thermosetting polymers.

(London)

4 Using pictures cut from magazines and newspapers, make a chart to show some of the uses of plastics. You could fix small samples of different types of plastic to your chart, or make a display of plastic objects to go with the chart.

.8 Rubber

Natural rubber comes from the sap of a tree, the Para rubber tree, which originally grew only in the tropical part of South America. Christopher Columbus first brought rubber to Europe in 1496, and almost up to the end of the nineteenth century all rubber came from Brazil. Then, in 1876, Sir Henry Wickham smuggled some seeds to England. They were cultivated in Kew Gardens and some of the seedlings were planted in Ceylon and Malaya. From these few plants grew the vast plantations which now supply most of the world's rubber.

The trees are TAPPED by making a spiral cut through the bark (Figure 9.52). The sap, or LATEX, is a white liquid rather like milk. It is a suspension of tiny particles of rubber in water, about 35 per cent rubber and 65 per cent water. If the latex is kept alkaline the rubber remains in suspension, but when acid is added it coagulates and solid rubber separates from the water.

Rubber is a natural hydrocarbon polymer with the structure:

$$\text{etc.}-\underset{H}{\overset{H}{C}}-\underset{H}{\overset{CH_3}{C}}=\underset{}{\overset{H}{C}}-\underset{H}{\overset{H}{C}}-\underset{H}{\overset{CH_3}{C}}=\underset{}{\overset{H}{C}}-\underset{H}{\overset{H}{C}}-\underset{H}{\overset{CH_3}{C}}=\underset{}{\overset{H}{C}}-\underset{H}{\overset{H}{C}}-\text{etc.}$$

(Only three of the repeating monomer units in the chain are shown. A rubber molecule contains on average about 11 000 of these units.) The long chain molecules are not normally stretched out straight, but are bent and tangled. Figure 9.53 shows a model of part of a rubber molecule.

Figure 9.52 (above) A tapper cutting away the bark of a rubber tree. A small cup is attached to the trunk ready to collect the latex.
Figure 9.53 (below) Model of part of a rubber molecule

Raw rubber is soft and pliable, rather like plasticine. It does not possess the main property which we associate with rubber: ELASTICITY (the ability to return to its original shape after stretching). To make rubber elastic it must be VULCANISED. Between 1 per cent and 3 per cent of sulphur is mixed with the raw rubber and the mixture is heated. Some of the double bonds in the rubber molecule break and the separate chains become joined by sulphur atoms:

$$\begin{array}{c} \sim\!\!C\!=\!C\!\sim\!\!\sim\!\!\sim\!\!C\!=\!C\!\sim \\ \\ \sim\!\!C\!=\!C\!\sim\!\!\sim\!\!\sim\!\!C\!=\!C\!\sim \\ \\ \sim\!\!C\!=\!C\!\sim\!\!\sim\!\!\sim\!\!C\!=\!C\!\sim \\ \\ \text{raw rubber} \end{array} \xrightarrow[\text{heat}]{\text{sulphur}} \begin{array}{c} \sim\!\!C\!-\!C\!\sim\!\!\sim\!\!\sim\!\!C\!-\!C\!\sim \\ |\quad\; |\quad\quad\quad\;\; |\quad\; | \\ S\quad S\quad\quad\quad S\quad S \\ |\quad\; |\quad\quad\quad\;\; |\quad\; | \\ \sim\!\!C\!-\!C\!\sim\!\!\sim\!\!\sim\!\!C\!-\!C\!\sim \\ |\quad\; |\quad\quad\quad\;\; |\quad\; | \\ S\quad S\quad\quad\quad S\quad S \\ |\quad\; |\quad\quad\quad\;\; |\quad\; | \\ \sim\!\!C\!-\!C\!\sim\!\!\sim\!\!\sim\!\!C\!-\!C\!\sim \\ \\ \text{vulcanised rubber} \end{array}$$

(Only the double bonds in the rubber molecules are shown, to make the process clearer.)

When raw rubber is stretched, the tangled chain molecules (Figure 9.53) tend to straighten out, and since they are not connected in any way they can slip past one another. When the stretching force is removed the molecules stay where they are, so there is very little tendency for the rubber to return to its original shape. When vulcanised rubber is stretched the tangled chains are partly straightened out and they cannot slip past one another because of the sulphur atoms holding the chains together. When the stretching force is removed the chains spring back to their original positions and the rubber returns to its original shape. Vulcanised rubber is almost the only substance which has this remarkable degree of elasticity. It is the presence of double bonds in the hydrocarbon chains which permits vulcanisation and makes this elasticity possible.

If a much larger proportion of sulphur is added to raw rubber (between 25 per cent and 40 per cent) a very different material, EBONITE, is produced, which is hard, black, and non-elastic. Ebonite is an exceptionally bad conductor of electricity, which makes it useful as an insulating material in the electrical industry.

Apart from sulphur, other substances are added to natural rubber to give it the properties required for different uses. To make it stronger and more resistant to wear, but still flexible, CARBON BLACK (finely-powdered graphite) is added. For car tyres rubber is mixed with up to half its own weight of carbon black. If flexibility is not important FILLERS such as clay or French chalk are added to make the rubber harder and stiffer. Rubber for floor tiles, carpet underlay, and mats contains cheap fillers of this type.

Synthetic rubber

The world's biggest producer of natural rubber throughout this century has been Malaya. In times of war, countries have been cut off from Malaya's rubber supplies, and this led chemists to look for an artificial substitute. German chemists first produced synthetic rubber during the First World War (1914–18). The method turned out to be too expensive for use in peacetime and, by 1939, 99 per cent of the rubber used throughout the world was still natural rubber. But during the Second World War (1939–45), when Japan invaded Malaya, supplies of natural rubber were interrupted. As a result, the manufacture of synthetic rubber on a large scale was started in the United States.

The problem was to produce a synthetic hydrocarbon polymer with chain-like molecules, rather like polythene or the other addition polymers (pages 293–5), but with double bonds in the chain to make vulcanisation possible. The basic monomer in synthetic rubber production is BUTADIENE:

$$\begin{array}{c} H \quad H \quad H \quad H \\ | \quad | \quad | \quad | \\ C=C-C=C \\ | \qquad\qquad | \\ H \qquad\qquad H \end{array}$$

an unsaturated hydrocarbon made by cracking butane (C_4H_{10}) obtained from petroleum. Just as ethylene can be polymerised with the aid of a catalyst to give polyethylene (polythene), butadiene polymerises to give POLYBUTADIENE:

$$\underset{H}{\overset{H}{\underset{|}{C}}}=\underset{H}{\overset{H}{\underset{|}{C}}}-\underset{|}{\overset{H}{C}}=\underset{H}{\overset{H}{C}} \;+\; \underset{H}{\overset{H}{\underset{|}{C}}}=\underset{H}{\overset{H}{\underset{|}{C}}}-\underset{|}{\overset{H}{C}}=\underset{H}{\overset{H}{C}} \;+\; \underset{H}{\overset{H}{\underset{|}{C}}}=\underset{H}{\overset{H}{\underset{|}{C}}}-\underset{|}{\overset{H}{C}}=\underset{H}{\overset{H}{C}}$$

$$\downarrow \text{catalyst}$$

$$\text{etc.}-C-C=C-C-C-C=C-C-C-C=C-C-\text{etc.}$$

Polybutadiene can be vulcanised just like natural rubber. 15 per cent of the world's synthetic rubber production is BUTADIENE RUBBER. It has particularly good resistance to wear, which makes it especially useful for making tyres.

Other types of synthetic rubber are made by mixing a proportion of another unsaturated hydrocarbon monomer with butadiene. The most important of these is STYRENE-BUTADIENE RUBBER (*SBR*). 75 per cent of the world's production of synthetic rubber is of this type. Styrene is an unsaturated hydrocarbon with the formula:

$$\underset{H\quad C_6H_5}{\overset{H\quad H}{\underset{|}{C}=\underset{|}{C}}}.$$

A mixture of six parts of butadiene with one part of styrene is polymerised:

$$B + B + B + B + B + B + S + B + B + B + B + B + B + S$$

$$\downarrow \text{catalyst}$$

$$\text{etc.}—B—B—B—B—B—B—S—B—B—B—B—B—B—S—\text{etc.}$$

(B = butadiene molecule, S = styrene molecule).

Questions

1 Try to find out more about the story of the rubber industry, from the time when Columbus brought rubber from America to the present day. You could, for instance, look under *Rubber* in an encyclopaedia. Write a short essay (illustrated, if you like) about what you discover.

2 (a) Why is rubber an important material?
  (b) What element is reacted with rubber to vulcanise it?
  (c) Explain what happens to rubber molecules during the process of vulcanisation. (Use diagrams if you think it will make your answer clearer.)
  (d) Why is vulcanised rubber more elastic than unvulcanised rubber?
  (e) Why can rubber, which is a hydrocarbon polymer, be vulcanised, whereas polythene (polyethylene), which is also a hydrocarbon polymer, cannot?

9 Detergents

The word *detergent* means *something which cleans*. The oldest detergent, which has been in use for at least 3000 years, is soap. From the time of the Phoenicians, who are believed to have discovered soap, up to the present day, it has been made in the same way. The basic materials are animal fats (such as lard) or vegetable oils (such as olive oil) and an alkali, usually sodium hydroxide.

Fats and oils are esters (page 288). They are compounds of glycerol (commonly called *glycerine*) with organic acids whose molecules contain a long chain of carbon atoms. Each glycerol molecule is combined with three acid molecules:

The main difference between the various fats and oils is the number of carbon atoms in the acid molecules, usually between 12 and 18. When the fat or oil is heated with sodium hydroxide solution the acids are broken away from the glycerol and are neutralised by the alkali to form a salt. Soap is the sodium salt of an organic acid with a long chain of carbon atoms in its molecule. A model of a typical soap molecule is shown in Figure 9.54.

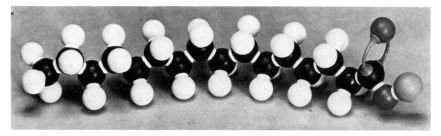

Figure 9.54 Model of a typical soap molecule

Experiment 9.15 Making soap

Put 2 cm^3 of castor oil in a beaker and add 10 cm^3 of concentrated sodium hydroxide solution. (**CARE: this solution is extremely caustic.**)

Stand the beaker on a tripod and wire gauze and warm it gradually until it boils; stir constantly with a glass rod. Continue boiling gently for 5 minutes.

Add 10 cm^3 of water and 3 spatulas-full of sodium chloride. Reheat the mixture to boiling. Filter through a Buchner funnel using a filter pump.

Use a spatula to transfer a little of the product left on the filter paper to a test tube. Half-fill the test tube with distilled water, put your thumb over the end, and shake the tube well.

What happens in the test tube? Does your product behave like soap? Explain the chemical reactions which take place when the oil is boiled with sodium hydroxide solution.

The manufacture of soap

You may have made soap by heating an oil with sodium hydroxide solution. This is the basis of the modern manufacturing process, but instead of making small batches the whole process is continuous. The oil and sodium hydroxide solution are fed into an enclosed reaction vessel under pressure. The high pressure enables the decomposition of the oil to be carried out at a temperature well above the normal boiling point of the mixture. At this temperature the reaction is complete in a few minutes.

The mixture of soap and glycerol from the reaction vessel is cooled, and concentrated sodium chloride (*salt*) solution is added. Glycerol dissolves easily in salt solution but soap does not, so solid soap separates out from the mixture and is removed by centrifuging. While still hot it is sprayed into a hot vacuum chamber to dry it. Perfume is added and the particles are compressed into a continuous block which is cut into bars.

How does soap clean?

Water by itself does not wash efficiently, for two reasons. First, it does not easily make things wet. If drops of water are put on to a piece of cloth they do not soak in straight away, but sit on the surface of the cloth in little globules. Secondly, water does not mix with oil and grease and will therefore not easily remove oily or greasy dirt.

Water does not easily wet things because of its unusually large SURFACE TENSION. Water molecules are strongly attracted to one another. In the centre of a drop of water (Figure 9.55) these forces of attraction operate equally in all directions, but at the surface of the water all the forces are inwards. The result is that water tends to pull itself into spherical drops, which prevents it from coming into contact with the material to be washed or with the dirt.

Soap is a salt. It is therefore ionic, and in solution in water the positive sodium ion separates, leaving a long hydrocarbon chain with a negative charge on one end. The charged 'heads' of the soap molecules are attracted to the water molecules and push in between them. The hydrocarbon 'tails' are not attracted and interfere with the forces between the water molecules. The effect of adding soap to water is to reduce the water's surface tension and break up the spherical droplets so that they spread and wet the surface (Figure 9.56).

The hydrocarbon 'tails' of soap molecules are attracted towards particles of grease or oil, which are also hydrocarbons. The 'tails' stick into the surface of the grease, leaving the charged 'heads' of the soap molecules protruding into the water,

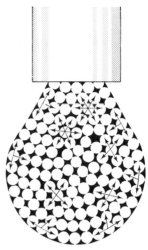

Figure 9.55

Figure 9.56 A drop of water on a surface tends to form a globule, as shown on the left. If a little detergent, such as soap, is added the drop breaks up and wets the surface, as shown on the right.

Figure 9.57

Carbon 307

to which they are attracted (Figure 9.57). Each of the particles of grease or oil becomes surrounded by a shield of negative charges (the ionic 'heads' of the soap molecules). The dislodged particles repel one another which keeps them dispersed in the washing water so that they do not re-attach themselves to the article being washed.

Soapless detergents

Animal fats and vegetable oils are important foodstuffs, and in a world where there is a shortage of food it is a waste to use edible material to make something which cannot be eaten, even something as useful as soap. It was the discovery that margarine could be made from vegetable oils (page 285) that prompted a search for an alternative way of making detergents.

The raw material must be a chemical with a long hydrocarbon chain molecule to which an ionic group can be attached. Petroleum provides the obvious source of hydrocarbons. It was found that certain hydrocarbons which could be obtained from petroleum could be reacted with concentrated sulphuric acid to produce a HYDROCARBON SULPHONIC ACID:

$$\sim\!\!\sim\!\!\sim\!\!\sim\!\!\sim\!\!\sim\!\!\sim\!\!\sim H + H_2SO_4 \longrightarrow \sim\!\!\sim\!\!\sim\!\!\sim\!\!\sim\!\!\sim\!\!\sim SO_3H + H_2O$$

long chain hydrocarbon　　　　　　　　　　　hydrocarbon sulphonic acid

The acid is neutralised with sodium hydroxide to produce a salt which has very similar properties to the sodium salts of organic acids used as soap:

$$\sim\!\!\sim\!\!\sim\!\!\sim\!\!\sim\!\!\sim SO_3H + NaOH \longrightarrow \sim\!\!\sim\!\!\sim\!\!\sim\!\!\sim\!\!\sim SO_3^- Na^+ + H_2O$$

hydrocarbon sulphonic acid　　　　　　　　　　soapless detergent

A model of a typical soapless detergent molecule is shown in Figure 9.58. Like the soap molecule in Figure 9.54, page 305, it is made up of a long hydrocarbon chain with an ionic group attached to it. Soapless detergents have the same properties as soap in reducing the surface tension of water to make it wet things better and in mixing with oily or greasy dirt.

Soapless detergents can be manufactured in solid form (for washing powders) or in liquid form (for washing-up liquids and shampoos). Washing powders contain a number of other components, apart from the soapless detergent, which usually makes up only about 20 per cent of the weight. Phosphates are added to help in the removal of the clay which may be present in dirt; they convert the clay to a soluble form. Sodium perborate gives the washing powder a mild bleaching action. Sodium sulphate and sodium silicate help to keep the powder dry and free-flowing. In addition most powders contain foam stabilisers to prevent excessive frothing, perfume, colouring, and fluorescent substances which counteract the gradual yellowing of white materials (the *whiter-than-*

Figure 9.58 Model of a typical detergent molecule

white ingredients). Some powders also contain enzymes (page 64) to digest organic dirt like food stains and blood.

The effect of soap and soapless detergents on hard water

Hard water contains dissolved salts, mainly calcium and magnesium salts. Soap reacts with these salts in a double decomposition (page 203) to produce the insoluble calcium and magnesium salts of the organic acid. These insoluble salts are precipitated as SCUM. The formation of scum is a waste of soap, and also makes washing more difficult, as the particles of scum tend to become lodged between the fibres of the cloth.

Soapless detergents have the advantage that they do not produce scum. The calcium and magnesium salts of the hydrocarbon sulphonic acids of which these detergents are composed are soluble in water, so there is no precipitate formed.

Experiment 9.16   Making a soapless detergent

Measure 4 cm^3 of concentrated sulphuric acid in a beaker (**take great care**). Add 2 cm^3 of castor oil, a little at a time, stirring constantly.

Pour the mixture into 10 cm^3 of water in another beaker. Stir the contents of the beaker well, then allow the oil to separate.

Use a teat pipette to transfer a few drops of the oil to half a test tube of water. Shake the tube well.

Describe the appearance of your product. What happens in the test tube? Does your product behave like a detergent?

Carbon

Experiment 9.17 Investigating the effect of detergents on the surface tension of water

(a) Lay a piece of dry cloth on the bench. Use a teat pipette to put drops of water on the cloth. Do the drops soak into the cloth? What shape are they?

Now add one drop of washing-up liquid to half a beaker of water. Put drops of this detergent solution on to the cloth near the drops of water. What happens? Why?

(b) Fill a crystallising dish with water. Float a piece of filter paper on the surface of the water and quickly lay a needle on the filter paper. Leave the dish to stand for a minute or so without disturbing it at all. If you have been successful, the filter paper will sink and the needle will be left floating on the surface of the water.

Use a teat pipette to run one drop of washing-up liquid carefully down the inside of the dish into the water. What happens to the needle? The needle is made of steel, which is much more dense than water. Why does the needle float at first? Why does it sink when detergent is added to the water?

(c) Half-fill a beaker with water. Add a spatula-full of powdered sulphur. Does the sulphur sink or float?

Stir the contents of the beaker vigorously. Allow it to settle for a few seconds. What happens to the sulphur now?

Using a teat pipette, add one drop of washing-up liquid to the beaker and stir again. What happens to the sulphur? Why?

Experiment 9.18 Investigating the effect of detergents on oil

Half-fill three test tubes with distilled water. Use a teat pipette to add 3 drops of olive oil to each tube.

To one tube add 2 cm^3 of soap solution and to another 2 cm^3 of washing-up liquid. Shake all three tubes vigorously.

Describe and explain what happens in each of the three tubes.

Experiment 9.19 Investigating the effect of detergents on hard water

Half-fill four test tubes with distilled water, and label them 1, 2, 3, and 4. Dissolve a spatula-full of magnesium sulphate in tubes 2 and 4 (this makes the water in these tubes 'hard').

To tubes 1 and 2 add 1 cm^3 of soap solution and to tubes 3 and 4 add 1 drop of washing-up liquid. Shake all four tubes vigorously.

Describe what happens in each tube. What do you see in tube 2? Why are the results in tubes 2 and 4 different?

Questions

1 (a) Describe how you would make a small quantity of soap in the school laboratory.
 (b) Explain why hard water wastes soap.
 (c) Say in a few sentences why modern detergents are more efficient than soap for most washing purposes. (*Wales*)

2 (a) Name *two* substances used in the preparation of soap.
 (b) Name *one* other useful substance formed in this reaction.
 (c) Name *one* substance present in natural hard water which prevents soap from lathering. (*Wales*)

3 Which of the following is the purpose of the salt (sodium chloride) used in the manufacture of soap?
A To convert the organic acid to its sodium salt.
B To catalyse the decomposition of the fat or oil.
C To purify the soap.
D To help to separate the soap from the glycerol.
E To react with the glycerol.

4 Explain briefly how detergents help to remove oily and greasy dirt.

For the teacher

Experiment 9.1 A 75 × 12 mm tube is best for the lime water as a small quantity is necessary to obtain a positive result. All reactants should be dried by heating in an evaporating basin for 5 minutes and cooling in a dessicator so that no condensation is seen in the class experiment. The experiment is not, of course, conclusive, but most pupils will not find the identity of these two allotropes unacceptable.

The identity of graphite and diamond is less easy to accept and it is worthwhile demonstrating a similar experiment with a small industrial diamond (available from the Industrial Diamond Company) as shown in Figure 9.59. The identity of the diamond can be established

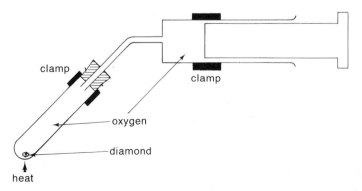

Figure 9.59

by scratching a piece of glass. The apparatus is filled with oxygen and the diamond is heated strongly until it burns. The syringe plunger should be rotated slowly during heating to prevent it sticking. The gas in the tube is tested as in experiment 9.1.

Models A ball-and-spoke model of graphite can be constructed from Linnell-type components (from Gallenkamp). 48 carbon atoms (black, 5-hole, trigonal bipyramidal), 57 38-mm springs, and 32 60-mm springs makes a useful size. 3 sheets of 16 atoms are made up as shown in Figure 9.60 using 38-mm springs. The layers are linked by 60-mm springs. A tangential model can be constructed from a convenient size of polystyrene spheres. Use a protractor template (page 244) to mark the spheres at 120° round the 'equator'. 16 spheres joined in a layer give 4 hexagons. Two or three layers should be made and rested on top of one another (not connected) to show the sliding effect.

Figure 9.60

A diamond model as shown in Figure 9.4, page 248, can be constructed from 60 carbon atoms (black, tetrahedral) and 60 springs. For a tangential model, polystyrene spheres must be marked tetrahedrally. First mark the equator at 120°. Then place the protractor template in line with an end 'pip' and one of the equatorial marks and mark at $109\frac{1}{2}°$ from the end 'pip'. Repeat in line with the other two 120° equatorial marks. The three $109\frac{1}{2}°$ marks and the end pip form the tetrahedral markings.

Properties of carbon allotropes (a) Scratch a piece of glass with a diamond (hardness). (b) Pass samples of graphite (from R. F. D. Parkinson or Lythe Minerals) round the class for pupils to mark paper and feel the slippery texture (use as pencil leads and as lubricant). (c) In a fume cupboard put 2 or 3 drops of bromine in a gas jar and cover with a lid. When the vapour has diffused through the jar drop in 1 or 2 large pieces of wood charcoal, replace the lid, and shake the jar. The bromine vapour is absorbed and the colour disappears (gas masks). (d) Dissolve 6 spatulas-full of dark brown Demerara sugar or some treacle in 50 cm³ of water in a beaker. Add 3 or 4 spatulas-full of activated charcoal, boil for two minutes, and filter. The filtrate is colourless (manufacture of sugar).

Experiment 9.2 As experiment 9.1. A selection from the following would be suitable: starch, flour, potato, rice, sugar, glucose, lard, butter, dried grass or leaves.

Sugar from starch Pupils can be given small pieces of bread. If bread is chewed for several minutes without swallowing, the taste becomes sweet, due to the breakdown of starch by enzymes in saliva. See also experiment 3.9.

Experiment 9.3 For control experiments the apparatus shown in Figure 9.61 is convenient.

Experiment 9.4 It is extremely difficult to obtain a realistic thread of cuprammonium rayon in the school laboratory and even more

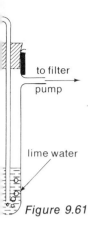

Figure 9.61

difficult to produce viscose rayon. This rather crude experiment does, however, illustrate the principle of solution and reprecipitation of cellulose. For (a) 0.5 M copper(II) sulphate and 2 M ammonia. For (b) 'ammoniacal copper(II) sulphate' is M copper(II) sulphate to which *just* sufficient 0.88 ammonia has been added to redissolve the precipitate. 2 M sulphuric acid and 2 cm³ (or larger) disposable plastic syringes (without needles).

Experiment 9.5 The lime water should have turned milky within an hour. By the end of the fermentation it will have cleared again due to the further reaction:

$$CaCO_3(s) + CO_2(g) + H_2O(l) \rightarrow Ca(HCO_3)_2(aq).$$

To get a reasonable sample of ethanol it is best to pool the class results from the fermentation experiment. Without disturbing the flasks decant the solution from the bulk of the yeast (alternatively, filter) into a 1000 cm³ round-bottomed flask and assemble for fractionation as in Figure 9.15, page 260. Add a few pieces of unglazed porcelain or anti-bumping granules. The fractionating column should be at least 15 cm long and packed with glass beads or $\frac{1}{2}$ cm lengths of glass tubing. Once the boiling point of the mixture has been reached, heating must be continued with great care to prevent excessive frothing. Collect the product distilling up to 90°C. This will be sufficiently pure ethanol to be ignited on a watch glass.

Brewing A simple beer recipe requires: 2 lb (900 g) sugar, 4 lb (1 800 g) malt extract, 2 oz (50 g) hops in a muslin bag, and $\frac{1}{4}$ oz (7 g) dried beer yeast. Make up the yeast according to the instructions on the packet. Bring 2 gallons (10 000 cm³) of water to the boil, add the sugar and hops and simmer for half an hour. Remove the hops and add the malt extract. Pour into a larger container, add 2 gallons (10 000 cm³) of cold water and stir in the yeast. A head will appear in a day or two; this should be removed every day. The beer is ready for bottling when all working has ceased (7–10 days). Pupils who are interested in pursuing brewing and wine-making can be recommended to Tritton, S. M. *A Guide to Better Wine and Beer Making for Beginners* (Faber, 1969), of which there is a paperback edition.

Experiment 9.6 It is extremely difficult to clean the tubes after this experiment. It is advisable to keep sets of apparatus specially for coal distillation. A display of some of the products made from chemicals produced by the fractionation of tar, based on Figure 9.22, page 268, is useful.

Experiment 9.7 Samples can be collected in 75 × 12 mm tubes. A small free sample of 'topped' crude is available from Shell and larger quantities from many laboratory suppliers; before use add 20 per cent of petrol to replace the missing low boiling point fractions.

The chemistry of petroleum: models The idea of the multiplicity of hydrocarbon molecules is best introduced by allowing pupils to make ball-and-spoke models for themselves. Each pupil will require a set of Linnell-type components (from Gallenkamp) comprising 4 carbon atoms (black, tetrahedral), 10 hydrogen atoms (white, one-hole), and 13 38-mm springs. For later work in section 9.6 the following can be added to each set: 2 oxygen atoms (red, tetrahedral, with two holes blocked e.g. with sealing wax), and 2 halogen atoms (green, one-hole).

Another type of relatively inexpensive pupil kit, which can be used to build both ball-and-spoke and space-filling models, is marketed by Spiring Enterprises. Tangential models can be built from polystyrene spheres (page 244).

Experiment 9.8 A large crystallising dish serves as a pneumatic trough. For bromine water use a saturated solution diluted x5.

For further experiments on petroleum see *Oil Experiments* (Shell, free). Another free teachers' booklet from Shell *Making an Oil Display* may also be useful, as well as the pupils' quiz *Find Out About Oil* (Shell, free).

Experiments 9.9, 9.10, 9.11 Industrial methylated spirits (66 O.P. or stronger). 0.01 M potassium manganate(VII). M sulphuric acid.

Natural polymer models Models of glucose, starch, and cellulose may be helpful. The following Linnell-type components (from Gallenkamp) are required. For glucose: 6 carbon atoms (black, tetrahedral), 6 oxygen atoms (red, tetrahedral), 12 hydrogen atoms (white, one-hole), 24 springs. For starch: 24 carbon atoms, 20 oxygen atoms, 40 hydrogen atoms, and 92 springs will make a 4-unit section of the chain. For cellulose: as for starch.

glucose

starch

cellulose

Polymerisation models Poppet beads (from Woolworths) provide a useful illustration. *Addition polymerisation:* Start with beads of one colour (preferably black) joined in pairs to represent the ethylene molecule. Link pairs into a chain to represent polythene. *Condensation polymerisation:* Start with chains of 6 beads (preferably black) with a different coloured bead (e.g. white) on each end to represent one monomer and chains of 6 beads (also preferably black) with a different coloured bead (e.g. red) on each end to represent the other monomer. Link several of each monomer unit alternately into a long chain, eliminating the red-white pairs.

Experiment 9.12 Pieces of polystyrene ceiling tiles or polystyrene spheres can be used.

The polymerisation of styrene can be demonstrated by refluxing 5 cm^3 of styrene with 5 cm^3 of paraffin for 45 minutes. Allow to cool and pour into 50 cm^3 of industrial methylated spirits (66 O.P. or stronger) in a beaker.

Experiment 9.13 Sebacoyl chloride solution: 5 per cent in carbon tetrachloride. 1,6-diaminohexane solution: 5 per cent in water. (On a commercial scale the acid, not the acid chloride, is used.)

Experiment 9.14 It is advisable to use tin cans or other disposable containers, since it is extremely difficult to clean the resins from beakers.

For other polymerisation experiments see *Experiments in Polymer Chemistry* (Shell, free) and *The Uses of Plastics in School Laboratories and Classrooms* (ICI, free).

A plastics demonstration kit *Plastics Explained*, containing 16 labelled samples, is available free from BP, and Shell supply a plastics display kit for £2. Further samples are available from Alan Griffiths and Partners. Samples of natural and synthetic textiles are available free from Courtaulds and of acrilan and nylon fibre, yarn, and fabric from Monsanto.

If identification of plastics is to be attempted the free ICI booklet *Identification of Plastics for Schools* will be found helpful, and a kit of plastics with equipment and instructions for identification is marketed by The Plastics Identification Service, Ltd.

Rubber A kit of samples, including natural latex, is available from the Natural Rubber Producers Research Association, and natural and synthetic latexes can be obtained from MacAdams and Co. The coagulation of latex can be demonstrated by adding 2 M acetic acid to latex in a beaker and stirring.

Vulcanisation, by heating coagulated latex with sulphur, is not easy to imitate in the school laboratory, but a cold-cure process, used commercially for vulcanising thin films (e.g. rubber-proofing on fabrics), can easily be demonstrated. Spread natural latex in a thin film on a piece of glass and leave to dry. Dip the film for 2 minutes in a $2\frac{1}{2}$ per cent solution of sulphur monochloride in carbon disulphide.

Thin films of vulcanised and unvulcanised rubber made as above can be compared by (a) stretching and (b) warming gently above a small flame.

Models of detergent molecules For a ball-and-spoke model of a typical soap molecule (sodium palmitate) the following Linnell-type components (from Gallenkamp) will be required: 16 carbon atoms (black, tetrahedral), 31 hydrogen atoms (white, one-hole), 2 oxygen atoms (red, tetrahedral), 1 sodium atom (silver, one-hole), and 50 springs. For a typical soapless detergent molecule the following components will be required: 16 carbon atoms, 28 hydrogen atoms, 3 oxygen atoms, 1 sodium atom, 1 sulphur atom (yellow, octahedral), and 57 springs.

$$CH_3-CH_2-CH_2-CH_2-CH_2-CH-CH_2-CH_2-CH_2-CH_3$$

with the CH group attached to a benzene ring bearing $-S(=O)(=O)-O^- Na^+$

Experiment 9.17 A woven fabric, such as denim, worsted, or tweed, is ideal.

Experiments 9.18 and 9.19 For soap solution dissolve 5 g of soap flakes or Castile soap in 100 cm³ of industrial methylated spirits (66 O.P.) and dilute to 1000 cm³.

For further experiments with detergents see *Experiments in Detergency* (Shell, free). Some other interesting experiments involving natural oils and synthetic detergents are included in *Cosmetics in the School Laboratory* (Shell, free).

10 Energy in chemistry

Energy is the ability to do work. Coal, for instance, contains stored energy. If it is burnt in a steam engine it can provide the work to pull a train along a track. If it is burnt in a power station it provides the work to turn dynamos to make electricity.

Types of energy

Energy can take many different forms. The energy stored in coal is called CHEMICAL ENERGY. The HEAT ENERGY of steam can be made to do the work of driving a steam engine. LIGHT ENERGY from the sun provides the work to make plants grow (page 252). The SOUND ENERGY of a sonic boom can do the work of breaking windows. The work to run all kinds of things, from vacuum cleaners to milk floats, is provided by ELECTRICAL ENERGY. A car running along the road has MECHANICAL ENERGY; it will do the work of

Figure 10.1 Battersea power station, an example of the transformation of chemical energy into electrical energy

climbing a hill or, if the driver is careless, of knocking down a wall.
NUCLEAR ENERGY is the energy stored in the nuclei of atoms; it is released to do work in an atomic power station.

Changing one sort of energy into another

Energy of one kind can be changed into another kind. In an electric fire electrical energy is changed into heat. In a light bulb it is changed into light. In the explosion of an atomic bomb, nuclear energy changes into heat, light, sound, and mechanical energy. In an electric motor electrical energy is changed into mechanical energy. In a dynamo mechanical energy is changed into electrical energy.

In changes like these the form of the energy is altered but the total amount of energy stays the same. This is called the *law of conservation of energy*.

Energy cannot appear or disappear.
It can only be changed from one form into another.

Chemical energy can be changed into almost any other kind of energy. You will have come across many examples of this.
1 When coal or oil is burnt, chemical energy is changed into heat.
2 When magnesium ribbon burns, some of the chemical energy changes into light.
3 When a mixture of hydrogen and oxygen explodes, chemical energy changes into heat, light, and sound.
4 In a battery, chemical energy is changed into electrical energy.
5 When a stick of dynamite explodes chemical energy is changed into mechanical energy.

Questions

In each of the following situations one form of energy is transformed into another form. Which form is being changed into which? (*Example* When town gas is burnt in a gas cooker. Answer Chemical energy → heat energy.)

1 An electric fire.   2 Playing a violin.   3 A microphone.   4 A loudspeaker.   5 An electric light.   6 An electric motor.   7 A dynamo.   8 An oil-fired central heating system.   9 Electrolysis.   10 Photosynthesis.   11 A battery.   12 Dehydrating copper(II) sulphate by heating.   13 An atomic reactor.

10.2 Heat energy from chemical energy: fuels

A fuel is a substance which reacts with oxygen, changing its stored chemical energy into heat. Some important fuels are listed in Figure 10.2.

Notice that all chemical fuels originate from the sun's energy. Sugar, starch, fat, and wood are all products of photosynthesis (page 252). Coal and petroleum, and all the fuels made from them, are also derived from living things (page 264 and page 270).

Gaseous fuels	Natural gas (methane) Other gases obtained from petroleum (propane, butane) Acetylene Coal gas
Liquid fuels	Petrol Paraffin Diesel oil Fuel oil
Solid fuels	Wood Peat Coal Coke
'Human fuels'	Sugar Starch Fat

Figure 10.2

In choosing the best fuel for a particular job, a number of points have to be considered. To heat your home, for instance, you might have the choice of coal, gas, or oil. You will want to know the cost, and how much heat each fuel gives. You will also have to decide which fuel needs the cheapest burner. If you live in a smokeless zone you may not be allowed to burn certain types of coal. Also, some fuels are more convenient than others: liquid and gaseous fuels are easier to feed continuously to the burner. The choice of a fuel for an industrial furnace will involve exactly the same kind of problems.

The amount of energy produced by burning a fuel is measured in *joules* (abbreviated *J*). The joule is rather a small energy unit so the *kilojoule* (abbreviated *kJ*) is often used also:

1 kilojoule = 1000 joules.

The amount of heat produced by burning a fuel, called the CALORIFIC VALUE of the fuel, can be measured by burning a known mass of the fuel and using the heat produced to heat up some water. The piece of apparatus used for measuring the calorific value of a fuel is called a CALORIMETER. There are many different types. Figure 10.3 shows one type of calorimeter.

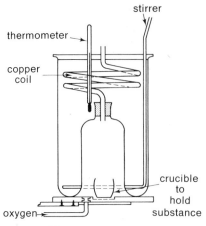

Figure 10.3 A calorimeter

A stream of pure oxygen gas is supplied from a cylinder. The fuel is ignited by passing a current through the coil of wire to make it red hot. Suction applied by a water pump draws the hot gases from the burning fuel through the copper coil to heat up the water in the glass container. The water is kept well stirred so that it heats up uniformly, and the rise in temperature is measured.

Calculation of the calorific value of a fuel

Before a calorimeter can be used to measure the calorific value of a fuel it is necessary to find out how much heat is required to heat that particular calorimeter through 1°C. Usually, pure carbon is burnt in the calorimeter, since it is known that 1 g of pure carbon produces 32.8 kJ when it is burnt.

Example In a particular calorimeter, burning 1 g of carbon produced a temperature rise of 4°C.
A temperature rise of 4°C is produced by 32.8 kJ of energy.
A temperature rise of 1°C would be produced by $32.8 \div 4 = 8.2$ kJ.

We now know that in this calorimeter every 1°C rise in temperature means that 8.2 kJ of energy have been produced by the burning fuel.

If 1 g of wood is burnt in the same calorimeter the temperature rises by 2.7°C.
A temperature rise of 1°C is produced by 8.2 kJ of energy.
A temperature rise of 2.7°C has been produced by $8.2 \times 2.7 = 22.14$ kJ.

The calorific value of any fuel can be measured in the same way. Figure 10.4 shows some typical results (though there will obviously be differences between, for example, different types of wood).

	Calorific value in kJ per gram
Town gas	80
Methane (natural gas)	56
Oil (petrol, fuel oil, etc.)	48
Fat	36
Coal	34
Coke	32
Wood	22
Carbohydrates (sugar, starch)	16

Figure 10.4

Questions

1 What is understood by the word 'fuel'?
Give the names and formulae of three hydrocarbon fuels. What are the products of combustion of such fuels?
Use one hydrocarbon of your own choice, write an equation to illustrate the combustion. *(North West)*

2 A particular calorimeter shows a rise in temperature of 1°C for every 6 kJ of energy released in it. When this calorimeter is used to burn various fuels the following results are obtained:
   (a) coal: 1 g gives a temperature rise of 6°C;
   (b) peat: 2 g gives a temperature rise of 9.5°C;
   (c) fat: 0.5 g gives a temperature rise of 3°C;
   (d) bread: 1.2 g gives a temperature rise of 3.2°C.
Calculate the calorific value of each of these fuels in kJ per gram.

3 When 1 g of carbon was burned in a particular calorimeter the temperature rose by 8°C. When 1 g of rice was burned in the same calorimeter the temperature rose by 3.5°C. What is the calorific value of rice in kJ per gram? (Calorific value of carbon = 32.8 kJ g^{-1}.)

4 When 0.5 g of town gas was burnt in a particular calorimeter the temperature rose by 3.3°C. Calculate the calorific value of town gas. (When 0.5 g of hydrogen, calorific value 120 kJ g^{-1}, was burnt in the same calorimeter the temperature rose by 5°C.)

5 Methane (CH_4) is a gaseous fuel. The heat energy released when ONE MOLE of the gas is completely burned in oxygen is 850 kJ.
   (i) To which class of organic compounds does methane belong?
   (ii) Name one other member of this family.
   (iii) Give the number of atoms of each element in a molecule of METHANE.
   (iv) If the atomic masses of carbon and hydrogen are 12 and 1 respectively, how much heat energy is released when 160 g of methane is completely burned in oxygen? *(West Yorkshire)*

Exothermic and endothermic reactions

All fuels give out heat when they react with oxygen. A reaction in which energy is given out is called an EXOTHERMIC REACTION. Most chemical processes are exothermic, for example: neutralising an acid with a base; displacing copper from copper(II) sulphate solution with zinc; precipitating silver chloride; diluting concentrated sulphuric acid; dissolving sodium hydroxide in water; adding water to white anhydrous copper(II) sulphate.

There are, however, a few chemical processes in which the reacting substances take in heat energy from their surroundings, so that the surroundings are cooled down. The heat energy taken in is changed into chemical energy, which is stored in the products of the reaction. This type of reaction is called an ENDOTHERMIC REACTION. Examples are: the reaction between carbon and sulphur to form carbon disulphide; dissolving ammonium nitrate in water; dissolving blue hydrated copper(II) sulphate in water.

Measuring the heat change in a chemical reaction

Fuel engineers usually measure the calorific value of a fuel in kilojoules per gram or kilojoules per tonne. Chemists prefer to make their measurements per mole of the reacting substance, so that they can compare energy changes for equal numbers of particles.

For reactions like the burning of a fuel the heat change per mole can be measured using a calorimeter like the one in Figure 10.3. But many chemical reactions take place in solution in water, and for these reactions there is a much simpler way of finding the heat change. If the reaction is exothermic the heat given out raises the temperature of the solution. If the reaction is endothermic the solution is cooled down. It is known that 1 cm^3 of water (or a dilute solution) is heated through 1°C by 4.18 joules. It is therefore quite simple to measure the temperature change of the solution in °C and calculate from this the amount of heat given out or taken in during the reaction in joules.

Example 1 When 1 mole (98 g) of concentrated sulphuric acid is poured into 1000 cm^3 of water, the temperature rises by 17°C.
Heating 1 cm^3 of water through 1°C takes 4.18 joules.
Heating 1000 cm^3 of water through 17°C takes $4.18 \times 1000 \times 17$ joules = 71060 joules = 71.06 kJ.
Dilution of concentrated sulphuric acid to a molar solution *produces* 71 kJ per mole.

Example 2 When 1 mole (80 g) of ammonium nitrate is dissolved in 1000 cm^3 of water, the temperature falls by 6°C.

$$\text{The heat taken in is } 4.18 \times 1000 \times 6 \text{ joules}$$
$$= 25080 \text{ joules} = 25.08 \text{ kJ}.$$

Dissolving ammonium nitrate to make a molar solution *uses up* 25 kJ per mole.

Example 3 When 100 cm^3 of 2 M sodium hydroxide solution is neutralised with 100 cm^3 of 2 M hydrochloric acid, the temperature rises by 2.6°C.

$$200 \text{ } cm^3 \text{ of solution have been heated through 2.6°C.}$$
$$\text{The heat produced is } 4.18 \times 200 \times 2.6 \text{ joules}$$
$$= 2173.6 \text{ joules.}$$
$$100 \text{ } cm^3 \text{ of a 2 M solution contains } \frac{100}{1000} \times 2 = \frac{1}{5} \text{ mole.}$$

Neutralisation of $\frac{1}{5}$ mole of sodium hydroxide with $\frac{1}{5}$ mole of hydrochloric acid produces 2173.6 joules.
Neutralisation of 1 mole of sodium hydroxide with 1 mole of hydrochloric acid will therefore produce $5 \times 2173.6 = 10368$ joules $= 10.368$ kJ.
Neutralisation of sodium hydroxide solution by hydrochloric acid *produces* 10.4 kJ per mole.

Energy level diagrams

Instead of writing *dilution of concentrated sulphuric acid produces 71 kJ per mole*, we can show the heat change in the equation. The equation representing the dilution of sulphuric acid is:

$$H_2SO_4(l) + aq \rightarrow H_2SO_4(aq).$$

(*aq* means *a lot of water*: remember that H_2O means 1 mole of water.)

Heat is given out in this reaction, so the solution of sulphuric acid left at the end contains *less* stored chemical energy than the concentrated sulphuric acid and water had between them at the beginning. We can show the loss of energy by writing the equation in the form:

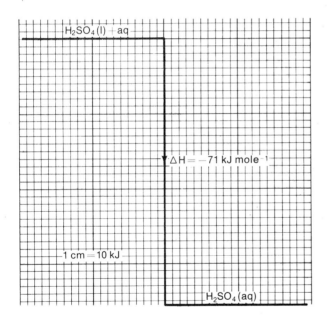

This is called an ENERGY LEVEL DIAGRAM. It is really a form of chemical equation in which an extra piece of information, the energy change in the reaction, is shown. ΔH means *heat change*, and the *minus* sign shows that energy is *lost* by the chemicals to their surroundings (the reaction is exothermic).

In dissolving ammonium nitrate 25 kJ per mole are taken in, so the

Energy in chemistry

solution of ammonium nitrate contains *more* stored chemical energy than the solid ammonium nitrate and water had between them at the beginning. The equation is:

$$NH_4NO_3(s) + aq \rightarrow NH_4NO_3(aq),$$

and the energy level diagram is:

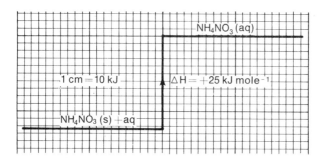

The *plus* sign shows that the chemicals have *gained* energy from their surroundings (the reaction is endothermic).

It is quite easy to draw an energy level diagram to represent any chemical reaction if the heat change in the reaction is known.

NB It is usual to draw energy level diagrams to scale. In each of the diagrams above 1 cm = 10 kJ, but any convenient scale can be chosen.

Experiment 10.1 Measuring the heat change in a displacement reaction

(a) Shake a spatula-full of iron filings with about 2 cm³ of copper(II) sulphate solution in a test tube. Explain what you see.

Write a word equation for the reaction which has taken place. Write an equation in symbols for the reaction.

(b) Measure 25 cm³ of 0.2 M copper(II) sulphate solution into a polythene bottle. Fit a bung with a thermometer into the bottle, turn the bottle upside-down and measure the temperature of the solution. Write it down.

Add 2 spatulas-full of iron filings to the copper(II) sulphate solution in the bottle (this is more than enough metal to react completely with all the solution). Shake the bottle well, fit the thermometer, turn the bottle upside down, and make a note of the maximum temperature reached.

What was the temperature rise of the solution?

How much heat energy was produced in the experiment? (Assume that it takes 4.2 joules to raise the temperature of 1 cm³ of the mixture in the bottle by 1°C.)

How many moles of copper(II) ions are present in 25 cm³ of 0.2 M copper(II) sulphate solution?

How many moles of copper must have been produced in the experiment?

How much energy would have been produced if sufficient copper(II) sulphate solution had been used to produce 1 mole of copper?

Draw an energy level diagram for the reaction.

(c) Repeat experiments (a) and (b) using powdered zinc in place of iron filings. Draw an energy level diagram for this reaction. Explain why the heat of reaction of zinc with copper(II) sulphate is greater than the heat of reaction of iron with copper(II) sulphate.

Industrial fuel gases

Producer gas and *water gas* are two important industrial fuels. Producer gas is a mixture of carbon monoxide and nitrogen, made by reacting air with coke (carbon). The reaction is exothermic:

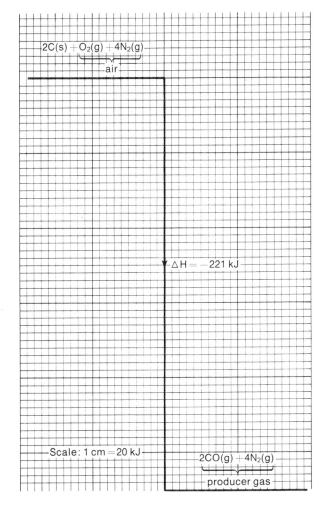

$2C(s) + O_2(g) + 4N_2(g)$ — air

$\Delta H = -221$ kJ

Scale: 1 cm = 20 kJ

$2CO(g) + 4N_2(g)$ — producer gas

Energy in chemistry

Water gas is a mixture of hydrogen and carbon monoxide, made by reacting steam with coke. This is an endothermic reaction:

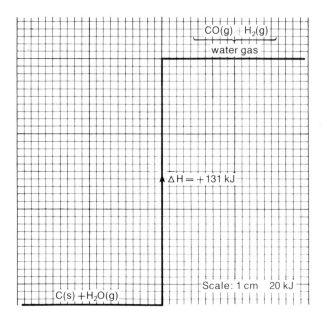

These two fuel gases are made in the same furnace. Air is blown through coke which is red-hot (about 1000°C). Producer gas is formed and, since the reaction is exothermic, the temperature of the coke rises. When it is white-hot (about 1400°C) the air supply is cut off and steam is blown through the coke. Water gas is formed and the endothermic reaction causes the temperature of the coke to fall. When it drops to about 1000°C the steam is shut off and air is blown through again. By alternating air and steam both fuel gases can be made without the need for any outside supply of heat.

In some industrial plants a modified process is used. A mixture of air and steam is blown through the coke in just the right proportions so that the heat given out in the reaction of coke with air is just balanced by the heat taken in by the reaction of coke with steam. A mixture of carbon monoxide, hydrogen, and nitrogen is produced, which is called *semi-water-gas*.

When producer gas is burnt as a fuel the carbon monoxide burns to carbon dioxide:

$$2CO(g) + O_2(g) \rightarrow 2CO_2(g).$$

Only about one-third is carbon monoxide (the rest is mainly nitrogen), so it does not have a very high calorific value (about 3.4 kJ per gram). It is always made on the industrial site where it is to be used, for instance to melt steel or to heat kilns for firing bricks or pottery.

Water gas is a much better fuel, with a calorific value of 17.5 kJ per gram, since both the hydrogen and the carbon monoxide burn:

$$\underbrace{CO(g) + H_2(g)}_{\text{water gas}} + O_2(g) \rightarrow CO_2(g) + H_2O(g).$$

It is used as an industrial fuel gas and also as a source of hydrogen (though nowadays it is less important as a source of hydrogen, because of the increase in the production of hydrogen from petroleum).

Carbon monoxide is a very poisonous gas. It combines with haemoglobin, the chemical in red blood cells which carries oxygen round the body. Once haemoglobin has combined with carbon monoxide it can no longer do its job properly. Town gas made from coal or oil contains a small proportion of carbon monoxide, and it is this which makes it poisonous. Natural gas contains no carbon monoxide, and is much safer.

There is a danger of carbon monoxide being formed when hydrocarbons burn. With plenty of oxygen all the carbon in a hydrocarbon molecule burns to carbon dioxide, but if there is not a sufficient supply of oxygen, combustion of the hydrocarbon is incomplete and some carbon monoxide is formed. Car engines do not burn the hydrocarbons in petrol completely, and their exhaust fumes contain carbon monoxide. This is why it is dangerous to run a car engine in a closed space, such as a garage. But even in the open air, pollution of the atmosphere by carbon monoxide from car exhausts is becoming a serious problem, especially in big cities.

Questions

1 The burning of coke and the digestion of fat are both said to be exothermic reactions. Say in one sentence what we mean by an exothermic reaction. *(Wales)*

2 When 6.5 g of zinc are reacted with 100 cm^3 of 5 M hydrochloric acid all the zinc dissolves and the temperature rises by 22°C.

(a) How many kilojoules were produced in this reaction? (1 cm^3 of 5 M hydrochloric acid is heated through 1°C by 4 joules.)

(b) How many kilojoules would have been produced by the reaction of 1 mole of zinc with excess hydrochloric acid? (Atomic mass of zinc = 65)

(c) Draw an energy level diagram for the reaction.

3 When 160 g of anhydrous copper(II) sulphate (CuSO$_4$) is dissolved in 1000 cm^3 of water the temperature rises by 15.7°C. When 250 g of hydrated copper(II) sulphate (CuSO$_4$.5H$_2$O) is dissolved in 1000 cm^3 of water the temperature falls by 2.7°C.

(a) Which of these reactions is exothermic and which is endothermic?

(b) Calculate the heat change for each reaction in kilojoules per

mole of copper(II) sulphate. (Atomic masses: Cu = 64, S = 32, O = 16, H = 1. 1 cm³ of water is heated through 1°C by 4 joules.)
  (c) Draw energy level diagrams for both reactions.
  (d) Why is the heat change for anhydrous copper(II) sulphate different from the heat change for hydrated copper(II) sulphate?

4 (a) Describe an experiment to determine the heat of reaction of some given metals with an acid.
  (b) How would these results enable you to place these metals in the activity series? (*Wales*)

5 Petrol is a hydrocarbon substance. A typical molecule present in petrol is Heptane C_7H_{16} which burns in oxygen according to the following equation.

$$C_7H_{16}(l) + 11O_2(g) \rightarrow 7CO_2(g) + 8H_2O(g)$$
$$\Delta H = -4430 \text{ kJ/mole}$$

  (a) Draw an energy level diagram for this reaction.
  (b) What does the symbol (g) after H_2O indicate?
  (c) The type of energy change that has occurred in the reaction is which of the following?
  A endothermic   D exothermic
  B calorific   E electrical
  C thermal
  (d) When a petrol engine is started from cold, a colourless liquid is seen to drip from the exhaust pipe. The liquid is best described as which of the following?
  A unburned petrol
  B water from the radiator
  C engine oil
  D water formed by the burning petrol
  E antifreeze
  (e) Explain your answer to part (d).
  (f) What is the weight of a mole of heptane? (C = 12, H = 1.)
  (g) How many kJ are produced by burning 10 g of heptane?
  (h) When petrol burns in an engine, a quantity of carbon is formed in the cylinders, and less energy is released than the equation states. Suggest a reason for these observations.
  (i) How might this be overcome? (*London*)

6 The sugar glucose burns in oxygen according to the equation

$$C_6H_{12}O_6(s) + 6O_2(g) \rightarrow 6CO_2(g) + 6H_2O(g)$$
$$\Delta H = -2834 \text{ kJ/mole}.$$

  (a) Draw an energy level diagram to represent this change.
  (b) This type of reaction is best described as which of the following?

A isothermal D endothermic
B electrothermal E calorific
C exothermic

(c) What is the weight of a mole of glucose? (C = 12, H = 1, O = 16)

(d) What is the heat change if 1 mole of glucose is burned?

(e) Glucose is often eaten as a food. How is the chemical energy stored in the glucose released?

(f) What effect does this produce in the animal that has eaten the glucose?

(g) Plants make glucose from carbon dioxide and water. Draw an energy level diagram to represent this change.

(h) Where has the energy stored in the glucose been obtained from?

(i) On a sunny day bubbles can be seen rising from pond weed. What gas is being produced? (*London*)

7 (a) Name the two gases obtained by passing air through a mass of hot coke. What is the name given to this mixture of gases?

(b) Name the two gases obtained by passing steam through white hot coke. What is the name given to the mixture of these gases?

(c) The two processes referred to in parts (a) and (b) are usually carried out alternately at seven-minute intervals in the same furnace on an industrial scale. Explain the reason for this. (*West Midlands*)

8 (a) Why is it dangerous to run a car engine in a closed garage?

(b) (i) Water gas is a mixture of equal volumes of carbon monoxide and hydrogen. Why is it called water gas?

(ii) When burnt, water gas gives out more heat than does an equal volume of producer gas. What is the reason?

(iii) Give *two* advantages of gaseous fuels as compared with solid fuels. (*East Anglia*)

.4 Chemical energy into electricity: cells

If a piece of zinc is dipped into a solution of copper(II) sulphate an exothermic reaction takes place. The more active metal displaces the less active one:

zinc + copper(II) sulphate → zinc sulphate + copper
$Zn(s)$ + $CuSO_4(aq)$ → $ZnSO_4(aq)$ + $Cu(s)$.

Both copper(II) sulphate and zinc sulphate are electrolytes:

$CuSO_4(aq)$ = $Cu^{2+}(aq)\ SO_4^{2-}(aq)$;
$ZnSO_4(aq)$ = $Zn^{2+}(aq)\ SO_4^{2-}(aq)$.

The sulphate ions do not play any part in the reaction. The ionic equation shows that this is simply a reaction between zinc atoms and copper(II) ions in which the positive charge from the copper ions is transferred to the zinc atoms:

$$Zn(s) + Cu^{2+}(aq) \rightarrow Zn^{2+}(aq) + Cu(s).$$

A simple cell

If the zinc and copper(II) sulphate solution are arranged as in Figure 10.5 the voltmeter registers that electrical energy is being produced. This energy is coming from the same reaction, even though the zinc and copper(II) sulphate are not in direct contact.

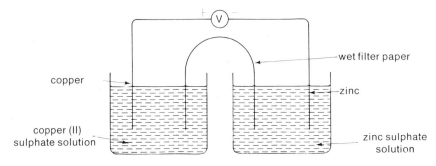

Figure 10.5 A simple cell

At the zinc rod zinc atoms become zinc ions:

$$Zn(s) \rightarrow Zn^{2+}(aq) + 2e.$$

The electrons produced at the zinc rod travel through the outside wire as an electric current to the copper rod, where they convert copper ions to copper atoms:

$$Cu^{2+}(aq) + 2e \rightarrow Cu(s).$$

This is called a CELL. A cell is an arrangement by means of which an exothermic chemical reaction produces its energy as electricity instead of as heat. (Cells are sometimes incorrectly called batteries. A battery is several cells joined up in series.)

The cell based on the reaction between zinc and copper(II) sulphate is called the *Daniell cell*, after its inventor. It produces a maximum voltage of 1.1 volts. Other similar cells can be made using different metals.

Experiment 10.2 Investigating some simple cells

(a) Half-fill one beaker with copper(II) sulphate solution and another with zinc sulphate solution. Place the beakers side by side and hang a strip of wet filter paper across the two beakers so that it dips into both solutions. Stand a piece of copper foil in the copper(II) sulphate solution and a piece of zinc foil in the zinc sulphate solution.

By means of two wires with crocodile clips on the end connect a voltmeter between the copper foil and the zinc foil. (Make sure the crocodile clips do not dip into the solution.) If there is no reading on the voltmeter you have probably connected it the wrong way round: try it the other way. Make a note of the voltage produced by this cell.

Which electrode is negative and which is positive? (You can check this from the terminals on the voltmeter. They will either be marked + and −, or they will be coloured red for positive and black for negative.)

Which is the negative electrode, the metal which is higher in the activity series or the metal which is lower?

What do you think is the function of the piece of wet filter paper?

(b) Repeat the experiment with magnesium sulphate solution and a piece of magnesium ribbon in one beaker and copper(II) sulphate solution and a piece of copper foil in the other. Clean the magnesium ribbon with sandpaper (why?).

Use your knowledge of the activity series to decide which electrode will be positive and which negative before you connect the voltmeter. Make a note of the voltage.

(c) This time use a solution of magnesium sulphate and magnesium ribbon (cleaned with sandpaper) in one beaker and zinc sulphate solution and a piece of zinc foil in the other. Decide which electrode will be positive and which negative before you connect the voltmeter. Make a note of the voltage.

Can you see any connection between your results for experiments (a), (b), and (c)? Try to explain this.

(d) If you have time, measure the voltages produced by two or three other cells, involving different metals.

The dry cell

The Daniell cell and other similar cells have the disadvantage that they are rather messy, since they have liquid electrolytes. The DRY CELL, the type used in torches, transistor radios, etc., is shown in Figure 10.6. It produces 1.5 volts.

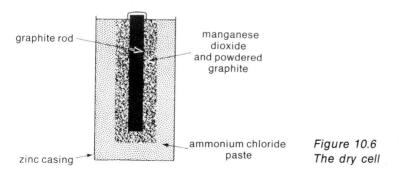

Figure 10.6
The dry cell

Energy in chemistry 331

The zinc casing is the cathode, where electrons are produced just as in the Daniell cell:

$$Zn(s) \rightarrow Zn^{2+}(aq) + 2e.$$

The electrolyte is a jelly-like paste of ammonium chloride. The anode is a graphite rod in contact with manganese(IV) oxide. At the anode manganese(IV) ions combine with electrons to become manganese(III) ions:

$$Mn^{4+}(s) + e \rightarrow Mn^{3+}(s).$$

The full ionic equation for the main reaction which takes place inside the dry cell is:

$$Zn(s) + 2Mn^{4+}(s) \rightarrow Zn^{2+}(aq) + 2Mn^{3+}(s).$$

Experiment 10.3 Investigating the dry cell

(a) Measure the voltage produced by a dry cell.

(b) Half-fill a beaker with ammonium chloride solution. Attach two wires with crocodile clips to a voltmeter. Fix a piece of zinc foil into one crocodile clip and a piece of manganese(IV) oxide into the other. Dip the zinc foil and manganese(IV) oxide into the ammonium chloride solution. Make a note of the voltage. (If the voltmeter does not register, reverse the leads.) Is this the same voltage as is obtained from a dry cell?

Experiment 10.4 Investigating the lead accumulator

(a) Fill an electrolysis cell (Figure 5.3, page 112) with dilute sulphuric acid. Pass a current through the cell until both tubes are full of gas. What do you think the gases in the tubes are? Carry out tests to check if you are correct.

(b) Half-fill a beaker with 4M sulphuric acid. (**CARE: this acid is quite concentrated.**) Electrolyse the acid for 5 minutes, using strips of lead foil as electrodes.

What do you see happening at the cathode? What gas do you think is being given off here?

What do you see happening at the anode? From your observations in experiment (a), what do you think the dark brown deposit on the anode might be?

(c) Connect a voltmeter between the two strips of lead foil. What voltage does this cell produce? Which is the positive electrode and which is the negative electrode?

(d) Connect a lamp bulb between the two strips of lead foil. For how long does the lamp remain lit? After the lamp has gone out, measure the voltage of the cell again.

Describe the appearance of the two electrodes.

(e) If you have time, electrolyse the acid for 5 minutes again, making the lead foil from which gas was produced the negative electrode and the lead foil which went brown the positive electrode. Describe what happens to the appearance of the electrodes. Measure the voltage and connect a lamp bulb between the electrodes once more.

What is the advantage of this cell over the others which you have studied?

The lead accumulator

The cathode in this cell is a plain lead plate, the anode is another lead plate coated with lead(IV) oxide, and the electrolyte is sulphuric acid.

At the cathode lead atoms lose electrons, forming lead(II) ions which appear as a deposit of insoluble white lead(II) sulphate on the cathode plate:

$$Pb(s) \rightarrow Pb^{2+}(s) + 2e.$$

At the cathode these electrons convert the lead(IV) ions in the lead(IV) oxide into lead(II) ions, changing the brown deposit of lead(IV) oxide on the cathode plate to a white deposit of lead(II) sulphate:

$$Pb^{4+}(s) + 2e \rightarrow Pb^{2+}(s).$$

The overall reaction in the lead accumulator can therefore be represented by the equation:

$$Pb(s) + Pb^{4+}(s) \rightarrow 2Pb^{2+}(s).$$

Since the products at the anode and cathode remain attached to the electrodes, the reaction in this cell is quite easy to reverse. When a lead accumulator has run down because the chemical reaction is complete, an electric current is passed through the cell in the opposite direction. This provides the energy to reverse the reaction:

$$2Pb^{2+}(s) \rightarrow Pb(s) + Pb^{4+}(s).$$

The lead(II) sulphate is removed from the cathode, the lead(II) sulphate on the anode is converted into lead(IV) oxide, and the cell is ready to produce electricity again.

The lead accumulator is therefore useful for storing electricity, hence its use in a car. Current is drawn from the battery to start the car, and it is then recharged by the dynamo (driven by the engine) while the car is running. A single lead accumulator cell produces 2.05 volts when fully charged. A normal 12 volt car battery consists of six of these cells connected in series.

Questions

1 (a) Draw a diagram to show how you would arrange a piece of magnesium ribbon, a piece of silver foil, some silver(I) nitrate

solution, and some magnesium(II) nitrate solution to form a simple cell.

(b) Mark on your diagram which metal will be the anode and which the cathode.

(c) Explain what happens at the anode when the cell is producing current.

(d) Explain what happens at the cathode when the cell is producing current.

(e) Write an ionic equation to describe the overall reaction which is taking place inside the cell.

2 Take an old dry cell to pieces. Where is the zinc? Where is the manganese(IV) oxide? Where is the ammonium chloride? What do you think is the function of the carbon rod in the middle of the cell?

Draw a diagram of the cell, labelling all of these main parts.

3 Lead-acid type cells have been in common use for many years but now their use is mainly restricted to car batteries.

(a) Why is lead a suitable metal for this purpose?

(b) What is a disadvantage of using lead for this purpose?

(c) When the cell is fully charged one plate has a brown colour. Why? *(East Anglia)*

10.5 Getting work out of fuels

Fuels are useful because they are convenient stores of energy which we can release for our own use. Sometimes we use a fuel simply for the heat which it provides when it is burnt, like coal on a living-room fire or gas in the kitchen cooker. But often we want to transform the stored chemical energy of a fuel into other forms. Petrol is burnt in a car engine to produce the mechanical energy to drive the car. Coal is burned in a power station to produce electricity. In these cases the energy may have to go through several transformations to reach the form we require.

The car engine

In a car engine the petrol burns in the cylinders and produces heat energy. Hot gases, mainly steam and carbon dioxide, are produced in the cylinders and their fast-moving molecules push the pistons down the cylinders. This motion is transmitted to the crankshaft which turns the wheels, making the car move. The sequence of energy transformation is:

$$\text{chemical energy} \rightarrow \text{heat energy} \rightarrow \text{mechanical energy.}$$

Although energy cannot disappear during a transformation from one form into another it does not necessarily all change into the form one wants. During the burning of the petrol most of the chemical energy is turned into heat, but a little is wasted. Some sound energy is produced (you can hear the explosion of the petrol/air mixture) and some light energy is also produced (when the petrol explodes there is a flash). In the second transformation (heat energy into mechanical energy) there is much more wastage. Many of the hot gas molecules produced in the explosion collide with the top and sides of the cylinder, heating up the metal. Even the gas molecules which collide with the piston head only transfer part of their heat energy into mechanical energy to move the piston. So only a part of the heat energy produced from the fuel is transformed into mechanical energy. A considerable amount remains as heat and is wasted. This is why a car engine needs to be cooled by circulating water or air around the cylinder block.

There is still more wastage. However well lubricated the engine may be, there will still be some friction between the pistons and the sides of the cylinders, between the gear wheels, between the wheels and the road, and so on. Friction causes part of the mechanical energy to be changed back into heat energy and wasted. Overall, even in the most efficient car engine, only about 20 per cent of the chemical energy stored in the petrol is transformed into useful work to drive the car. The other 80 per cent is wasted, mainly as heat.

Power stations

In an electric power station there are even more energy transformations than in a car engine. The fuel (coal, oil, or gas) is burnt, changing chemical energy into heat energy. The heat is used to turn water into steam. The heat energy of the steam molecules is converted to mechanical energy in the turbines. The mechanical energy is transformed into electrical energy in the dynamos:

chemical energy → heat energy →
                    mechanical energy → electrical energy.

Just as in the car engine, at every stage in this process some energy is wasted by not being transformed into the kind that is required. A lot of the heat energy is not changed into mechanical energy in the turbines, but remains as wasted heat. Friction in the turbines and dynamos wastes some of the mechanical energy by changing it back into heat energy. The maximum efficiency of a power station is about 30 per cent. The other 70 per cent of the chemical energy stored in the fuel is wasted.

The human body

The human body is also a form of engine. Its fuel is food, particularly carbohydrates, which react with oxygen to produce energy (page 253). Some of the chemical energy is transformed into heat energy to keep the body warm. Some is transformed into mechanical energy for movement. And some is transformed into electrical energy for the electrical circuits of the brain and nervous system. The human body, however, runs at about the same efficiency as a car engine or power station. For every one kilojoule of mechanical energy produced in the muscles, between 3 and 4 kilojoules of extra heat energy are produced, so the body is less than 25 per cent efficient. This is why we get hot when we exert ourselves.

Electricity direct from chemical reactions

An electric cell, as we have seen, changes chemical energy direct into electrical energy. In the Daniell cell the reaction

$$Zn(s) + Cu^{2+}(aq) \rightarrow Zn^{2+}(aq) + Cu(s)$$

produces a maximum 1.1 volts. This means that when the cell is operating at maximum voltage 1.1 joules of energy are produced for every coulomb of electricity that passes.

This reaction involves the transfer of 2 faradays (192 000 coulombs) of positive charge from 1 mole of copper ions (Cu^{2+}) to 1 mole of zinc atoms (Zn). So when 1 mole of zinc reacts completely the amount of energy produced is:

$$1.1 \times 192\,000 = 211\,200 \text{ joules} = 211.2 \text{ kJ}$$

On page 324 we saw that if the same reaction is carried out in a test tube so that the chemical energy is transformed into heat, the energy produced is 217 kJ per mole of zinc. The amount of electrical energy produced in the cell is slightly less than the amount of heat energy produced in the test tube because in the cell some energy is lost as heat. The amount of energy wasted in the cell as heat is

$$217 - 211 = 6 \text{ kJ},$$

which is about 3 per cent of the total energy. The Daniell cell is therefore a 97 per cent efficient way of turning chemical energy into electrical energy, whereas a power station is only 30 per cent efficient. In practice it is not usually possible to run a cell at its maximum voltage and if it is run at a lower voltage the wastage of energy as heat is greater. Even so, the rate of conversion of chemical energy into electrical energy in a cell is much better than in a power station.

Ordinary cells, however, have one great disadvantage. The chemicals involved in their reactions are relatively expensive. These cells are useful where a small portable supply of electricity is needed and cost is not very important, as in torches, transistor radios, hearing aids, etc. But they could never compete as suppliers of electricity on a large scale, even though they are more efficient than power stations. It would cost a fortune to provide all the electricity you use in your home from batteries.

Fuel cells

A real advantage would be obtained if it were possible to react ordinary fuels such as coal, oil, and gas with oxygen in a form of cell so that electricity could be produced directly. Electricity could then be produced cheaply and efficiently.

This was first successfully achieved with the reaction between hydrogen and oxygen to form water. Hydrogen, obtained as a by-product of the refining of petroleum (page 280), is a reasonably cheap fuel. Hydrogen and oxygen (or air) are passed through separate porous tubes in a tank through which potassium hydroxide solution is circulating.

At the positive electrode (a platinum wire in contact with the oxygen gas):

$$O_2(g) + 2H_2O(l) + 4e \rightarrow 4OH^-(aq).$$

At the negative electrode (a platinum wire in contact with the hydrogen):

$$2H_2(g) + 4OH^-(aq) \rightarrow 4H_2O(l) + 4e.$$

The overall reaction in this fuel cell is:

$$2H_2(g) + O_2(g) \rightarrow 2H_2O(l) + \text{electrical energy}.$$

Other fuel cells have now been developed using, for instance, methane as a fuel, so that the overall cell reaction is:

$$CH_4(g) + 2O_2(g) \rightarrow CO_2(g) + 2H_2O(l).$$

Fuel cells are still in the early stages of development for use on a commercial scale, though they are used as a source of energy in spacecraft.

Figure 10.7 An experimental fuel cell using methanol as fuel. The overall reaction is $2CH_3OH + 3O_2 \rightarrow 2CO_2 + 4H_2O$.

Energy in chemistry

Questions

1 Indicate the energy changes that occur in a car engine when it drives a car along, starting with the chemical energy in the petrol.
(London)

2 What is the sequence of energy transformations in each of the following: (a) a steam locomotive, (b) a battery-operated electric toothbrush, (c) a nuclear power station, (d) a diesel lorry?
In each case say in which stage you think the biggest energy wastage occurs.

3 What are the advantages and disadvantages of producing electricity from dry cells as opposed to power stations?

For the teacher

The joule The Association for Science Education have recommended that the joule should be adopted as the energy unit used in school science, and that the use of other units such as the calorie and kilowatt-hour should be progressively abandoned (*SI Units, Signs, Symbols, and Abbreviations* ASE, 1970).

If pupils are not familiar with the joule from their work in physics, a short demonstration may be helpful. A 12-volt, 50-watt immersion heater is wired as shown in Figure 10.8, and used to heat 100 cm^3 of water in a copper calorimeter. The calorimeter should be insulated by standing in a block of expanded polystyrene or polyurethane foam or by lagging with cotton wool. The time taken to heat the water through 30°C is measured. The number of joules produced is calculated:

$$\text{joules} = \text{volts} \times \text{coulombs} = \text{volts} \times \text{amps} \times \text{seconds},$$

and hence the number of joules required to heat 1 cm^3 of water through 1°C. For the purposes of this experiment the thermal capacity of the calorimeter may be neglected.

Heat of combustion apparatus A form of the apparatus shown in Figure 10.3, page 320 is available from most manufacturers under the name of food calorimeter or Thiemann calorimeter. An oxygen cylinder is also required. Following the calculation on page 320, the apparatus is first calibrated using 0.5 g of wood charcoal dried by heating in a test tube and cooled in a dessicator. The calorific value of a fuel such as wood or coal can then be measured.

Demonstration of exothermic and endothermic changes (a) To 950 cm^3 of water in a 2000 cm^3 beaker add 1 mole (53.5 cm^3) of concentrated sulphuric acid, stirring with a thermometer, and record the rise in temperature (about 17°C). This produces approximately 1000 cm^3 of M solution.

(b) Dissolve 1 mole (80 g) of ammonium nitrate in 950 cm^3 of water

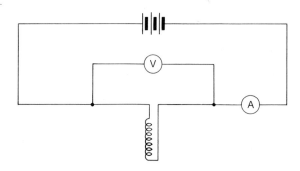

Figure 10.8

in a 2000 cm³ beaker, and record the fall in temperature (about 6°C). This produces approximately 1000 cm³ of M solution.

(c) To 500 cm³ of 2 M hydrochloric acid in a 2000 cm³ beaker add 500 cm³ of 2 M sodium hydroxide solution, and record the rise in temperature (about 13°C). This produces approximately 1000 cm³ of M sodium chloride solution. In each experiment it is helpful to let pupils feel the temperatures of the beakers before and after the reactions. The heat of reaction per mole of reactants can be calculated in each case and energy level diagrams drawn.

Experiment 10.1 Polythene bottles of about 100 cm³, fitted with bungs carrying − 10 to 110°C × 1°C or − 5 to 55°C × 1°C thermometers with the bulbs just protruding through the bungs.

Daniell cell demonstration A simple Daniell cell can be constructed from a small porous pot (about 10 × 3 cm) standing inside a 250 cm³ beaker. The porous pot is filled with M zinc sulphate solution and the beaker with M copper(II) sulphate solution. A strip of zinc foil stands in the porous pot and a strip of copper foil in the beaker. Ideally the maximum voltage should be measured with an electronic voltmeter or on a potentiometer so that the cell is producing zero current, but for the purposes of this experiment an ordinary voltmeter is accurate enough and should record between 1.0 and 1.1 volts.

Experiment 10.2 M solutions of copper(II) sulphate, zinc sulphate and magnesium sulphate. The experiment can be extended to other metal foils and molar solutions of sulphates if desired. One 0–3 V or 0–5 V DC voltmeter per bench is adequate for this experiment and for experiments 10.3 and 10.4.

Experiment 10.3 M ammonium chloride solution. Lumps of manganese dioxide (*pyrolusite*) can be obtained from Lythe Minerals or from R. F. D. Parkinson.

Experiment 10.4 For experiment (a) 2M sulphuric acid. For (b) a supply of 3 or 4 volts DC is required. For (d) a 2.5 V 0.3 amp bulb in a holder is suitable.

Fuel cell demonstration M sodium hydroxide solution is electrolysed at about 6 volts in an electrolysis cell (Figure 5.3, page 112) until both tubes are filled with gas. The supply is disconnected. A voltmeter is then used to identify the positive and negative terminals and to measure the voltage.

Energy in chemistry 339

11 Silicon

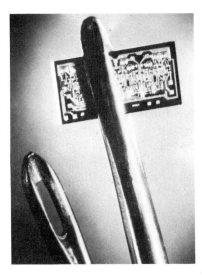

Figure 11.1 A silicon chip on which extremely complicated electrical circuits are made passes through the eye of an ordinary sewing needle

Silicon is the second most common element in the earth's crust (25.7 per cent by weight). It is never found as the free uncombined element, but only in the form of compounds with oxygen, together with other elements. Compounds of silicon have been used for thousands of years to make bricks, pottery and glass.

Because it is very difficult to separate from its compounds, the element itself was discovered only in 1823. In the last 30 years two new uses have been found for it. Very pure silicon crystals are manufactured for use as transistors. They have the property of conducting electricity in one direction, but not in the other. The earliest transistors were made of germanium, an element in the same group of the periodic classification as silicon, but most transistors are now made of silicon, which is very much cheaper. The other new use for silicon is in the manufacture of a remarkably useful group of chemical compounds called silicones.

11.1 Silicon compounds in the earth

Granite is a common type of rock, formed when molten material inside the earth cooled and solidified millions of years ago. It is a tough building stone, and it can be polished for use as headstones for graves or to decorate shop-fronts. Like all rocks in the earth's crust, granite is made up of several minerals, mainly *quartz*, *mica*, and *felspar*. These three minerals are all compounds of silicon. In addition to granite, three-quarters of all the rocks of the earth's crust contain silicon; it is one of the most important rock-forming elements.

Silica

The simplest compound of silicon in the earth is silicon oxide (SiO_2), commonly called *silica*. Silica occurs in crystalline form as *quartz*, which forms long hexagonal crystals. They are usually about 5 cm long, but in 1959 a quartz crystal weighing 13 tonnes was found in Siberia. It was as tall as a two-storey building.

Silica is also found in forms with no definite crystalline shape. *Flint* is an example of this type of silica. It was used for weapons and tools by Early Man, and also in flint-lock guns. Another non-crystalline form of silica, called *rock-crystal*, is a clear glass-like material, used for fortune-tellers' crystal balls.

Chemical impurities in silica can produce many different colours. For example, traces of iron compounds in crystalline silica produce the deep purple mineral called *amethyst*, which is a semi-precious stone used in jewellery. Other examples of semi-precious forms of silica are *cornelian*, *onyx*, *agate*, *jasper*, *rose quartz*, *citrine*, and *opal*. Opal can be coloured blue, red, orange, green, or black. This variation in colour is caused by varying amounts of water of crystallisation.

Silicates

Mica and felspar, the other two minerals in granite, look very different from each other. *Mica* splits into very thin leaves, and may be black or transparent. *Felspar* is more solid, and may be white or pink in colour. Both these minerals contain silicon, combined with oxygen and various metals, mainly aluminium. They are two members of a family of chemical compounds called SILICATES.

Differences between silicates are due partly to their chemical composition and partly to the way the atoms are arranged. The basic arrangement is one silicon atom and four oxygen atoms in the shape of a tetrahedron (Figure 11.2). In *mica* the tetrahedra are arranged in sheets (Figure 11.3), which are only weakly connected together, rather like the sheets of atoms in graphite (page 248). Because of the arrangement of the atoms, mica will easily split along its lines of weakness. It can grow into very large sheets

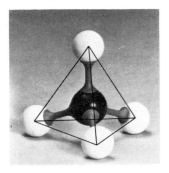

Figure 11.2 A model of the basic tetrahedron of silicon and oxygen atoms from which the complex silicates are built up

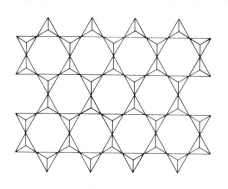

Figure 11.3 Part of a sheet of SiO$_4$ tetrahedra in mica

and in Northern India colourless mica is mined to make fire-proof windows for furnaces and insulators for electrical apparatus. Mica is also used as Christmas tree snow.

Felspar has a more complicated three-dimensional arrangement of its tetrahedra and it also contains atoms of other elements such as aluminium, sodium, and potassium. This more complicated structure makes felspar a stronger mineral than mica. It is used in pottery glazes and enamel and as a mild abrasive.

Mica and felspar are just two of the hundreds of different silicates found in the earth's crust. A few of the others are listed in Figure 11.4. These minerals have all been solidified from molten rock while the earth was cooling down; they are called IGNEOUS ROCKS.

Name	Uses
Asbestos	Building materials, fire-proofing
Felspar	Abrasives, enamel, pottery glazes
Garnet	Abrasives, jewellery, sand-blasting
Mica	Electrical insulators, furnace windows
Olivine	Firebricks, pastes for blast furnace repairs
Palygorskite	Paints, putty
Talc	Cosmetics, crayons, wallpaper
Vermiculite	Insulation, paints.

Figure 11.4

The weathering of rocks

When igneous rocks become exposed at the surface of the earth they are WEATHERED. This means that they are broken up by water, ice, and wind. The hardest minerals, like quartz, do not break up as much as the softer minerals, such as mica and felspar.

Many of the rocks of the Sahara Desert are granite. The sand dunes are made of tiny pieces of broken quartz which were left behind as the granite weathered. In wetter lands the hard quartz is carried by rivers to the sea where it forms the sand of our beaches. If sand collects on the sea bed in great thicknesses it may be compressed to form a rock called *sandstone*.

The felspar in granite has sometimes been broken down by hot gases produced inside the earth when the granite was formed millions of years ago. When felspar is broken down in this way *kaolin*, sometimes called *china clay*, is formed. This is a white clay which is mined mainly on the granite moors of Devon and Cornwall for use in the pottery industry of Staffordshire. The china clay is washed out by high-powered hoses, and the unwanted white quartz and shiny mica are left in tall conical waste heaps. Other types of soft silicate minerals weather to form other, less pure, kinds of clay.

Questions

1 How was granite formed in the earth? What is the name given to rocks like granite which were formed in this way? Give an example of a type of rock which was formed in a quite different way.

2 With the exception of diamond, most precious and semi-precious stones contain silicon, together with other elements. Make a list of some of these minerals and their colours (looking in a jeweller's window may help).

3 Carbon and silicon are in the same group of the periodic classification. Why do you think silicon dioxide is a solid whereas carbon dioxide is a gas at room temperature?

Glass

Glass is a most unusual material. One of the few other substances like it is toffee. Toffee is made by heating sugar and other ingredients together until they melt. When the melted toffee is cooled it slowly becomes stiffer and stiffer, but there is no point at which one can say it has actually solidified. Glass is also made by melting together several solid ingredients. The hot liquid glass is then cooled slowly and crystallisation never occurs. The molecules never arrange themselves in a regular pattern, so glass is described as a SUPERCOOLED LIQUID. It appears to be solid, but a very old glass window pane may be thicker at the bottom than at the top; the glass is still flowing like a liquid, but very, very slowly.

The manufacture of glass

The basic ingredient of most glass is silicon oxide (*silica*), in the form of sand. Glass can be made by melting pure silicon oxide, but this kind of glass is expensive to make because silicon oxide needs a temperature of at least 1700°C to melt it.

Glass can be made at a lower temperature by adding a FLUX to the sand. The commonest flux is sodium carbonate. A mixture of sand and sodium carbonate melts at about 800°C. The two substances react together to form sodium silicate, sometimes called *waterglass*:

$$SiO_2(s) + Na_2CO_3(s) \rightarrow Na_2SiO_3(l) + CO_2(g).$$

Waterglass, as the name suggests, is soluble in water. It is used for preserving eggs, fire-proofing wood, and in washing powders (page 308).

To make ordinary, insoluble glass for bottles and jars a STABILISER is added, usually calcium carbonate (*limestone*). A mixture of 50 per cent sand, 15 per cent sodium carbonate, 10 per cent limestone, and 25 per cent scrap glass, is heated to about 1500°C. Gradually the ingredients melt and react to form a mixture of sodium silicate and calcium silicate, which is glass. At one time the chemicals were melted together in small clay pots. Nowadays a TANK FURNACE is used which may have a capacity of up to 2000 tonnes. The mixture of ingredients is fed in continuously at the top and a continuous flow of molten glass is tapped off from the bottom.

Types of glass

Glass made from sand, sodium carbonate, and limestone is the cheapest type, used for bottles, tumblers and window panes. It is called SODA-GLASS.

LEAD GLASS is made in the same way as soda-glass but uses lead oxide in place of sodium carbonate. It is much more expensive than soda-glass but it has a higher refractive index: it bends light more and is therefore an attractive glass for making decorative glassware such as wine glasses, fruit bowls, flower vases, and chandeliers. Lead glass windows are used in nuclear power stations and radioactivity laboratories because the large lead atoms absorb radiation.

BOROSILICATE GLASS is made from a mixture of silicon oxide and boron oxide. This type of glass will withstand rapid heating and cooling without breaking. It is used for kitchen-ware and laboratory apparatus. The best-known kind goes under the trade-name *Pyrex*.

COLOURED GLASS is made by adding small quantities of other chemicals to the basic ingredients. Blue glass contains traces of cobalt oxide or copper oxide, green glass contains chromium oxide, and so on.

Working with glass

The traditional way of forming glass is by mouth-blowing. A *gob* of glass is gathered by dipping the blowing-iron into a pot of molten glass. The glassmaker blows down the blowing-iron and shapes the glass by controlling his breath, turning the blowing-iron, and shaping with tools. Some glassware, especially decorative objects and complicated laboratory apparatus, is still made by mouth-blowing.

Simple articles, like bottles, are now made by a completely automatic process. The bottle machine cuts off a gob of molten glass of the correct size, drops it into a mould to form the basic shape, transfers it to the automatic blower, and after it has been blown carries it off on a conveyor belt for cooling (Figure 11.5).

Sheet glass may be made by continuously drawing molten glass upwards from a tank. Patterned sheet glass is made by rolling the molten glass between water-cooled rollers. The rollers are embossed to give a

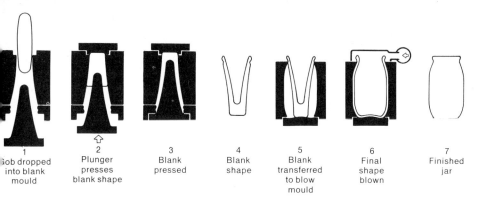

1	2	3	4	5	6	7
Gob dropped into blank mould	Plunger presses blank shape	Blank pressed	Blank shape	Blank transferred to blow mould	Final shape blown	Finished jar

Figure 11.5 The stages in the manufacture of a jar

surface pattern to the glass. A modern method of making sheet glass is to allow the molten glass to float across the surface of a bath of molten tin. This produces sheets with very flat and parallel surfaces, suitable for plate glass windows.

Fibreglass

Glass fibres were first made by the ancient Egyptians. But it was not until about 1930 that these fibres were produced on a commercial scale, when the usefulness of fibreglass as an insulating material was discovered. To make the tangled mass of glass fibres used for insulation, molten glass direct from the furnace flows into a rapidly rotating steel spinner. The molten glass is thrown outwards through small holes in the side of the spinner and the fibres fall at random to form a tangled mat on the conveyer belt (Figure 11.6). Glass wool is used for heat, cold, and sound insulation. It is also used in air filters.

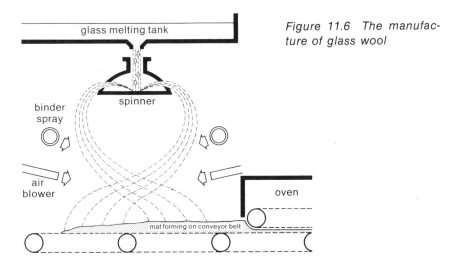

Figure 11.6 The manufacture of glass wool

Silicon 345

Glass fibre can also be produced in the form of a continuous thread instead of in a tangled mat. Molten glass falls through a series of fine holes in the bottom of the furnace and as the glass threads cool they are wound on to bobbins. Glass thread can be woven into fabric which is strong, will not shrink or stretch, and is resistant to fire, chemical attack, and moisture. But the biggest single use for glass thread is to reinforce plastics and rubber to make car bodies, boat hulls, furniture, and prefabricated building units.

Enamels

ENAMEL, sometimes called VITREOUS ENAMEL or PORCELAIN ENAMEL, is a form of glass fused on to metals, particularly iron and steel. It is very hard and scratch-proof and resists corrosion and staining. At home you will find the kitchen cooker enamelled, both outside and inside the oven. You may also have an enamelled bath, refrigerator, or washing machine.

Enamel, like glass, is a mixture of silicates. It is made by melting together sand, sodium carbonate, and boron oxide, together with small quantities of other materials. The hot molten glass is poured into water, which shatters it into small pieces known as FRIT. The frit is ground to powder and mixed with felspar and water to a thin cream. This is applied to the cleaned metal surface either by spraying or dipping. The enamel is fused to the metal by passing it through a furnace at about 700°C. The final coating is usually between $\frac{1}{1000}$ cm and $\frac{1}{100}$ cm thick.

Questions

1 Soda-glass is made from sand (silicon oxide), sodium carbonate, and limestone (calcium carbonate). What is the purpose of (i) the sodium carbonate and (ii) the limestone?

2 (a) What are the main differences in properties between soda-glass, lead glass, and borosilicate glass (*Pyrex*)?
 (b) Which of these types of glass would be most suitable for making (i) a cut-glass flower vase, (ii) a milk bottle, (iii) a chemical flask, (iv) a window through which an experiment involving radioactive chemicals is to be watched, (v) glass for a picture frame?

3 Explain briefly how each of the following glass objects could be made: (a) a milk bottle, (b) a plate-glass shop window, (c) a jug.

4 Make a chart using pictures cut out of old magazines, newspapers, etc. to illustrate some of the things that can be made from fibreglass.

3 From skyscrapers to porcelain

Bricks

Bricks are made from various types of *clay* (metal silicates from weathered rocks). The clay is sometimes obtained as a by-product of other mining operations, or it may be specially mined. It is crushed between rollers and mixed with water to a suitable consistency for moulding. At one time moulding was carried out by hand but nowadays most bricks are made by fully automatic processes. Either the wet clay is fed by machines into individual moulds or it is squeezed into a huge block which is then cut by wires into bricks of the required size. The wet bricks are stacked on cars running on rails. They are first carried through a drying tunnel where they are heated gently to dry out most of the water. Then they run through the firing kiln, in which the temperature is gradually raised to over 1000°C. In the firing process water molecules which were trapped between the silicon-oxygen tetrahedra of the silicates are driven out of the clay. Once removed, this water cannot be replaced: firing is a chemical reaction. The colour of the final bricks depends on the chemical composition of the clay and on the firing temperature.

Bricks are the oldest manufactured building material. They have been found on the sites of the ancient cities of Babylonia, some of which are estimated to be over 6000 years old. This shows how resistant they are to corrosion and decay. In Britain about 10 000 000 000 bricks are made each year, in various shapes and sizes for all kinds of constructional purposes.

Cement

Cement is manufactured by heating together clay and calcium carbonate (*limestone*) in a CEMENT KILN (Figure 8.13, page 237). Some cement is also obtained as a by-product in the manufacture of sulphur dioxide from *anhydrite* (page 237).

Clay is a mixture of complex aluminium silicates, which are composed of aluminium oxide (Al_2O_3) and silicon oxide (SiO_2), combined chemically in various proportions. Heating the clay with calcium carbonate produces a mixture of calcium silicate and calcium aluminate:

$$2CaCO_3(s) + \underbrace{SiO_2(s) + Al_2O_3(s)}_{\text{clay}} \rightarrow$$
$$\underbrace{CaSiO_3(s) + CaAl_2O_4(s)}_{\text{cement}} + 2CO_2(g).$$

Cement is a mixture of calcium silicate and calcium aluminate, together with smaller quantities of other silicates. In use it is mixed to a paste with sand and water. As a result of chemical reactions between the water, calcium silicate, and calcium aluminate the mixture sets into a hard mass.

Concrete

Concrete is a mixture of cement with an AGGREGATE. Various kinds of aggregate are used, such as gravel, crushed brick or stone, limestone, or blast furnace slag. When the cement sets it binds the aggregating material together to give a strong, cheap building material.

Concrete may be further strengthened by forming it over a network of steel rods. This is called FERRO-CONCRETE or REINFORCED CONCRETE. PRE-STRESSED CONCRETE is even stronger. It is formed round flexible steel rods which are stretched or twisted so that the whole concrete block is held under compression.

Mortar

Mortar, used for joining bricks together, is of various types. The simplest kind, called CEMENT MORTAR, is a mixture of cement and sand. The greater the proportion of cement the stronger is the mortar, but the more easily it cracks due to slight movement of the brickwork. The most frequently used type is LIME MORTAR, made with a mixture of cement, sand, and calcium hydroxide (*slaked lime*). This type gives a much stronger bond between the bricks and makes the brickwork more resistant to penetration by rain. Mortar sets in the same way as plain cement, as a result of a chemical reaction between the cement and the water.

Pottery and china

Chemically there is very little difference between pottery, china, and bricks. All are composed mainly of a mixture of metal silicates from which the trapped water is driven by firing, giving a hard product, resistant to chemical attack. The main difference is in the quality of the clay used. For pottery and china various kinds of pure white clay are used, particularly *kaolin* or *china clay* (page 342).

Pottery is made from a mixture of 65 per cent of fine clay and 35 per cent of crushed *flint*, blended to a paste with water. The paste is squeezed into slabs in a press and then kneaded to drive out air bubbles. It is shaped in a mould or on the potter's wheel and the articles are loaded into trucks for running through the kiln, where they are fired for two or three days at over 1000°C. Decoration is added by spraying on paint, with transfers, or by hand painting.

At this stage the pottery is still porous to water. The final stage of manufacture is coating with a GLAZE, which seals the clay and prevents water seeping into it. Glazing is a process rather like enamelling. The article is coated with a paste of crushed frit, felspar and water and it then goes through a second kilning for about 30 hours. This dries out the glazing paste and melts the frit on to the surface of the pottery.

The best quality china is called BONE CHINA. This is also made from fine clay, but mixed with 50 per cent of burnt animal bones instead of flint. The manufacturing process is the same as for pottery.

Questions

1 Bricks are made by removing the water from a paste of clay and water. Why does rain not turn bricks back into a clay paste?

2 (a) Cement is a mixture of calcium and aluminium silicates. What two substances are heated together to manufacture cement?
  (b) What is used to supply the heat?
  (c) Fill in the blanks. Concrete is made by mixing cement and _____, then folding in _____. If a mesh of steel rods is held in the mould before the concrete is poured in, it is then known as _____ concrete. (London)

3 A mixture of calcium hydroxide, sand and water is called which of the following?

A cement B mortar C quicklime D limestone E Plaster of Paris
(North West)

4 What difference is there between the raw materials used in the brick industry and those used in the pottery industry?

4 Silicones

Silicones are a group of man-made compounds in which the element silicon is combined with carbon, hydrogen, and oxygen in big polymer molecules. The name *silicone* for this group of compounds was coined by Professor F. S. Kipping, who discovered them accidentally in 1910. For over 30 years no one showed much interest in Kipping's discovery, until two American companies began to take an interest in silicones as a bonding material for fibreglass. During their investigations the research chemists of these companies found that silicones had all sorts of unexpected properties. Nothing would stick to them, they stood up to great extremes of heat and cold, they repelled water, they suppressed foam, and they were excellent electrical insulators. Gradually these remarkable properties were exploited and today all sorts of new uses are being found for these substances.

The chemistry of silicones

The starting materials for the manufacture of silicones are silicon and methyl chloride. Silicon is manufactured by heating silicon oxide (*sand*) with carbon (*coke*) at 2000°C:

$$SiO_2(s) + 2C(s) \rightarrow Si(s) + 2CO(g).$$

The raw materials are reasonably cheap but electric furnaces are required to reach the necessary temperature, so silicon is manufactured in countries where cheap hydro-electric power is available (Sweden, France, Canada, and the United States). Methyl chloride is manufactured from the methane in natural gas and chlorine:

$$CH_4(g) + Cl_2(g) \rightarrow CH_3Cl(g) + HCl(g).$$

Silicon and methyl chloride are heated together to produce a gas which is a mixture of the three methylchlorsilanes:

$$\underset{Cl}{\underset{|}{Cl-\underset{|}{\overset{CH_3}{Si}}-Cl}} \qquad \underset{CH_3}{\underset{|}{Cl-\underset{|}{\overset{CH_3}{Si}}-Cl}} \qquad \underset{CH_3}{\underset{|}{CH_3-\underset{|}{\overset{CH_3}{Si}}-Cl}}$$

The mixture of methylchlorsilanes is condensed and the liquid mixture fractionated (page 259) to separate them.

When methylchlorsilanes are treated with water the chlorine atoms in their molecules are replaced by hydroxyl (—OH) groups:

$$\underset{OH}{\underset{|}{HO-\underset{|}{\overset{CH_3}{Si}}-OH}} \qquad \underset{CH_3}{\underset{|}{HO-\underset{|}{\overset{CH_3}{Si}}-OH}} \qquad \underset{CH_3}{\underset{|}{CH_3-\underset{|}{\overset{CH_3}{Si}}-OH}}$$

These compounds are easily polymerised. With $\underset{CH_3}{\underset{|}{HO-\underset{|}{\overset{CH_3}{Si}}-OH}}$ long chain polymers are formed:

$$\underset{CH_3}{\underset{|}{HO-\underset{|}{\overset{CH_3}{Si}}-OH}} + \underset{CH_3}{\underset{|}{HO-\underset{|}{\overset{CH_3}{Si}}-OH}} + \underset{CH_3}{\underset{|}{HO-\underset{|}{\overset{CH_3}{Si}}-OH}} + \underset{CH_3}{\underset{|}{HO-\underset{|}{\overset{CH_3}{Si}}-OH}}$$

↓ loss of water

$$etc.-\underset{CH_3}{\underset{|}{\overset{CH_3}{\overset{|}{Si}}}}-O-\underset{CH_3}{\underset{|}{\overset{CH_3}{\overset{|}{Si}}}}-O-\underset{CH_3}{\underset{|}{\overset{CH_3}{\overset{|}{Si}}}}-O-\underset{CH_3}{\underset{|}{\overset{CH_3}{\overset{|}{Si}}}}-etc.$$

Silicone fluids

By blending the three methylchlorsilanes in various proportions an almost limitless range of different silicones can be produced. If a proportion of

$$\text{CH}_3-\underset{\underset{\text{CH}_3}{|}}{\overset{\overset{\text{CH}_3}{|}}{\text{Si}}}-\text{OH}$$ is mixed in before polymerisation it acts as an end-stop to the polymer chain:

$$\text{CH}_3-\underset{\underset{\text{CH}_3}{|}}{\overset{\overset{\text{CH}_3}{|}}{\text{Si}}}-\text{OH} \;+\; \text{HO}-\underset{\underset{\text{CH}_3}{|}}{\overset{\overset{\text{CH}_3}{|}}{\text{Si}}}-\text{OH} \;+\; \text{HO}-\underset{\underset{\text{CH}_3}{|}}{\overset{\overset{\text{CH}_3}{|}}{\text{Si}}}-\text{OH} \;+\; \text{HO}-\underset{\underset{\text{CH}_3}{|}}{\overset{\overset{\text{CH}_3}{|}}{\text{Si}}}-\text{OH}$$

↓ loss of water

$$\text{CH}_3-\underset{\underset{\text{CH}_3}{|}}{\overset{\overset{\text{CH}_3}{|}}{\text{Si}}}-\text{O}-\underset{\underset{\text{CH}_3}{|}}{\overset{\overset{\text{CH}_3}{|}}{\text{Si}}}-\text{O}-\underset{\underset{\text{CH}_3}{|}}{\overset{\overset{\text{CH}_3}{|}}{\text{Si}}}-\text{O}-\underset{\underset{\text{CH}_3}{|}}{\overset{\overset{\text{CH}_3}{|}}{\text{Si}}}-\text{etc.}$$

According to the proportions used, chains of different lengths can be produced. Silicones with chain-like molecules of various lengths are liquids: the longer the chain the thicker the liquid.

They are water-repellent, and are used for shower-proofing overcoats, umbrellas, tents, shoes, etc. (Figure 11.7). The silicone coating can be applied during manufacture or with a do-it-yourself aerosol spray. They are also used to waterproof masonry. Silicone coatings on brickwork or concrete last for more than ten years.

They are good lubricants. Used in car and furniture polishes they help to take the work out of polishing by acting as lubricants for the wax. Long-chain silicones are useful greases. They have the advantage over petroleum-based greases that they remain thick and retain their lubricating properties up to a much higher temperature.

They can be used to quell foam. Excessive foaming is troublesome, messy, wasteful of time and material, and restricts production capacity. Minute traces of silicones dramatically reduce foam in sewage works, dyeing vats, and food processing.

They are bad conductors of electricity. High voltage insulators on overhead power lines are coated with a silicone grease. No short-circuiting occurs even when the insulators are wet because of the water-repellent properties of the silicone.

They are not poisonous and can be used in protective skin creams, lipsticks, and sun-tan lotions where their water-repellent and lubricating properties are useful.

They will not stick to anything. Self-adhesive plastics like *Fablon*, self-adhesive wallpapers, labels, and surgical plasters will stick to almost anything except their silicone-treated backing papers. In tyre manufacture the inside of the mould is sprayed with a silicone fluid to ensure that the tyre can be removed easily from the mould (Figure 11.8).

Figure 11.7 A suede leather jacket waterproofed with silicone fluid

Figure 11.8 A tyre is quickly and cleanly released when the mould has been sprayed with a silicone fluid

Silicone resins

When HO—Si(CH₃)(OH)—OH is polymerised, the three hydroxyl groups link up in three dimensions to produce a giant-molecule solid silicone, called a SILICONE RESIN. Large bakeries use bread tins coated with a silicone resin. The non-stick properties of the silicone enable the bread to be removed easily from the tin after baking and the tin can be used over 200 times before the silicone coating has to be renewed.

Silicone resins are also used in the insulation of electric locomotive power units and in the transformers of television sets, as well as to insulate the coils of electric furnaces used in the manufacture of steel.

Silicone rubbers

By the blending together of all three of the basic monomers used in making silicones, polymers can be obtained whose molecules consist of long chains linked together at intervals. These polymers are called SILICONE RUBBERS, as their structure and elastic properties are similar to those of vulcanised rubber (page 302).

Silicone rubbers are used in oven door seals and the crankshaft seals in cars. They not only stand up to high temperatures, which ordinary rubber would not, but are also much more resistant to oil and grease. The electric cables in the *Snowcats* used by the British Antarctic Expedition were insulated with silicone rubber which remains flexible to below −50°C. Silicone rubbers are also used for the seals on aeroplane doors, windows, and flaps which are subjected to constantly varying temperatures.

Questions

1 The elements carbon and silicon belong to the same group of the periodic table (Group 4).
(a) What does the term *group* mean in this context?
(b) What can you say about the properties of compounds of elements in the same group?
(c) Magnesium burns in air and produces a white ash. It also burns in carbon dioxide producing black specks and a white ash.
Name these substances.
Explain how the reaction with carbon dioxide happens.
(d) Sand (silicon dioxide) is a source of silicon. How can silicon be extracted from sand? In your answer give the starting materials and the products and the conditions used. (*London*)

2 Silicones have the chain backbone
$$-\underset{|}{\overset{|}{Si}}-O-\underset{|}{\overset{|}{Si}}-O-\underset{|}{\overset{|}{Si}}-O-$$

(i) Why are they *not* organic polymers?
(ii) How do they differ structurally from organic polymers?
(iii) From a study of the periodic table, what would lead you to expect that silicon might be capable of forming polymers?
(iv) To what common inorganic substance are the silicones structurally related?
(v) Silicone resins are used as insulating covers for electrical machinery. What property apart from electrical resistivity enables them to be put to this use?
(vi) Give one other use for silicones. (*London*)

3 Explain briefly the differences between the molecular structures of silicone fluids, resins, and rubbers. How are the properties of these groups of silicones related to their molecular structures?

4 Make a list, write a short essay, or make an illustrated chart, to show as many everyday uses of silicones as you can think of or find out about.

For the teacher

Rocks and minerals A collection of appropriate rocks and minerals to pass round the class is helpful. Samples of granite (*microgranite*), quartz, mica (*muscovite*), felspar (*orthoclase felspar*), asbestos (*actinolite*), and most of the silicates listed in Figure 11.4 are available from Lythe Minerals or R. F. D. Parkinson. A useful introduction to the identification of rocks and minerals to which interested pupils may be referred is Zinn, H. S and Shaffer, P. R. *Rocks and Minerals* (Paul Hamlyn, 1971).

Extraction of silicon Silicon can be extracted from sand by heating with magnesium. It can be pointed out that the industrial process is based upon the same activity series principle but using carbon, which is much cheaper. Thoroughly mix 2 parts by volume of white sand (previously dried by heating strongly in an evaporating basin for 5 minutes and cooling in a dessicator) with 1 part by volume of magnesium powder. Put about 3 g of this mixture in a hard-glass test tube and clamp the tube horizontally. **Place a safety screen between the tube and the class.** Commence heating the mixture at the end nearest the mouth of the tube; then, when the mixture begins to glow, follow the reaction to the bottom of the tube with the bunsen. Allow the tube to cool and tip the contents into 25 cm^3 of 2M hydrochloric acid in a 100 cm^3 beaker and stir. The magnesium oxide and any excess magnesium dissolve and a few harmless (but impressive) explosions occur due to the reaction of magnesium silicide with hydrochloric acid to form silicon hydrides which ignite spontaneously on contact with the air. Heat the mixture to boiling and filter to recover the silicon.

Water-repellency of silicones A 2 per cent solution of silicone fluid MS 1107 (obtainable from Hopkin and Williams) in carbon tetrachloride or acetone is suitable. Materials to be tested are dipped in this solution. Any excess is allowed to drain off, the article is allowed to dry in the air, and it is then heated in an oven for 15 minutes at 150°C (or 2 hours at 100°C). (a) Treated cotton wool floats on water; untreated cotton wool becomes soaked and sinks. (b) Ink poured on to treated cotton cloth will wash off; untreated cloth is stained. (c) An untreated brick absorbs water; a treated brick does not. Many other materials can also be tested.

The advantage of silicone lubricants Half-fill a test tube with a silicone grease (e.g. the silicone stopcock grease obtainable from most apparatus suppliers) and another tube with a petroleum-based grease such as vaseline. Immerse both tubes in boiling water. The vaseline becomes much less viscous whereas the viscosity of the silicone grease is hardly affected.

Anti-foaming properties of silicones About 5 drops of Antifoam RD (obtainable from Hopkin and Williams) is sufficient to dispel a good head of foam on a 1 000 cm^3 beaker of $\frac{1}{2}$ per cent detergent solution.

Non-stick properties of silicones Sellotape will not stick to paper treated with MS 1107 solution as above.

12 Nitrogen and other elements important to life

1 Nitrogen in nature

One of the most important groups of chemicals in living things is PROTEINS. These are the main 'building materials' of which animals are made. Skin, muscle, and hair are all composed chiefly of proteins, and so are the catalysts, called enzymes (page 64), which control the chemical reactions in animals' bodies.

Many animals, from lions to spiders, get their proteins by eating other animals. Others (vegetarians) get all their proteins by eating plants. Cows, for instance, get most of the proteins they need from grass. In rich countries, like Britain, we get most of our proteins by eating animal products. Meat, fish, eggs, cheese, and milk all contain a large proportion of proteins. In poorer countries, where these foods are not available, people have to get the protein their bodies need by eating plant foods. All kinds of beans and peas are good sources of vegetable protein.

Plants have to make their own protein. Proteins are complicated polymer molecules but they are composed mainly of only four elements: carbon, hydrogen, oxygen, and nitrogen. Plants get the carbon and oxygen they need in the form of oxygen and carbon dioxide gases from the air. They obtain hydrogen from water, absorbed through their roots. Supplies of nitrogen in a form that plants can use to make proteins are more of a problem. Plants cannot use nitrogen from the air; they are not capable of turning it into the nitrogen compounds they need. They can only take in nitrogen in the form of soluble nitrogen compounds in the soil which pass into the plants, in solution in water, through their roots.

The nitrogen cycle

The supply of these nitrogen compounds in the soil is not inexhaustible. If life is to continue the nitrogen compounds taken out of the soil by one generation of plants must be returned in some way for the use of the next generation. There must be a natural NITROGEN CYCLE, just as there is a natural carbon cycle (page 253).

The nitrogen compounds which a plant has taken out of the soil may eventually be returned when the plant dies. If the plant is eaten by animals some of the nitrogen is returned to the ground in the animals' excreta, and some when the animal dies.

There is some wastage in this process. Not all the nitrogen taken out of the soil by plants is eventually returned. When plants or animals decay some nitrogen gas escapes into the air. Some nitrogen compounds may be washed into the sea by rain. There is also a type of bacteria, called DENITRIFYING BACTERIA, which changes nitrogen compounds in the soil into nitrogen gas which escapes into the atmosphere.

These losses of nitrogen from the soil are made up in two ways. Flashes of lightning cause reaction between nitrogen, water, and oxygen in the air, producing nitrogen compounds which are carried into the soil by rain. Also some kinds of plants, including peas, beans, and clover, have little lumps, or nodules, on their roots in which another type of bacteria, called NITRIFYING BACTERIA, live. This type of bacteria makes nitrogen compounds out of the nitrogen gas in the air. Nitrifying bacteria are the only living things which are capable of doing this.

The need for fertilisers

The natural nitrogen cycle is shown in Figure 12.1.

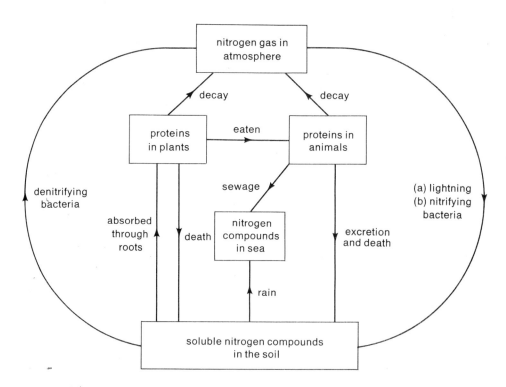

Figure 12.1 The nitrogen cycle

356 Chemistry today

There is a perfect balance in nature between nitrogen compounds removed from the soil and nitrogen compounds returned to the soil. Man, however, has interfered with the natural nitrogen cycle. The concentration of population in cities, particularly in the last 150 years, has caused a major disturbance. Huge quantities of plants are picked or dug up for food and are transported to the cities. In the fields where these plants grew, nitrogen compounds are not being returned to the soil for the use of the next generation of plants. Human waste products are dumped into the sea as sewage. The land becomes poorer in nitrogen compounds and eventually nothing will grow on it.

Up to about 100 years ago, farmers had a simple solution to the problem, called CROP ROTATION. One year a cereal crop such as barley or wheat was grown. This was harvested and sent to market, thus removing nitrogen compounds from the soil of that field. The next year a crop such as clover would be planted. The nitrifying bacteria in the roots of the clover restored nitrogen compounds to the soil so that the field was ready for growing cereal crops again.

Nowadays the demand for food is so great that fields cannot be allowed to rest for a year while the nitrogen compounds are being restored. The only alternative is to spread nitrogen compounds artificially on the soil in the form of fertiliser. The earliest fertilisers were *compost* (rotted

Figure 12.2 In this maize plot in Venezuela all the seeds were planted at the same time, but no fertiliser was applied to the central strip

vegetable remains) or *manure* (decaying animal excreta or remains), but these are not available on a large enough scale. In the first half of the nineteenth century huge quantities of *guano*, hardened birds' droppings from islands off the coast of Peru, were imported into Europe for use as fertilisers. The guano deposits were, however, soon used up. Towards the

Nitrogen and other elements important to life   357

end of the nineteenth century guano was replaced by *saltpetre*, a mineral found in Chile which is composed mainly of sodium nitrate, a nitrogen compound which plants can absorb and use. At the same time ways were being found to convert coal's nitrogen compounds, which collect in the ammoniacal liquor during the manufacture of coal gas (page 267), into fertilisers. But by 1900 it was becoming obvious that within 25 years at the most supplies of saltpetre and fertilisers made from coal would not be sufficient to produce enough crops to feed the increasing population, even in Europe.

Fertiliser from air

Nitrogen is not a very common element. It forms only $\frac{3}{100}$ per cent by weight of the earth's crust. The atmosphere, however, contains 78 per cent of nitrogen by volume, and this could be an almost inexhaustible supply out of which to make nitrogen compounds suitable for use as fertilisers.

Unfortunately, nitrogen is an extremely unreactive gas. It is colourless, odourless, almost insoluble in water, and neutral to indicators. The only elements with which it combines easily are a few of the more reactive metals in Groups I and II of the periodic classification. Magnesium, for instance, burns in nitrogen gas to form magnesium nitride:

$$3Mg(s) + N_2(g) \rightarrow Mg_3N_2(s).$$

These metals are all too expensive to be used in manufacturing fertilisers on a large scale.

In 1913 a German chemist, Fritz Haber, solved the problem when he discovered that under certain conditions nitrogen could be reacted with hydrogen to form AMMONIA (NH_3), from which fertilisers could easily be made. In fact it was not a search for fertilisers which led to Haber's discovery. Ammonia can also be used to make the nitric acid which is required for the manufacture of explosives. In the shadow of the First World War (1914–1918) the need was for a process to make ammonia for explosives. Only after the war was it seen that Haber's discovery was vital in the production of fertilisers.

Questions

1 When a burning splint is inserted in the mouth of a jar containing a certain gas, the splint is immediately extinguished. The gas has no effect on limewater. Which of the following could the gas be?
A   carbon monoxide   D hydrogen
B   carbon dioxide   E producer gas.
C   nitrogen
                                                                 (*East Anglia*)

2 Using the apparatus shown in Figure 12.3, about 100 cm³ of air from syringe A was passed over the heated copper by very slowly pushing in the plunger of syringe A.

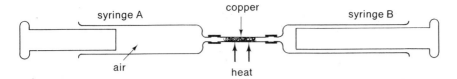

Figure 12.3

After cooling, with all the air expelled from A, syringe B was seen to contain about 80 cm³ of gas.
(a) What is the name of the gas present in the largest proportion by volume in syringe B after the experiment?
(b) Explain the disappearance of about 20 cm³ of gas during the experiment. (*West Midlands*)

3 (a) Why is it necessary continually to supply nitrogen to soil used for crop cultivation?
(b) Explain any three ways in which soil receives nitrogen by natural means.
(c) For what purpose do plants require nitrogen?
(*West Midlands*)

4 Explain why the manufacture of artificial fertilisers has only become important in the twentieth century.

5 Amino-acids are a group of chemicals which, joined together, form proteins. The simplest amino-acid is glycine which has the formula:

$$\text{H}-\underset{\underset{\text{NH}_2}{|}}{\overset{\overset{\text{H}}{|}}{\text{C}}}-\text{C}\overset{\displaystyle\nearrow\text{O}}{\searrow\text{OH}}$$

It is thought that glycine was first formed from a primeval 'soup' and the first step to life as we know it was taken. Suggest from which simple chemicals glycine could have been formed. (*London*)

2 The Haber process

The basic reaction of the Haber process is:

nitrogen + hydrogen ⇌ ammonia.
$N_2(g)$ + $3H_2(g)$ ⇌ $2NH_3(g)$.

This reaction is very difficult to bring about. In order for ammonia to be produced reasonably quickly the conditions under which the reaction is carried out must be chosen very carefully. Conditions vary slightly from one ammonia plant to another, but the biggest Haber plant in Britain, the ICI plant at Billingham, uses a pressure of 250 atmospheres, a temperature of 400°C, and an iron catalyst.

Nitrogen and hydrogen are mixed in the ratio 1:3 by volume, as required by the equation, and the gas mixture is compressed to a pressure of 250 times atmospheric. At 400°C the gases pass through the catalyst chamber, a huge reinforced steel cylinder containing about $7\frac{1}{2}$ tonnes of pea-sized pieces of iron.

When the gases leave the catalyst chamber they contain between 10 per cent and 20 per cent of ammonia. They are cooled by passing through pipes around which cold water is circulated. Under the high pressure, liquid ammonia condenses out of the gas mixture and is collected. The unreacted hydrogen and nitrogen are passed back through the catalyst chamber. Ammonia is usually stored and transported liquefied, under pressure, in steel cylinders by road and rail tankers.

Nitrogen for ammonia

The nitrogen required for the manufacture of ammonia is obtained from the air. Air is first cooled sufficiently to freeze out water (freezing point 0°C) and carbon dioxide (freezing point −56°C). It is then compressed. Compression causes it to heat up (a bicycle pump gets hot during use for the same reason). The hot compressed air is cooled by passing it through pipes around which cold water is circulated. The cool compressed air is then rapidly expanded. Just as compression causes heating up, expansion causes cooling. With repeated compression, cooling, and rapid expansion the air is cooled to −200°C and liquefies.

The liquid air is fractionated, in much the same way as crude oil is fractionated (page 273). Liquid air is fed into the middle of the column (Figure 12.4). Nitrogen, which has the lower boiling point (−196°C), boils off at the top of the column. Oxygen, with the higher boiling point (−183°C), collects as a liquid at the bottom of the column. Fractionation of liquid air, in addition to providing nitrogen for the Haber process, also provides oxygen and carbon dioxide for all kinds of uses. By more elaborate fractionation the noble gases helium, neon, argon, krypton, and xenon can also be obtained. Some of their uses are shown in Figure 12.5.

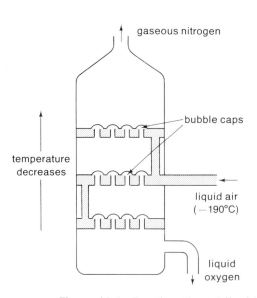

Figure 12.4 Fractionation of liquid air

Helium	In meteorological balloons (low density, non-inflammable). Mixed with oxygen for deep-sea divers to breathe (unlike nitrogen, helium does not dissolve in the blood under pressure, and so does not cause 'bends'). Liquid helium (boiling point −269°C) as a coolant in low temperature research.
Neon	To fill discharge tubes for advertising signs.
Argon	To fill household light bulbs. As an inert atmosphere for welding.
Krypton Xenon	To fill certain types of electronic valves.

Figure 12.5

Hydrogen for ammonia

In Britain up to about 1960 almost all hydrogen for the Haber process came from water gas, made from coke (page 326). Just as in the last few years town gas manufacture has switched almost completely from coal to oil (page 280), so also the hydrogen for ammonia manufacture is now almost all obtained from oil or natural gas (page 280). One method is to react methane with steam at a high temperature, using a nickel catalyst:

$$CH_4(g) + H_2O(g) \rightarrow CO(g) + 3H_2(g).$$

Experiment 12.1 Some properties of ammonia

(a) Place a spatula-full of ammonium chloride and a spatula-full of calcium hydroxide in a hard-glass test tube and mix them together well. Hold the tube in tongs and warm it *gently* while you conduct the following tests.

(i) **Very cautiously** smell the ammonia gas which is produced. Do you recognise the smell? Where have you come across it before?

(ii) Hold a piece of moist universal indicator paper in the mouth of the tube. What pH is registered? Is ammonia an acid or a base?

(iii) Hold a lighted splint just inside the mouth of the tube. What happens? Does ammonia burn? Does it support burning?

(iv) Dip a glass rod in concentrated hydrochloric acid and hold it at the mouth of the tube. Describe what you see.

(b) Fit the test tube with a bung and delivery tube as shown in Figure 12.6.

Warm the mixture until a piece of moist universal indicator paper, held at the mouth of the second tube, turns dark blue. Keeping the

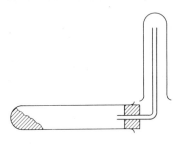

Figure 12.6

tube containing the ammonia gas vertical, dip the open end quickly into half a beaker of water.

What happens? What does this tell you about ammonia gas? Test the water in the test tube with universal indicator paper. Does this confirm your conclusions?

The tube supported mouth downwards in Figure 12.6 became filled with ammonia gas. What does this tell you about ammonia?

Experiment 12.2 Investigating some foodstuffs

Put a small piece of meat (about the size of a pea) in a test tube and cover it to a depth of 2 cm with *sodalime*. (Sodalime is a mixture of sodium hydroxide and calcium hydroxide.)

Heat the tube strongly and test the fumes evolved with:
(i) moist universal indicator paper;
(ii) a glass rod dipped in concentrated hydrochloric acid.

Make a note of the results of these two tests. What can you deduce from the results?

If you have time you could repeat the experiment with other foodstuffs, for instance a spatula-full of milk powder or a pea-sized piece of cheese. Are the results of the two tests the same every time? What do the foodstuffs you tested have in common?

The properties of ammonia

Ammonia is a colourless gas with a sharp, penetrating smell. It is about half as dense as air and extremely soluble in water. One volume of water will dissolve about 800 volumes of ammonia at room temperature and pressure. For laboratory use ammonia is usually supplied as a saturated solution in water. The solution is often known as 880 ammonia, because its density is 0.880 g cm^{-3}. This solution contains about 35 per cent of ammonia by weight.

As with most gases which are very soluble in water, the great solubility of ammonia is due to reaction of the gas with water. Ammonia is a base and removes protons from water to produce ammonium ions and hydroxide ions:

$$NH_3(g) + H_2O(l) \rightleftharpoons \underset{\text{ammonium ion}}{NH_4^+(aq)} + OH^-(aq).$$

Ammonia solution is therefore an alkali (a soluble base). It changes the colour of indicators and reacts with acids to form salts. With hydrochloric acid, for instance, the salt ammonium chloride is produced:

$$NH_3(aq) + HCl(aq) \rightarrow NH_4Cl(aq).$$

(It is important to distinguish between the words *ammonia* and *ammonium*. Ammonia is a compound, formula NH_3, which is a pungent-smelling gas at room temperature. Ammonium is the name of a cation, formula NH_4^+, which exists in solid salts such as ammonium chloride, NH_4Cl, which have no smell.)

Ammonium salts are not very stable. They are all decomposed by heat. Also, since ammonia is not a very strong alkali, the salts are decomposed by heating with a stronger alkali, such as sodium hydroxide or calcium hydroxide. For example:

$$2NH_4Cl(s) + Ca(OH)_2(s) \rightarrow CaCl_2(s) + 2NH_3(g) + 2H_2O(g).$$

This provides a convenient method for preparing a small quantity of ammonia gas in the laboratory.

Experiment 12.3 Investigating the action of heat on some ammonium salts

(a) Place a spatula-full of ammonium chloride in a hard-glass test tube. Hold the tube in tongs and heat gently at first, then more strongly.
Describe what you see happening.

(b) Arrange a hard-glass test tube as shown in Figure 12.7. Holding the tube in tongs heat gently, then more strongly. What happens to the pieces of moist indicator paper?

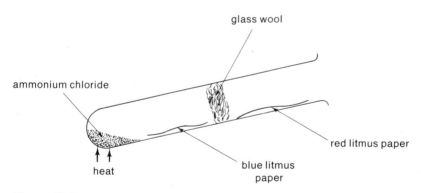

Figure 12.7

Try to work out what ammonium chloride (formula NH_4Cl) decomposes into when heated. Write an equation for the decomposition.
Can you now explain the result of experiment (a)?
Can you also explain why dense white fumes are produced when a glass rod dipped in concentrated hydrochloric acid is held in ammonia gas (Experiment 12.1)?

(c) Repeat Experiment (a), using ammonium sulphate in place of ammonium chloride. Describe and explain what you see.

Nitrogen and other elements important to life

Some important ammonium salts

AMMONIUM CHLORIDE, NH_4Cl, is the electrolyte in dry cells (page 331). It is also used as a flux to clean the surface of metals before soldering. On heating it decomposes into ammonia and hydrogen chloride gases:

$$NH_4Cl(s) \rightarrow NH_3(g) + HCl(g),$$

but the ammonia and the hydrogen chloride recombine when the gases cool, producing a deposit of ammonium chloride at the top of the test tube. The ammonium chloride appears to be subliming, but in fact it is decomposing and reforming.

AMMONIUM SULPHATE, $(NH_4)_2SO_4$, is one of the most important of the nitrogenous fertilisers.

AMMONIUM NITRATE, NH_4NO_3, is another important fertiliser. It is also used as an explosive. On heating it decomposes to produce dinitrogen oxide, the dentist's anaesthetic gas, which used to be called *laughing gas*:

$$NH_4NO_3(s) \rightarrow N_2O(g) + 2H_2O.$$

AMMONIUM CARBONATE, $(NH_4)_2CO_3$, is the most unstable of all the ammonium salts. It decomposes slowly at room temperature into ammonia, carbon dioxide, and water:

$$(NH_4)_2CO_3(s) \rightarrow 2NH_3(g) + CO_2(g) + H_2O(l).$$

It was once popular as *smelling-salts* to bring Victorian ladies out of fainting fits. A little was carried in a tightly-stoppered bottle. When the bottle was opened under the patient's nose the pungent ammonia produced by the decomposition of the ammonium carbonate had a stimulating effect.

Ammonia for fertilisers

About 85 per cent of all the ammonia manufactured in the world is used to make fertilisers. Ammonia gas can be injected directly into the soil, but it is difficult to avoid some escaping into the air and being wasted. It is more usual to convert ammonia into one of its solid salts, which is much easier to apply to the ground, either as the solid or dissolved in water.

AMMONIUM SULPHATE is the most popular fertiliser. It is made by the reaction between ammonia and sulphuric acid:

$$2NH_3(aq) + H_2SO_4(aq) \rightarrow (NH_4)_2SO_4(aq),$$

or by reacting ammonia and carbon dioxide with calcium sulphate (*anhydrite*):

$$CaSO_4(s) + 2NH_3(aq) + CO_2(g) + H_2O(l) \rightarrow$$
$$CaCO_3(s) + (NH_4)_2SO_4(aq).$$

The calcium carbonate is filtered off and the solution of ammonium sulphate evaporated down.

AMMONIUM NITRATE is made by reacting ammonia with nitric acid:

$$NH_3(aq) + HNO_3(aq) \rightarrow NH_4NO_3(aq).$$

It has the disadvantage that it has a tendency to explode, so it is usually mixed with *chalk* (calcium carbonate) to reduce the risk of accidental explosion. The mixture is called NITROCHALK.

A nitrogenous fertiliser which is increasing in popularity is UREA, made by reacting ammonia with carbon dioxide under pressure:

$$2NH_3(g) + CO_2(g) \rightarrow \underset{\text{urea}}{CON_2H_4(s)} + H_2O(l).$$

The most useful nitrogenous fertiliser is one containing a high proportion of nitrogen by weight. This gives the maximum effect with the lowest transport costs. The nitrogen content is expressed as a percentage by weight, for example:

1 mole of ammonium nitrate (NH_4NO_3) is $14+4+14+48 = 80$ g.

Of this 28 g is nitrogen.

Percentage of nitrogen in ammonium nitrate is $\frac{28}{80} \times 100 = 35$ per cent. Figure 12.8 shows the percentages by weight of nitrogen in various fertilisers. These figures explain the increasing popularity of urea, though ammonium sulphate is still the cheapest fertiliser to produce and it has the advantage that it also provides the soil with sulphur which, like nitrogen, is necessary for the growth of plants.

	% nitrogen by weight
Ammonium sulphate	21%
Ammonium nitrate	35% (less in nitrochalk)
Urea	47%

Figure 12.8

Some other uses of ammonia

1 In the manufacture of nitric acid.
2 To make urea for use in the manufacture of thermosetting plastics (page 300).
3 To make one of the monomers required for nylon manufacture (page 297).
4 To adjust the pH of latex to increase its stability (page 301).
5 As the cooling liquid in some refrigerators.
6 In the manufacture of a wide range of chemicals, including drugs and dyestuffs.

Questions

1 The industrial manufacture of ammonia is carried out on a very large scale, and it is essential to obtain as much ammonia as quickly as possible.
The equation for this reaction is:

$$N_2(g) + 3H_2(g) \rightleftharpoons 2NH_3(g).$$

(a) How does temperature affect the speed of a reaction?
(b) This is a reaction in which all the starting materials, and the products are gases. How would pressure affect the speed of this reaction? Explain your answer.
(c) Mention one other way in which the speed of this reaction may be increased.
(d) What does the sign $\rightleftharpoons$ indicate about the reaction?
(e) In such a reaction as this what will be happening after the reaction has been proceeding for a long time in a closed vessel?
(f) A mole of any gas occupies the same volume under the same conditions of pressure and temperature. What can you say about the pressure change that occurs as the reaction proceeds in the closed vessel?

The graphs shown in Figure 12.9 indicate the amounts of ammonia present in the reacting gases under different conditions.

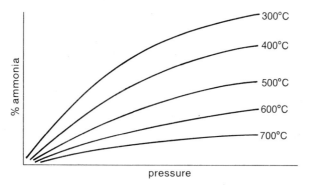

Figure 12.9

(g) From the graphs state how the amount of ammonia changes as the pressure increases.
(h) From the graphs state under what conditions the amount of ammonia in the reacting gases is the greatest.
(i) Suggest reasons for the fact that the commercial process operates at about 500°C. (*London*)

2 Oxygen is obtained from air on a large scale by which of the following processes?
A reduction   D sublimation
B decomposition   E liquefaction and distillation
C oxidation
(*West Yorkshire*)

3 Each of the following statements about ammonia is true except one. Which one is false?
A   It is a gas at room temperatures and pressures.
B   It is an electrovalent (ionic) compound.
C   It dissolves readily in water.
D   It has the formula NH_3.
E   It combines readily with hydrogen chloride.   (London)

4 Read the following passages and then answer the questions asked.
   A salt (A) was heated with some calcium hydroxide. A colourless gas (B) with a characteristic smell was given off and it was seen to turn a piece of damp red litmus paper blue. A large amount of the gas (B) was then passed through an inverted filter funnel into some water and a colourless solution was obtained. To the colourless solution an equivalent amount of dilute sulphuric acid was added and the solution evaporated. A white solid (C) was obtained.
   (a) Name the salt A.
   (b) Give the name and formula of gas B.
   (c) Explain the purpose of the 'inverted filter funnel'.
   (d) Name the solid C and give a use for it on a large scale.
   (e) Write an equation for the reaction by which C was formed and name the type of reaction involved.
   On an industrial scale gas B is obtained by direct combination of two gases (E and F) by passing them under high pressure at 450°C over a catalyst (G).
   (f) What are the two gases E and F?
   (g) Explain what you understand by the word 'catalyst'. Name the catalyst G.   (West Midlands)

5 Solutions of ammonia and sodium hydroxide can be distinguished by holding a piece of moist red litmus paper above each solution. Explain why the litmus paper above the ammonia solution turns blue but that above the sodium hydroxide solution does not.
(North West)

6 A powdered foodstuff was heated with soda-lime. The powder was seen to blacken and smoky vapours were given off, which had a sharp irritating odour and turned damp universal indicator paper purple. (pH = 12)
   (a) What does the information tell us about the vapours?
   (b) Only one common gas has these properties. Name it.
   (c) What element is suggested as being present in the foodstuff by the blackening?
   (d) The type of foodstuff used in the experiment was which of the following?
   A   fat   C   sugar   E   mineral salt
   B   starch   D   protein

(e) Name one foodstuff of this type.

(f) The change brought about by heating with soda-lime is best described as which of the following?

A changing small molecules into large
B changing large molecules into small
C changing it into ions
D fermentation
E electrolysis

(g) Which one of the following processes involves a similar change?

A polymerisation
B digestion
C boiling
D condensation
E sublimation

(h) A pure form of the gas produced above was heated with copper(II) oxide. A colourless liquid was collected which turned anhydrous copper(II) sulphate blue. What do you think the liquid was?

(i) Another gas was also collected after the heating with copper(II) oxide described in (h). This was found to be neutral to litmus, it did not burn or support combustion. It did not turn lime water milky. Give the name of this gas.

(j) What elements are shown to be present in the original gas which you have named in (b)?

(k) Briefly explain the importance of this conclusion in the growing of food.

(l) Soil often needs liming before it will grow good crops. A sample of soil was tested and found to have a pH of 5. After treatment with lime the pH eventually rose to 7. What can you then say about the properties of lime?

(m) The effect that the lime had on the soil can be described as which of the following?

A fertilising
B oxidising
C neutralising
D acidifying
E manuring

(n) It is said that some artificial fertilisers and lime (calcium hydroxide) must not be applied to the soil together. Suggest a reason for this. *(London)*

7 *Fill in the blanks.* (i) Ammonium sulphate is important as a fertiliser because it contains _____ which plants can use to build compounds called _____.

(ii) How can ammonium sulphate be prepared in the laboratory?

(iii) Write an equation for the reaction.

(iv) Ammonium nitrate is another important fertiliser, which has to be stored mixed with chalk, and the mixture is called 'nitrochalk'. Why has it to be stored in this way?

(v) What are the products when ammonium nitrate is heated?

(vi) Write the equation for the reaction. *(London)*

3 The manufacture of nitric acid

The modern process for nitric acid production uses as its raw material ammonia produced by the Haber process. There are two stages.
1 Oxidation of ammonia, using oxygen from the air, to nitrogen oxide gas:

$$4NH_3(g) + 5O_2(g) \rightarrow 4NO(g) + 6H_2O(g).$$
nitrogen oxide

2 Reaction between nitrogen oxide, more oxygen, and water to form nitric acid:

$$4NO(g) + 3O_2(g) + 2H_2O(l) \rightarrow 4HNO_3(aq).$$
nitric acid

Stage 1   The catalytic oxidation of ammonia

Ammonia gas from the Haber plant is mixed with air in the proportion 10 per cent ammonia to 90 per cent air. The mixture passes through the converter, where it meets the catalyst at 900°C. The converter is a large cylindrical vessel, about 3 metres in diameter.
Across the centre is the catalyst, a pad of gauzes made of an alloy of 90 per cent platinum and 10 per cent rhodium. As the gas mixture passes through the catalyst over 95 per cent of the ammonia is converted into nitrogen oxide gas.

Figure 12.10   Nitric acid converters

Nitrogen and other elements important to life   369

Stage 2 The absorption of nitrogen oxide

Before the second stage of the process, the gas mixture from the converter must be cooled almost to room temperature. The cooled gases are mixed with more air and compressed to a pressure of 4 atmospheres. They then pass up a tower down which water is trickling. The water reacts with the nitrogen oxide and oxygen in the gas mixture, giving nitric acid of about 65 per cent strength.

For many industrial purposes, such as the manufacture of ammonium nitrate fertiliser, 65 per cent nitric acid is concentrated enough. For some applications, particularly the manufacture of explosives, concentrations up to 99 per cent are required. It has been found cheaper to manufacture the more dilute acid and then concentrate it than to manufacture very concentrated acid directly.

The properties of nitric acid

The normal laboratory strength for concentrated nitric acid is about 65 per cent. It is an oily liquid, usually slightly yellow in colour.

Like all acids, nitric acid reacts with water to produce hydronium ions:

$$HNO_3(l) + H_2O(l) \rightarrow H_3O^+(aq) + NO_3^-(aq).$$
$$\text{nitrate ion}$$

In solution in water it has all the normal properties of an acid (page 212). It forms a series of salts, called *nitrates*:

e.g. $NaOH(aq) + HNO_3(aq) \rightarrow NaNO_3(aq) + H_2O(l).$
$$\text{sodium nitrate}$$

Apart from being an important industrial acid, nitric acid is also a powerful oxidising agent, and also a nitrating agent (page 374).

Experiment 12.5 Investigating the action of heat on nitrates

(a) Put a spatula-full of potassium nitrate in a small hard-glass test tube. Holding the tube in tongs, heat it gently at first, then more and more strongly. When you are heating it very strongly indeed, hold a glowing splint just inside the mouth of the tube.
Describe all the changes you see taking place inside the tube.
What gas is given off?
The residue inside the tube at the end of the experiment is potassium nitrite.

(b) Repeat experiment (a) with sodium nitrate in place of potassium nitrate. Describe what happens. What do you think the residue in the tube is this time?

(c) Repeat experiment (a) with copper(II) nitrate in place of potassium nitrate.
Describe all the changes you see taking place inside the tube.

Is oxygen gas produced this time?

Describe the appearance of the new gas, called nitrogen dioxide, which is produced from the copper(II) nitrate.

What do you think the residue in the tube is?

(d) Repeat experiment (a) with lead nitrate in place of potassium nitrate. Describe what happens. Write a word equation to describe the reaction.

Some important nitrates

POTASSIUM NITRATE, KNO_3, is used in gunpowder. On heating, it decomposes into potassium nitrite and oxygen:

$$2KNO_3(s) \rightarrow 2KNO_2(s) + O_2(g).$$

SODIUM NITRATE, $NaNO_3$, occurs in nature as the mineral *saltpetre*, which was at one time an important fertiliser (page 358). Before the Haber process provided a cheap source of ammonia from which to make nitric acid, sodium nitrate was used to manufacture nitric acid, by heating it with concentrated sulphuric acid:

$$NaNO_3(s) + H_2SO_4(l) \rightarrow NaHSO_4(s) + HNO_3(g).$$

Although sodium nitrate is much cheaper than potassium nitrate, it cannot be used in the manufacture of gunpowder because it is deliquescent. On heating it decomposes into sodium nitrite and oxygen (just like potassium nitrate):

$$2NaNO_3(s) \rightarrow 2NaNO_2(s) + O_2(g).$$

AMMONIUM NITRATE, NH_4NO_3, is an important fertiliser and explosive. Most of the nitric acid manufactured is used to make this salt.

LEAD NITRATE, $Pb(NO_3)_2$, is a useful laboratory chemical because it is one of the few compounds of lead which is soluble in water. On heating, it decomposes into lead oxide, oxygen, and the brown gas nitrogen dioxide:

$$2Pb(NO_3)_2(s) \rightarrow 2PbO(s) + 4NO_2(g) + O_2(g).$$

Questions

1 Describe *one* method for the industrial preparation of nitric acid, giving the equations of the chemical reactions involved.
Name *two* important uses of nitric acid. (*Wales*)

2 A laboratory experiment was set up as in Figure 12.11.
It was observed that the platinum wire continued to glow red hot, and brown fumes could be seen forming in the flask.
(i) Write down the name and formula of the gas produced.
(ii) Explain the reaction which results in the formation of the brown gas, commenting on the purpose of the platinum.

(iii) Suggest a reason why the platinum wire continued to glow red hot.

(iv) What is the commercial importance of the reaction illustrated by the experiment? (*West Midlands*)

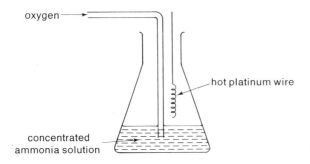

Figure 12.11

3 Give one important use for each of the following substances: (a) nitric acid, (b) ammonia, (c) ammonium sulphate, (d) potassium nitrate. (*Middlesex*)

4 Some lead nitrate crystals were heated in a clean, dry test tube. A brown gas (X) and an invisible gas (Y) which relit a glowing splint were given off. A yellow residue (Z) remained.
  (a) Identify the brown gas (X) and the colourless gas (Y).
  (b) What was the yellow residue (Z)?
  (c) Gas (X) will form a liquid if surrounded by a freezing mixture but gas (Y) will not liquefy under such conditions. Gas (Y) can be collected over water. Using this information draw a labelled diagram to show how you would prepare and collect samples of X and Y by heating the lead nitrate crystals. (*West Midlands*)

12.4 Explosives

An explosion is a very fast chemical reaction. For a reaction to take place explosively the following conditions must be satisfied.
1 It must be very exothermic (produce a lot of heat).
2 It must travel very fast through the reacting substances (in commercial explosives the reaction moves at a speed of several thousand metres per second).
3 The products of the reaction must be gases.
The effect of an explosion is that a very large quantity of gas is produced very quickly, and due to the heat produced the gas expands violently.

A number of quite common chemical reactions can take place explosively under certain conditions. The reaction of hydrogen with oxygen to form water is a familiar example. When hydrogen is reacted with a large volume of oxygen it burns quietly and steadily; the excess oxygen moderates the reaction. But if the gases are mixed in the proportions required by the equation:

$$2H_2(g) \; + \; O_2(g) \; \rightarrow \; 2H_2O(g),$$
$$\text{2 volumes} \quad \text{1 volume}$$

the reaction is explosively fast.

WARNING: UNDER NO CIRCUMSTANCES SHOULD ANY ATTEMPT BE MADE TO MIX OR MAKE EXPLOSIVES OR FIREWORKS. THE PRODUCTION OF EXPLOSIVES IS A HIGHLY COMPLEX AND DELICATE OPERATION REQUIRING GREAT SKILL AND KNOWLEDGE. EVEN EXPERIENCED CHEMISTS DOING RESEARCH INTO EXPLOSIVES HAVE BEEN SERIOUSLY INJURED AND NEWSPAPER REPORTS OF TERRIBLE INJURIES TO YOUNG PEOPLE WHO HAVE EXPERIMENTED WITH EXPLOSIVES ARE ALL TOO COMMON.

Gunpowder

Gunpowder, sometimes called *blackpowder*, was the earliest explosive known to man. It was discovered by the ancient Chinese, who used it mainly to make fireworks, but was not made in Europe until the thirteenth century. This led to the invention of guns of all kinds, and it was not until the nineteenth century that explosives began to find peaceful uses.

Gunpowder is a mixture of an oxygen-rich compound, usually potassium nitrate, with substances that react easily with oxygen, usually powdered carbon and sulphur. When gunpowder is ignited it burns vigorously, without exploding, producing carbon dioxide and sulphur dioxide gases. If it is packed tightly inside a container, as in a bullet or cartridge, the gases formed are compressed and their rapid expansion produces an explosion.

When gunpowder is compressed in this way it can be ignited by a flame, as it used to be in cannons and flint-lock pistols. Alternatively, the reaction can be started by detonation. A percussion cap is attached to the gunpowder container. The percussion cap contains a chemical such as mercury fulminate which explodes when it is hit, for instance by the hammer of a gun. The shock wave from this little explosion sets off the reaction of the gunpowder.

Ammonium nitrate

When ammonium nitrate is heated gently it decomposes steadily into dinitrogen oxide and water:

$$NH_4NO_3(s) \; \rightarrow \; N_2O(g) \; + \; 2H_2O(g).$$

But at higher temperatures, or when a solid mass of ammonium nitrate is detonated, a different reaction takes place which is completely uncontrollable, and there is an explosion:

$$2NH_4NO_3(s) \rightarrow 2N_2(g) + 4H_2O(g) + O_2(g)$$

Since extra oxygen is produced in this reaction an even more effective explosive can be produced by mixing the ammonium nitrate with another substance which will react with this oxygen. AMMONAL is a mixture of ammonium nitrate and powdered aluminium. The aluminium is oxidised to aluminium oxide.

Nitroglycerine

Concentrated nitric acid will react with many organic compounds containing hydroxyl (—OH) groups, replacing them by nitrate (—NO$_3$) groups. This reaction is known as NITRATION. With glycerol:

$$\begin{array}{l} CH_2\text{—}OH \\ | \\ CH\text{—}OH \\ | \\ CH_2\text{—}OH \end{array} + 3HNO_3 \longrightarrow \begin{array}{l} CH_2\text{—}NO_3 \\ | \\ CH\text{—}NO_3 \\ | \\ CH_2\text{—}NO_3 \end{array} + 3H_2O$$

glyceryl trinitrate
(nitroglycerine)

The reaction does not take place easily, unless the water produced is removed. Concentrated sulphuric acid, a powerful dehydrating agent (page 226), is used for this purpose.

Glycerol, obtained as a by-product in the manufacture of soap (page 306), is mixed with concentrated nitric and sulphuric acids. The mixture runs continuously through a reaction vessel which is kept at a temperature between 15°C and 20°C. The temperature must be very carefully controlled because the mixture is likely to explode if it reaches 25°C. An automatic safety system in the plant floods the reaction vessel with water if the temperature reaches a dangerous level.

Nitroglycerine explodes by reaction between the oxygen in its molecules with the carbon and hydrogen. There is enough oxygen to convert all the carbon to carbon dioxide and all the hydrogen to water. All of the products of the reaction are gases:

$$4C_3H_5N_3O_9(l) \rightarrow 12CO_2(g) + 10H_2O(g) + 6N_2(g) + O_2(g)$$

Nitroglycerine is an extremely dangerous explosive. It is a liquid which detonates with even a slight bump. Alfred Nobel discovered a safe way of using it in 1866. He absorbed the liquid nitroglycerine in sticks of a type of clay, called *kieselguhr*, to make DYNAMITE, which is harmless until a detonator is attached.

Guncotton

When cellulose (page 289) is reacted with a mixture of nitric and sulphuric acids all the hydroxyl groups in the molecule (three in each of the monomer units in the polymer chain) are replaced by nitrate groups. The product is called NITROCELLULOSE, and is also known as GUNCOTTON because cotton is the form of cellulose used in its manufacture. It is used in cartridges and to make GELIGNITE and CORDITE, both of which are mixtures of nitrocellulose and nitroglycerine.

TNT

Concentrated nitric acid will also nitrate hydrocarbon molecules, replacing hydrogen atoms with nitro ($-NO_2$) groups. With toluene it gives TRINITROTOLUENE (TNT). TNT can be used as an explosive on its own, but it is usually used in a mixture with other compounds to provide extra oxygen. A mixture of TNT with ammonium nitrate is called AMATOL.

Other explosives, such as LYDDITE and CYCLONITE, are also nitrated hydrocarbons.

Questions

1 What is an explosion?

2 In general, ammonia can be prepared in the laboratory by heating a mixture of an alkali and an ammonium salt. Nevertheless, it might be preferable to avoid using ammonium nitrate. Why?   (*East Anglia*)

3 Use an encyclopedia or other reference book to find out something about the life and work of Alfred Nobel.

5 Life-giving elements

Phosphorus

Phosphorus makes up 0.1 per cent of the earth's crust, mainly in the form of metal PHOSPHATES, which are salts of phosphoric acid, H_3PO_4. Phosphorus compounds are very important to all kinds of living things. Enzymes containing phosphorus act as catalysts in respiration, the reaction between carbohydrates and oxygen which provides living things with their energy (page 253). Our bones, and those of other animals, are composed chiefly of CALCIUM PHOSPHATE, $Ca_3(PO_4)_2$.

Animals get their phosphorus by eating plants or other animals. Plants get their phosphorus by taking in phosphates, dissolved in water, through their roots. Just as nitrogen compounds in the soil are depleted by intensive farming methods, so it is necessary to return phosphorus to the soil in the form of phosphatic fertilisers.

The commonest mineral containing phosphorus is *rock phosphate*, which is composed mainly of calcium phosphate. Calcium phosphate is insoluble in water and is therefore not much use as a fertiliser, since plants can only take in soluble compounds from the soil. It is therefore converted into a soluble compound, calcium hydrogen phosphate, by reaction with concentrated sulphuric acid:

$$Ca_3(PO_4)_2(s) + 2H_2SO_4(l) \rightarrow Ca(H_2PO_4)_2(s) + 2CaSO_4(s).$$

The mixture of calcium hydrogen phosphate and calcium sulphate is used directly as a fertiliser, called SUPERPHOSPHATE.

Another useful phosphatic fertiliser is blast furnace slag (page 155) which contains phosphorus compounds which were present as impurities in iron ore.

Some other essential elements

At least 24 elements are now known to be essential for the normal functioning of our bodies. CARBON, HYDROGEN, and OXYGEN are the components of carbohydrates and fats. Proteins, in addition to carbon, hydrogen, and oxygen, also contain NITROGEN and SULPHUR.

IRON is important as a component of the compound haemoglobin, which gives red blood cells their colour and enables them to carry oxygen around the body. An adult's body contains a total of about 5 g of iron and we need about $\frac{1}{100}$ g per day in our diet to maintain the manufacture of new cells.

We also need CALCIUM and PHOSPHORUS to make the calcium phosphate for our bones. An adequate supply of these elements, for instance from milk, is especially important for children whose bones are still being formed. It was because of the prevalence of rickets, a disease in which the bones do not harden properly and often bend, that free milk for children was introduced in Britain.

IODINE is needed in our diet to form the hormone thyroxine which controls our growth, and COBALT is needed to make vitamin B_{12}.

Most enzymes, the catalysts which control all the chemical reactions in our body, contain PHOSPHORUS, and certain enzymes also contain COPPER, IRON, or MOLYBDENUM. Apart from the enzymes themselves, a number of substances called CO-ENZYMES are necessary to enable the enzymes to function properly. So far it has been shown that no less than 16 metallic elements are necessary as co-enzymes: ALUMINIUM, CADMIUM, CAESIUM, CALCIUM, CHROMIUM, COBALT, COPPER, IRON, MAGNESIUM, MANGANESE, MOLYBDENUM, NICKEL, POTASSIUM, RUBIDIUM, SODIUM, and ZINC.

Questions

1 Calcium phosphate is a common artificial fertiliser. It is treated with

sulphuric acid and sold as 'superphosphate of lime'. What effect does the sulphuric acid have on the phosphate?   (London)

2 What are the functions of the following elements in the human body?

(a) calcium   (b) iron   (c) nitrogen   (d) phosphorus   (e) sulphur

3 Try to find out which items in our diet are good sources of compounds of the elements calcium, iodine, and iron.

4 It is said that the following elements are necessary for the proper growth of plants: nitrogen, phosphorus, potassium, and sulphur. Devise an experiment using cress seeds, which can be grown on wet blotting paper, to test the effects of sodium nitrate, sodium phosphate, potassium chloride, and sodium sulphate on the growth of this plant. Your teacher may be able to provide you with the materials to try out your experiment either at school or at home.

For the teacher

Haber process   Two syringes (100 cm^3 or 50 cm^3, glass or plastic) are connected to a 3-way stopcock and a 150 × 7 mm silica combustion tube as shown in Figure 12.12. The combustion tube is filled with steel wool. Flush the whole apparatus with nitrogen from a cylinder (obtainable from BOC) two or three times. Put 60 cm^3 of hydrogen and 20 cm^3 of nitrogen from cylinders into the syringe next to the stopcock (half these quantities for a 50 cm^3 syringe). Connect the two syringes with the stopcock, heat the combustion tube to redness, and pass the gases backwards and forwards across the steel wool two or three times. Allow the apparatus to cool and eject the gases through the stopcock on to a piece of moist universal indicator paper. Sufficient ammonia will have been formed to detect with the indicator.

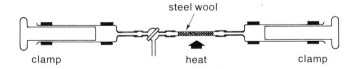

Figure 12.12

The preparation of nitrogen   If required, the preparation of nitrogen (contaminated with noble gases) from air can be demonstrated using the apparatus in Figure 12.13. Jars can be tested with indicator paper, for solubility, with a lighted splint, and with burning magnesium.

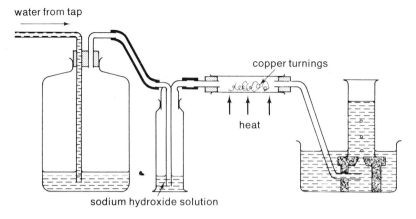

Figure 12.13

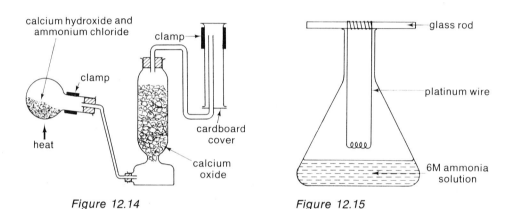

Figure 12.14

Figure 12.15

The preparation of ammonia If desired, the laboratory preparation may be demonstrated using the apparatus in Figure 12.14. Jars can be tested with universal indicator paper, for solubility, for density, with a lighted splint, and with a glass rod dipped in concentrated hydrochloric acid.

Catalytic oxidation of ammonia A very simple demonstration that a reaction does take place between ammonia and oxygen and that it is highly exothermic can be performed with the apparatus in Figure 12.15. The platinum wire is heated over a bunsen until it glows red then quickly lowered into the flask. It continues to glow, showing that an exothermic reaction is taking place.

A fuller demonstration of the first stage of the industrial process can be carried out with the apparatus in Figure 12.16. The platinised asbestos should be thoroughly dried by heating for a few minutes in a bunsen flame and cooling in a dessicator. 6M ammonia solution is placed in the conical flask and air is drawn through the apparatus by means of a filter pump. The calcium chloride tube is to absorb excess ammonia. When the platinised asbestos is heated to redness, brown

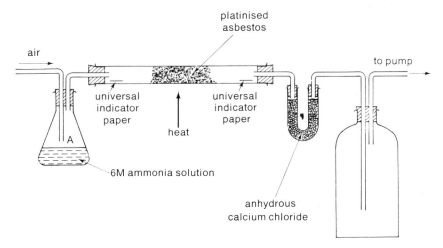

Figure 12.16

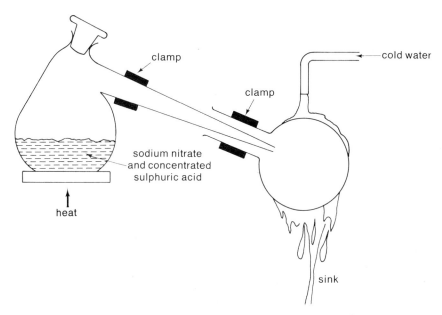

Figure 12.17

fumes of nitrogen dioxide appear in the aspirator bottle. For additional clarity pieces of moist universal indicator paper may be placed at either end of the combustion tube. The one at the ammonia inlet end turns blue and the one at the other end red.

The preparation of nitric acid If desired, the laboratory preparation may be demonstrated using the apparatus in Figure 12.17.

Nitrogen and other elements important to life 379

13 Limestone

Limestone is a rock composed mainly of calcium carbonate, $CaCO_3$. It is one of the most important raw materials of chemical industry. 50 000 000 tonnes a year are used in Britain alone. Calcium carbonate itself finds many uses, but it can also be converted into other important chemicals. When it is heated it loses carbon dioxide gas to give calcium oxide, known as *quicklime*:

$$CaCO_3(s) \rightarrow CaO(s) + CO_2(g).$$

Calcium oxide reacts with water to produce calcium hydroxide, commonly called *slaked lime*:

$$CaO(s) + H_2O(l) \rightarrow Ca(OH)_2(s).$$

It is important to remember the chemical names of these substances.

limestone	calcium carbonate
quicklime	calcium oxide
slaked lime	calcium hydroxide.

Unfortunately the word *lime* is often used to describe any or all of them. This can be very confusing since calcium carbonate, calcium oxide, and calcium hydroxide are all important industrial chemicals.

Experiment 13.1 Investigating limestone and some other substances

(a) To a small piece of limestone in a test tube add half a tube of dilute hydrochloric acid.
   Describe what you see happening. Carry out suitable tests to discover what gas is given off. What is the gas? What does this tell you about the elements present in limestone?
   (b) Dip a nichrome wire fitted into a glass rod into half a test tube of concentrated hydrochloric acid and heat the tip in the edge of a bunsen flame. Repeat the dipping and heating until the wire does not give any colour to the flame.

Now dip the wire into the hydrochloric acid, touch it on to a piece of limestone, and heat the tip in the flame. Describe what you see.

Clean the wire by repeatedly dipping in hydrochloric acid and heating as before. Then dip the wire into hydrochloric acid and into a little solid calcium chloride on a watch glass. Heat the wire in the flame. Describe what you see.

This procedure is called a *flame test*. Certain metallic elements impart a characteristic colour to a flame by which they can be identified. What element can you conclude must be present in limestone?

From experiments (a) and (b) of what three elements must limestone be composed?

(c) Repeat experiments (a) and (b) with as many of the following substances as you have time for: chalk, marble, calcite, eggshell, a sea shell. What can you conclude about all these substances?

Experiment 13.2 Further investigation of limestone

(a) Shake a spatula-full of powdered limestone with half a test tube of water. Does limestone appear to be soluble? Test the mixture with universal indicator paper. What pH is registered?

(b) Heat a small piece of limestone strongly in a hard-glass test tube. Do you see any evidence to suggest that the limestone is being decomposed by heating in this way?

Try heating a piece of limestone more strongly by holding it in tongs in a roaring bunsen flame for at least 5 minutes. Has the appearance of the limestone changed this time?

(c) Put the piece of limestone you have heated strongly into a dry test tube and add 2 or 3 drops of water from a teat pipette. Describe what happens. Test the mixture in the tube with universal indicator paper. What pH is registered?

What evidence is there that limestone has been decomposed by very strong heating?

What gas is given off when limestone is heated strongly like this? (The results of experiment 13.1 may help you to answer this.)

What is the residue remaining after limestone has been heated strongly? What is formed when water is added to this residue?

(d) Hold a piece of calcium metal in tongs in a roaring bunsen flame until it catches fire. What colour is the flame? Describe the appearance of the residue. What is the residue?

Drop the residue into a test tube and add half a tube of water. Shake well and filter into another test tube. What has the residue from burning the calcium been converted into?

Bubble carbon dioxide into the clear solution that passed through the filter paper. Describe what happens. What substance has been formed?

13.1 Limestone in nature

The oldest IGNEOUS ROCKS, the first to solidify when the earth cooled down, are composed mainly of silicates (page 341), including calcium silicate. Weathering by ice, wind, and water over millions of years changed some of the calcium silicate into soluble calcium salts. These were washed into the sea by rivers and used by small sea creatures to form their shells and skeletons, which are composed chiefly of calcium carbonate. As these animals died deposits of calcium carbonate were laid down on the ocean floor. Eventually the deposits were compressed into SEDIMENTARY ROCK. *Chalk*, the softest calcium carbonate rock, has been subjected to the least pressure during its formation. *Limestone* is a more highly compressed form of calcium carbonate. *Marble* is calcium carbonate that has been subjected to such high pressures and temperatures that it has been melted and then solidified again. Rocks like marble which have been changed in this way are called METAMORPHIC ROCKS. Sometimes calcium carbonate which has been melted inside the earth has cooled sufficiently slowly to crystallise. This is called *calcite*.

Calcium carbonate deposits are still being laid down on the ocean floor today. *Corals* are particularly spectacular deposits of crystalline calcium carbonate, often beautifully coloured by traces of impurities.

Of the three most important calcium carbonate minerals (chalk, limestone, and marble) limestone is by far the commonest. It contains between 65 per cent and 98 per cent of calcium carbonate together with impurities such as silicon oxide, iron oxide (which may give it a brownish colour), and magnesium carbonate. The purest forms of limestone were deposited on ocean beds far away from land. Where deposits were laid down near the shore they were more likely to be contaminated with silt and sand deposited where rivers meet the sea.

The British Isles are particularly rich in limestone rock (Figure 13.1). The earliest deposits were laid down about 350 million years ago. They are called *carboniferous limestone*, because they were deposited at about the same time that coal was being formed. These are the purest limestones in Britain. 'Younger' limestone, laid down between 270 and 70 million years ago, forms, for instance, the Chilterns, the Cotswolds and the Sussex Downs, and is much less pure.

The production and uses of limestone

Limestone is often found near the surface. After the topsoil has been removed the rock is quarried by blasting. Vertical holes are drilled behind the quarry face and filled with explosive. The charge is detonated electrically and a single blast can bring down as much as a quarter of a million tonnes of rock. If the charge is correctly laid the rock is blasted into fragments of a convenient size for handling by mechanical shovels. It is transported to the crushing plant and graded into lumps of suitable size for various purposes.

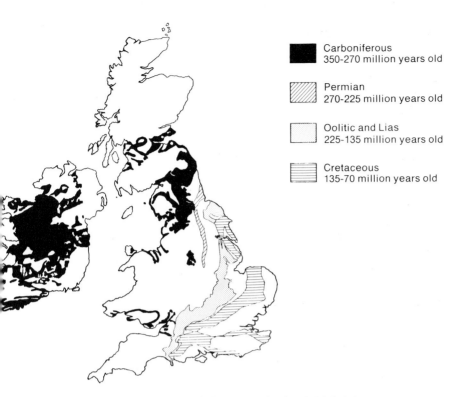

Figure 13.1 The distribution of limestone in the British Isles

Of the 50 million tonnes of limestone quarried each year in Britain about four-fifths is used directly in various industries. The remaining one-fifth is converted into calcium oxide or calcium hydroxide.

The biggest single use of limestone is in the manufacture of cement (page 347). It is also used as an aggregate for concrete (page 348). The iron and steel industry uses limestone as a flux to produce slag from the silicon oxide and other impurities in iron ore (page 153). Limestone is also used in the manufacture of glass (page 344) and sodium carbonate (page 417), and as a filler for rubber (page 303).

Questions

1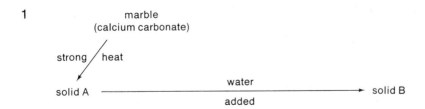

(i) Give the chemical name of A. (iii) Give the chemical name of B.
(ii) Give the common name of A. (iv) Give the common name of B.

Limestone 383

(v) Describe how, starting with B, a dry sample of calcium carbonate could be obtained.
(*East Anglia*)

2 The following equation represents the action of strong heat on calcium carbonate:

$$CaCO_3 \rightarrow CaO + CO_2.$$

The atomic mass of carbon is 12, of oxygen is 16 and of calcium is 40. Use the above information to answer the following questions:
   (i) What is the weight of a mole of calcium carbonate?
   (ii) If 200 g of calcium carbonate are heated until completely decomposed, what would be the weight of solid residue?
   (iii) What is this reaction used for producing industrially?
(*South East*)

3 A solid when treated with dilute hydrochloric acid gave off a gas which turned lime water milky. This tells us which of the following about the solid?
A It contained a carbonate.
B It was zinc carbonate.
C It contained powdered carbon.
D It was an element.
E It was a mixture of carbon and a metal oxide. (*West Midlands*)

4 In industrial areas old buildings constructed of limestone are badly pitted and worn. Explain the chemical changes involved.
(*South East*)

13.2 The manufacture of quicklime

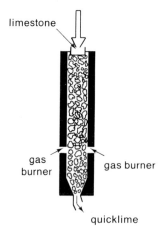

Figure 13.2 Lime kiln

Quicklime is calcium oxide. It is made by heating *limestone* (calcium carbonate), which decomposes giving off carbon dioxide gas:

$$CaCO_3(s) \rightarrow CaO(s) + CO_2(g).$$

This reaction requires a temperature of about 1200°C to make it proceed at a reasonably fast rate. The limestone is heated in a *kiln*. There are two common types of lime kiln. In the vertical kiln (Figure 13.2) limestone is fed in at the top and gradually falls through to be drawn off as lumps of calcium oxide at the bottom. Fuel is provided either by mixing the limestone with coke or by feeding powdered coal, gas, or oil into burners half-way down the kiln. The other type of kiln is the rotary kiln, similar to that used in the manufacture of cement (Figure 8.13, page 237). Limestone is charged into the upper end and the fuel is introduced through burners at the lower end. The limestone slowly falls through the length of the cylinder under gravity and quicklime falls out of the bottom into coolers.

Calcium oxide is a very reactive compound, and a strong base. It reacts violently with water, producing calcium hydroxide:

$$CaO(s) + H_2O(l) \rightarrow Ca(OH)_2(s).$$

Calcium hydroxide, being an alkali, can then absorb acidic carbon dioxide gas from the air, producing calcium carbonate:

$$Ca(OH)_2(s) + CO_2(g) \rightarrow CaCO_3(s) + H_2O(l).$$

Quicklime therefore needs careful transport and storage, and must be used as soon as possible after manufacture or it will be converted back to the calcium carbonate it was made from. The ease with which calcium oxide combines with water makes it a good drying agent and it is used for this purpose in the manufacture of many organic chemicals.

Quicklime in agriculture

Quicklime is used in agriculture for two purposes. As an alkali it neutralises soil which is too acidic for a particular crop (crushed limestone is also sometimes used for this purpose). It is also used to break up heavy clay soil to give it a finer texture ready for sowing. It does this by reacting with some of the water in the clay.

Quicklime for acetylene

When calcium oxide and *coke* (carbon) are heated together in an electric furnace at a temperature of 2000°C calcium carbide is formed:

$$CaO(s) + 3C(s) \rightarrow CaC_2(s) + CO(g).$$

Calcium carbide reacts with water to produce the welding gas, acetylene (page 285):

$$CaC_2(s) + 2H_2O(l) \rightarrow C_2H_2(g) + Ca(OH)_2(s).$$

.3 Slaked lime

Calcium hydroxide, produced by the reaction of calcium oxide with water, is known as *slaked lime*:

$$CaO(s) + H_2O(l) \rightarrow Ca(OH)_2(s).$$

Slaked lime is not very soluble in water. Its saturated solution contains about 0.1 g per 100 g of water at room temperature. This is the familiar *lime water* used as a laboratory reagent in testing for carbon dioxide gas. Lime water reacts with carbon dioxide to form calcium carbonate, which is very insoluble in water and appears as a fine white precipitate. We say the lime water turns *milky*:

$$Ca(OH)_2(aq) + CO_2(g) \rightarrow CaCO_3(s) + H_2O(l).$$

Calcium hydroxide is an alkali, and reacts with acids to form salts. For example:

$$Ca(OH)_2(s) + 2HCl(aq) \rightarrow CaCl_2(aq) + 2H_2O(l).$$

Since it is the cheapest alkali available to industry, most of its uses depend on this property. For ease of transport much calcium hydroxide is produced in the form of *milk of lime*, a suspension of solid calcium hydroxide in water made by reacting calcium oxide with a large excess of water.

Slaked lime is used in the manufacture of mortar (page 348), in which a proportion of calcium hydroxide gives a much stronger bonding material. Soaking in milk of lime is the first stage in the tanning of hides to make leather. Bleaching powder is manufactured from calcium hydroxide (page 415), and it is also used in the purification of sugar (page 254). Milk of lime is used in water purification (page 396).

Questions

1 Calcium oxide is made by strongly heating lumps of limestone. This process is called 'lime burning'.
   (i) Why is 'lime burning' a bad term to use?
   (ii) What is a by-product of the process?
   (iii) Give two large-scale uses of calcium oxide.   (*London*)

2 A farmer loses the labels from his lime and fertiliser bags. The fertiliser he uses is SULPHATE of AMMONIA.
   (a) Suggest how he could distinguish between samples of the lime and the ammonium sulphate by appearance.

His son has a simple chemistry set and wished to confirm the results. He performed two simple tests: first he warmed a sample of each substance with a little caustic soda (sodium hydroxide) and he found that the ammonium sulphate gave off a gas.
   (b) What gas was this?
   (c) Describe its smell.
   (d) What was its effect on damp litmus paper?
   (e) What would be observed in the lime/caustic soda experiment?

Secondly, the boy took a sample of each substance, shook it with water, filtered and then blew through the filtrate.
   (f) What happened in each case?

(g) Why would the farmer not add his lime and sulphate of ammonia to his land at the same time? (*West Yorkshire*)

3 A weighed lump of a shiny white solid is strongly heated in a bunsen flame for about five minutes. The lump loses weight, becomes dull white and is easily crushed to a white powder. When drops of cold water are put on the powder it swells up and becomes hot. A small amount of the powder is completely dissolved in water and a straw is used to blow into the solution until it turns milky. The milkiness is removed by filtration and is dried to give a white powder.

(a) What do you think the shiny white solid is?
(b) Explain the loss in weight after it has been heated.
(c) Name the white solid obtained after the heating and state one use for it.
(d) How do you know a chemical reaction is taking place when water is dropped on to the white powder?
(e) Explain the formation of the milkiness when the solution is blown into with a straw.
(f) Name the white powder obtained at the end.
(g) Describe in detail tests you would carry out to prove that the shiny white solid lump and the white powder obtained at the end are chemically identical. (*Lancashire*)

4 (a) What are the chemical names for chalk, quicklime and slaked lime? Describe *two* simple tests (you must use all three substances in each test) which could enable you to distinguish between samples of these substances.
(b) How would you prove, without the use of any other *laboratory* chemical, that the lime water in the bottle in your laboratory was fit for use in a test for carbon dioxide? (*South East*)

4 Carbon dioxide

Experiment 13.3   Some properties of carbon dioxide

To 2 or 3 small marble chips in a test tube add one-third of a tube of dilute hydrochloric acid. Describe and explain what happens.

(a) Hold a lighted splint just inside the mouth of the tube. What happens?
(b) Fit a delivery tube into the test tube as shown in Figure 13.3 and allow the gas to bubble through one-third of a tube of distilled water. Drop a piece of universal indicator paper into the water.
What pH is registered? What *two* properties of carbon dioxide can you deduce from this test?
(c) Hold one-third of a test tube of lime water under the tube containing marble chips and hydrochloric acid as shown in Figure

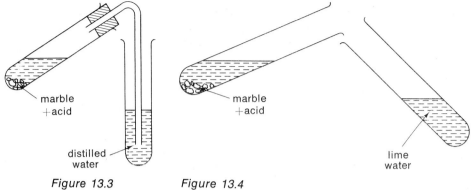

Figure 13.3   Figure 13.4

13.4 for about half a minute. Put your thumb over the end of the lime water tube and shake it.

What happens to the lime water? What **two** properties of carbon dioxide can you deduce from this test?

Some uses of carbon dioxide

Carbon dioxide, CO_2, is produced when any compound containing carbon is burnt. The atmosphere contains about 0.03 per cent of carbon dioxide. This concentration is kept reasonably constant by the balance of the carbon cycle (page 253). There is, however, real concern nowadays that our tremendous consumption of fuels will soon, if it has not already done so, upset this natural balance since all fuels, including coal, gas, and oil, produce carbon dioxide when they are burnt. No one knows for certain what the effect of this may be.

Although carbon dioxide is an important industrial chemical it does not have to be specially manufactured. It is obtained as a by-product in several manufacturing industries:
1 in making quicklime from limestone (page 384),
2 in fermenting carbohydrates to alcohol in the brewing and whisky industries (page 263), and
3 in the fractionation of liquid air to produce nitrogen and oxygen (page 360).

Carbon dioxide is a colourless, odourless gas with a density about $1\frac{1}{2}$ times that of air.

It is a weak acid and dissolves in water (1 volume of gas in 1 volume of water at room temperature and pressure) to produce a weakly acidic solution:

$$CO_2(g) + 2H_2O(l) \rightleftharpoons 2H_3O^+(aq) + CO_3^{2-}(aq).$$
$$\text{hydronium ion} \qquad \text{carbonate ion}$$

Carbon dioxide dissolved in the water provides sea-plants with the material for photosynthesis (page 252) and so makes life in the sea possible. The solution of carbon dioxide in water has a pleasant, slightly acid taste. Under pressure more than usual will dissolve and when the pressure is released the extra gas bubbles out. Soda water, lemonade, and other fizzy drinks contain carbon dioxide dissolved under pressure.

Carbon dioxide will not support burning. Cylinders of liquid carbon dioxide are used as fire extinguishers. When the gas is released it smothers the fire. This type of extinguisher has the advantage that it makes no mess at all, but it can only be used for fairly small fires.

Solid carbon dioxide is used as a refrigerant, for example in ice cream vans and in lorries for transporting frozen food. The solid sublimes at −78°C (passes directly from solid to gas without melting). Solid carbon dioxide is sometimes called *dry ice*.

Questions

1 (a) Figure 13.5 shows the apparatus which a student set up to prepare and collect a sample of carbon dioxide.

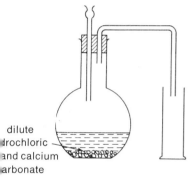

Figure 13.5

(i) What *two* errors did the student make?

(ii) Why is hydrochloric acid used in this experiment in preference to sulphuric acid?

(b) Carbon dioxide is denser than air.

(i) Describe a laboratory experiment which would demonstrate this property of carbon dioxide.

(ii) What important use is made of this property of carbon dioxide?

(*East Anglia*)

2 (a) Why is carbon dioxide used as a fire extinguisher?

(b) How would you deal with burning fat in a frying pan if no carbon dioxide were available?

(c) Why should a pupil not run from the laboratory into the playground if his clothing was on fire?

(d) What would you do to extinguish the fire on a pupil's clothing if no chemical extinguisher was available? (*Wales*)

3 Describe briefly the way in which carbon dioxide reacts with water, both PHYSICALLY and CHEMICALLY. (*West Yorkshire*)

4 Carbon forms two oxides. Make a table showing the difference (if any) between these two gases in respect of the following properties:

(i) action on litmus solution, (iv) the effect of a burning taper,
(ii) action of lime water, (v) the poisonous nature of the gas.
(iii) the effect of a glowing splint,
(The table should show the five properties of carbon dioxide and of the monoxide.) *(Middlesex)*

13.5 Hard water

Experiment 13.4 Examining tap water

(a) Put a single drop of tap water on the centre of a microscope slide. Holding the slide in tongs, wave it slowly back and forward about 10 cm above a medium bunsen flame until all the water has evaporated.

Examine the slide carefully, with a hand lens if necessary, and describe what you see.

(b) Repeat experiment (a) using a drop of distilled water in place of tap water.

How do you think the substances left on the microscope slide in experiment (a) got into the tap water?

How were they removed when the distilled water was made?

Experiment 13.5 What ions are responsible for hardness in water?

As rain runs across and through the ground before reaching our reservoirs it may dissolve substances from the rocks. Some common rocks in this country are: silicates, limestone (calcium carbonate, often with magnesium carbonate as an impurity), gypsum (calcium sulphate), rock salt (sodium chloride), and iron ore (iron oxides). Among possible ions which might be dissolved in tap water and which might cause hardness are:
cations—sodium, calcium, magnesium, iron;
anions—chloride, carbonate, sulphate, silicate.

You will be provided with solutions of sodium chloride, calcium chloride, magnesium chloride, sodium sulphate, iron(III) sulphate, magnesium sulphate, sodium carbonate, and sodium silicate.

To one-quarter of a test tube of soap solution add one-quarter of a test tube of one of these solutions. Put your thumb over the end of the tube and shake it well. If a good lather is produced which remains for at least half a minute, the water can be described as *soft*. If a lather which lasts for half a minute is not formed the water may be described as *hard*. Test each of the above solutions in turn and note down in each case whether they are hard or soft.

From the results of your experiments:
(i) Do any of the anions cause hardness?
(ii) Which cations cause hardness and which do not?

The causes of hardness

Before rainwater reaches reservoirs and then our taps at home it trickles across rocks and through the ground. As it does so small amounts of various compounds are dissolved. You have probably evaporated a drop of tap water on a microscope slide and seen the whitish deposit left behind. The main substances dissolved are various compounds of calcium and magnesium.

Not all tap water contains the same amount of dissolved solids. In Britain the concentration varies up to a maximum of about $\frac{1}{20}$ per cent by weight, depending on the type of rocks over which the rainwater has trickled. Water containing a lot of dissolved solids is known as HARD WATER. Pure water, containing hardly any dissolved material, is called SOFT WATER. Figure 13.6 (overleaf) shows the hardness of water in various parts of Britain, expressed in parts of dissolved solids per million parts of water by weight.

Types of hardness

For convenience, hardness of water is divided into two types according to the particular salts dissolved in it. Most tap water contains both types of hardness in various proportions.

1 TEMPORARY HARDNESS   Rainwater contains dissolved carbon dioxide from the air through which it has fallen. Carbon dioxide solution is weakly acidic (page 388) and all acids react with carbonates (page 212). The carbon dioxide in rainwater attacks the calcium carbonate in *limestone* to produce calcium hydrogen carbonate, which is soluble in water:

$$CaCO_3(s) + CO_2(g) + H_2O(l) \rightarrow Ca(HCO_3)_2(aq).$$

In the same way magnesium hydrogen carbonate is formed by the reaction between rainwater and magnesium carbonate in the rocks:

$$MgCO_3(s) + CO_2(g) + H_2O(l) \rightarrow Mg(HCO_3)_2(aq).$$

The hardness of water caused by calcium hydrogen carbonate and magnesium hydrogen carbonate is called *temporary hardness* because these salts are not very stable. When the water is heated they decompose by a reversal of the reaction by which they were formed. Calcium hydrogen carbonate is decomposed into calcium carbonate:

$$Ca(HCO_3)_2(aq) \rightarrow CaCO_3(s) + CO_2(g) + H_2O(l),$$

and magnesium hydrogen carbonate into magnesium carbonate. Calcium and magnesium carbonates are not soluble in water and are precipitated out.

2 PERMANENT HARDNESS   Calcium sulphate occurs in nature as *anhydrite*. It is not very soluble in water, but small quantities do dissolve. Water also picks up calcium chloride, magnesium sulphate, and magnesium chloride as it trickles through various types of rock. The chlorides and sulphates of

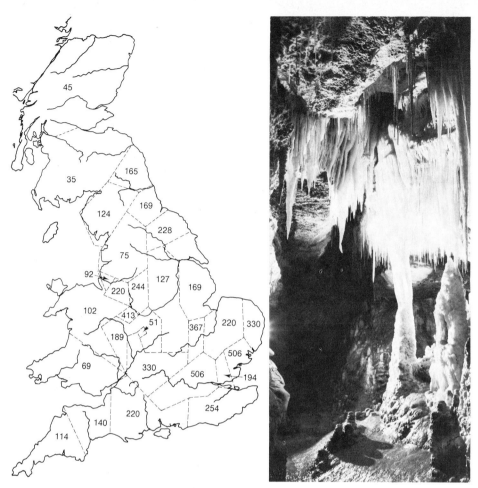

Figure 13.6 The hardness of water in the British Isles

Figure 13.7 Stalactites and stalagmites

calcium and magnesium cause hardness, but unlike the bicarbonates are stable and unaffected by heat. Hardness caused by these salts is therefore called *permanent hardness*.

Stalagmites and stalactites

In underground limestone caves, such as those found in the Cheddar Gorge, water seeping into the cave through the rock contains dissolved calcium hydrogen carbonate, produced by the action of rainwater containing dissolved carbon dioxide on the calcium carbonate (limestone):

$$CaCO_3(s) + CO_2(g) + H_2O(l) \rightarrow Ca(HCO_3)_2(aq).$$

As this temporarily hard water drips off the roof some evaporation takes place and because calcium hydrogen carbonate is rather unstable the reaction by which it was formed is partly reversed:

$$Ca(HCO_3)_2(aq) \rightarrow CaCO_3(s) + CO_2(g) + H_2O(l).$$

Calcium carbonate is gradually deposited on the roof of the cave. Over thousands of years this temporarily hard water dripping constantly from the same place produces a STALACTITE which 'grows' downwards from the roof (Figure 13.7). In the same way, at the point where the drops fall, a STALAGMITE is built up from the floor. Eventually the stalactite and stalagmite may link up to produce a pillar of calcium carbonate, often beautifully coloured by impurities in the water.

Experiment 13.6 Measuring hardness

(a) Measure 10 cm^3 of distilled water into a small conical flask. Use a teat pipette to add one drop of soap solution and shake the flask well. Do you get a lather which lasts for at least one minute?

(b) Measure 10 cm^3 of permanently hard water into a flask. Add soap solution, *one drop at a time*, shaking after each addition, until you get a lather which lasts for one minute.

How many drops of soap solution were required?

What did you notice forming in the flask when soap was added before a lather was produced?

(c) Measure 10 cm^3 of temporarily hard water into a flask and find out how many drops of soap solution are required to produce a lather lasting for one minute as in experiment (b).

How many drops of soap solution were required?

What did you notice forming in the flask when soap was added before a lather was produced? Was the appearance the same as in experiment (b)?

(d) Using the same method as in experiment (b) measure the total hardness of your tap water. How many drops of soap solution were required to remove all the hardness from 10 cm^3 of tap water?

Assuming that the temporarily hard water used in experiment (c) contains 750 parts per million by weight of dissolved solids, how many parts per million does your tap water contain? Compare your result with the average given for your area in Figure 13.6.

(e) Repeat experiment (d) using a sample of your tap water that has been boiled for 15 minutes.

Do you notice any difference in appearance after the tap water has been boiled?

How many drops of soap solution are required to remove the permanent hardness in 10 cm^3 of your tap water?

Using the results of this experiment and of experiment (d), calculate how many drops of soap solution are required to remove the temporary hardness in 10 cm^3 of your tap water.

Experiment 13.7 Removing hardness

For this experiment you will use the same samples of temporarily hard and permanently hard water as in experiment 13.6. Try each of the following tests on 10 cm³ of each type of water and then measure the hardness with soap solution as in experiment 13.6.

(a) Add 2 cm³ of *lime water* (a solution of calcium hydroxide) to 10 cm³ of each of the samples of hard water and shake.

Do either of the samples change in appearance?

Measure the hardness of each sample. Has the addition of calcium hydroxide affected the hardness of (i) the temporarily hard water, (ii) the permanently hard water?

(b) To 10 cm³ of each of the samples add a spatula-full of sodium carbonate and shake.

Do the samples change in appearance?

Measure the hardness of each sample. Has the addition of sodium carbonate affected the hardness of either sample?

(c) Repeat experiment (b) using a spatula-full of sodium polyphosphate.

(d) Fill a burette with ion-exchange resin. Run some of the temporarily hard water through the burette. Collect a 10 cm³ sample from the bottom of the burette and measure its hardness.

Does the water change in appearance after passing through the resin?

Is its hardness affected?

Repeat this experiment with some of the permanently hard water.

(e) If you have time you could also carry out experiments (a), (b), (c), and (d) with tap water.

The disadvantages of hard water

Figure 13.8 A section of a hot water pipe which has been almost blocked by deposits from temporarily hard water

1 When temporarily hard water is boiled, calcium and magnesium carbonates are precipitated. These usually take the form of a hard deposit known as *fur* or *boiler scale* which sticks to the side of the vessel. The carbonate deposit is a bad conductor of heat. An electric kettle whose heating element has become covered with fur can require twice as much electricity to boil the water. This may not make much difference to your electricity bill at home, but in industry where millions of gallons of water are being heated every day it can cost thousands of pounds a year.

Pipes can be gradually blocked up by these deposits from hard water (Figure 13.8). The flow of water is reduced and this may lead to loss of efficiency, for example in a central heating system. If pipes from a boiler become severely blocked steam pressure may build up inside the boiler and cause an explosion.

2 Soap is a mixture of the sodium salts of organic acids with long carbon-chain molecules (page 305). The calcium and magnesium salts of these acids are insoluble in water so soap forms a precipitate in hard water, known as *scum*. For example, with sodium stearate, one of the commonest salts in soap, and calcium sulphate from hard water:

sodium stearate + calcium sulphate →
(soap)
          calcium stearate + sodium sulphate.
          (scum)

In very hard water, such as London water, four-fifths of the soap goes to forming scum with the calcium and magnesium salts in the hard water, and only one-fifth for the actual washing. This may not make a lot of difference to the average household's shopping bill but in the laundry industry it would be very expensive. Scum is a nuisance, too. It gets trapped in the fibres of cloth during washing and is difficult to rinse out. It can discolour the material and even damage it as a result of the rubbing of the hard particles of scum against the fibres of the cloth.

The advantages of hard water

Absolutely pure water has a very dull, flat taste. Try drinking some distilled water. The presence of some dissolved salts improves the flavour of the water, although most people find very hard water unpleasant to drink.

Both calcium and magnesium are required by our bodies (page 376) and a substantial proportion of our requirements of these elements is obtained from drinking water. For drinking purposes, therefore, we do not want to remove hardness, but for most other purposes a great deal of money is saved by softening water before use.

Water softening

1 DISTILLATION If water is boiled and the steam condensed, all the dissolved salts are left behind. *Distilled water* is absolutely pure, but a great deal of heat is needed to make distilled water so this is not an economic purification process except for small quantities of water for special purposes where absolute purity is essential. Distilled water is used in the laboratory and also for topping up car batteries (page 333).

2 BOILING Boiling removes temporary hardness by decomposing the calcium and magnesium hydrogen carbonates to calcium and magnesium carbonates (page 391). This is also an expensive way of removing

hardness because of the amount of heat required. Also, of course, it does not remove hardness completely, but leaves the salts responsible for permanent hardness in the water.

3 ADDITION OF CALCIUM HYDROXIDE If calcium hydroxide (*slaked lime*) is added to water containing calcium bicarbonate, calcium carbonate is precipitated and the temporary hardness is removed:

$$Ca(HCO_3)_2(aq) + Ca(OH)_2(s) \rightarrow 2CaCO_3(s) + 2H_2O(l).$$

Exactly the right amount of calcium hydroxide must be added to react with the quantity of calcium hydrogen carbonate present. If too much calcium hydroxide is used a new source of hardness is introduced.

This method is cheap, since it uses slaked lime made from limestone (page 385). Only temporary hardness is removed, but some water authorities in limestone areas, where most of the hardness is temporary, use this method (for example, Canterbury and Southampton).

4 ADDITION OF SODIUM CARBONATE Sodium carbonate precipitates all the calcium and magnesium ions in hard water as calcium and magnesium carbonates, e.g.

$$Ca(HCO_3)_2(aq) + Na_2CO_3(s) \rightarrow CaCO_3(s) + 2NaHCO_3(aq);$$
$$MgSO_4(aq) + Na_2CO_3(s) \rightarrow MgCO_3(s) + Na_2SO_4(aq).$$

Sodium salts are left behind in the water, but these are harmless. They cannot react with soap, since soap is also a sodium salt, and they cannot form fur.

Sodium carbonate (*washing soda*) is used to soften water on a small scale, though it is not cheap enough for use in industry. Before detergents, which are not so much affected by hard water as soap (page 309), became popular, it was usual to add some washing soda to the home wash. *Bath salts* are sodium carbonate crystals, usually coloured and perfumed, and other bath preparations often contain sodium carbonate.

A modification of the sodium carbonate process for softening water is the use of *Calgon*. This is a trade name (CALcium GONe) for another sodium salt, sodium polyphosphate. Instead of precipitating the calcium and magnesium ions as calcium and magnesium carbonates, Calgon reacts with them to form complex calcium and magnesium phosphates which remain dissolved in the water but cannot form fur or react with soap.

5 ION-EXCHANGE PROCESSES Sodium aluminium silicate, though insoluble in water, has the property of exchanging other metal ions for its sodium ions. With calcium sulphate for instance:

calcium sulphate + sodium aluminium silicate $\rightarrow$
    calcium aluminium silicate + sodium sulphate.

When hard water is run through a tube containing pieces of sodium aluminium silicate all the calcium and magnesium ions are removed and replaced by sodium ions which do not cause hardness. This is called the ION-EXCHANGE process for softening water. The earliest forms of ion-exchange column used a natural form of sodium aluminium silicate (a type of clay) under the trade-name *Permutit*. Nowadays man-made organic polymers, called ION-EXCHANGE RESINS, are usually used.

The advantage of this process is that when all the sodium ions in the ion-exchange resin have been used up, the column can be regenerated by pouring a strong solution of sodium chloride through it. This reverses the ion-exchange reaction:

sodium chloride + calcium aluminium silicate → calcium chloride + sodium aluminium silicate,

and the column is ready for further use. The overall effect is that sodium chloride (*common salt*) is being used to soften the water: a very cheap method. On a large scale some water authorities soften their water by this method (for example, Henley and Cambridge) and it is often used in industry.

Absolutely pure water can also be made by an ion-exchange process. Two columns are usually used. The first contains an ion-exchange resin which removes all metal ions, replacing them by hydronium ions. The second column contains another type of resin which removes all the anions, replacing them by hydroxide ions. The hydronium ions from the first column and the hydroxide ions from the second combine together to form water:

$$H_3O^+(aq) + OH^-(aq) \rightarrow 2H_2O(l),$$

so pure water comes out of the second column. The first column is regenerated by pouring concentrated hydrochloric acid through it to replace its hydronium ions. The second column is regenerated with sodium hydroxide solution which replaces its hydroxide ions. This type of ion-exchange process is often used in laboratories nowadays to produce water which is just as pure as distilled water and cheaper to make.

Questions

1 Describe what you would expect to see when a small quantity of soap is shaken with (a) distilled water, (b) soft water, (c) hard water.
(Middlesex)

2 Explain why:
   (a) Some water supplies are hard whilst some are soft.
   (b) Hard water does not readily form a lather with soap.
   (c) The cleansing action of liquid detergents is unaffected by hard water. *(North West)*

3 Hard water does not lather readily and leaves scum when treated with soap. This is because hard water contains which of the following?
A   particles of insoluble calcium carbonate
B   soluble salts of calcium and magnesium
C   dissolved carbon dioxide

D potassium compounds in solution
E particles of insoluble sulphates (*West Midlands*)

4 Look carefully at Figure 13.1, page 383, and at Figure 13.6, page 392.
 (a) What connection is there between the two maps?
 (b) In what areas of the country do you see a particularly clear connection?

5 25 cm³ of water from three different districts are treated with soap solution until a permanent lather forms in each case. The same volume of water from each district is boiled and re-treated with soap. Three further samples are tested with soap, but this time after they have passed through a good water softener. The results are shown in Figure 13.9.

District	Volumes of soap used (cm³)		
	No treatment	Boiled	Water softener used
A	10.0	1.4	1.5
B	13.3	13.3	1.3
C	19.9	16.9	1.9

Figure 13.9

 (i) Which district has the hardest water?
 (ii) What type or types of hardness are present in (a) District A, (b) District B, (c) District C?
 (iii) What compound could cause the hardness found in the water from District B?
 (iv) What type of rock was responsible for the formation of the hardness in the water from District A?
 (v) If tests were carried out with rain water approximately what volume of soap would be required to give a lather with 25 cm³ of rain water? (*West Yorkshire*)

6 Say why it is best for a domestic water supply to be (i) not too hard (ii) not too soft. (*Wales*)

7 (a) Describe *two* everyday uses of carbon dioxide.
 (b) Explain why limestone is very slightly soluble in rain water but insoluble in pure water. (*Wales*)

8 A small lump of calcium carbonate was heated very strongly for

about five minutes, leaving a residue of a white powder A. This powder was allowed to cool and then placed in a test tube. A little cold water was added and the mixture was shaken. The test tube became very hot. A white solid B was filtered off, leaving an alkaline solution C. Carbon dioxide was passed into this alkaline solution, producing a white suspension D, which redissolved on passage of further carbon dioxide, forming a colourless solution E. When this solution was heated a colourless gas F was evolved and the white suspension D reappeared.

Identify the substances A to F and write equations for the reactions occurring. *(North West)*

9 Temporary hardness of water can be removed by each of the following methods EXCEPT one. Which one?
A adding soap solution
B boiling the water
C using ion-exchange resins
D adding washing soda
E adding sodium chloride.
(London)

For the teacher

Samples of limestone, chalk, and marble are available from Griffin and George. Calcite may be obtained from Lythe Minerals or R. F. D. Parkinson.

Experiment 13.1 2 M hydrochloric acid. 3 cm of 26 s.w.g. platinum wire fused into a 10 cm glass rod makes an ideal flame test wire. 5 cm of 18 s.w.g. nichrome wire provides a much cheaper alternative, but has the disadvantage that it fuses into glass rod much less satisfactorily.

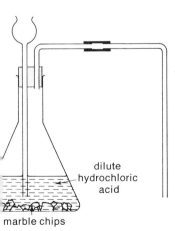

Figure 13.10

Experiment 13.2 A cylinder of carbon dioxide is ideal. Alternatively a Kipp's generator can be set up with marble chips and 4 M hydrochloric acid or the generator shown in Figure 13.10 can be used.

Experiment 13.3 2 M hydrochloric acid.

Laboratory preparation of carbon dioxide If desired, the full laboratory preparation can be demonstrated using the apparatus shown in Figure 13.10 and collecting the gas in jars by the upward displacement of air or over water. Jars can be tested with universal indicator paper; a lighted splint; for density; for solubility in water; with lime water; and with burning magnesium ribbon.

Dry ice If a cylinder of carbon dioxide with an expansion nozzle is available, a specimen of dry ice can be produced by wrapping a piece of cloth around the nozzle and turning the cylinder full on.

Experiment 13.5 Standard soap solution (also required in experiments 13.6 and 13.7) can be made by dissolving 10 g of soap (proprietary soap flakes or Castile powder) in 1000 cm^3 of distilled water and adding 1000 cm^3 of industrial methylated spirits. Salt solutions should all be approximately 0.1 M in distilled water.

Demonstration of formation of temporarily hard water Pass carbon dioxide into 100 cm^3 of lime water diluted with an equal volume of distilled water until calcium carbonate is just precipitated. The precipitate of calcium carbonate corresponds to the limestone rock. Continue passage of carbon dioxide until the precipitate redissolves (the effect of rain water containing dissolved carbon dioxide on limestone). Test a little of the solution with soap solution to show that it is hard. Divide the remainder into two parts. Stand one part on one side for a few days; a slight precipitate of calcium carbonate will form on the surface (formation of stalactites and stalagmites). Boil the other half for about 15 minutes in a conical flask with a delivery tube dipping into a little lime water; carbon dioxide is evolved. Filter off the precipitate of calcium carbonate and test it with 2 M hydrochloric acid and lime water to show that it is a carbonate. Test the filtrate with soap solution to show that it is no longer hard.

Experiment 13.6 Soap solution as for experiment 13.5. For temporarily hard water dilute lime water with an equal volume of distilled water and pass carbon dioxide until the precipitate of calcium carbonate redissolves. For permanently hard water use a saturated solution of calcium sulphate diluted with an equal volume of distilled water. If tap water is also to be examined it saves time to provide pupils with tap water that has been boiled for 15 minutes rather than having them boil their own.

Experiment 13.7 Any sodium zeolite or cation-exchange resin is suitable for (d). For complete de-ionisation, CLEAPSE have a pamphlet describing a simple de-ioniser which can be cheaply and easily made.

14 Salt

As rain water trickles through rocks, forming streams and then rivers, it dissolves soluble compounds and carries them into the sea. These compounds have been accumulating in the seas of the world ever since they were formed hundreds of millions of years ago. On average sea water contains $3\frac{1}{2}$ per cent by weight of dissolved substances, though the composition varies considerably from one sea to another. The Baltic Sea contains less than 1 per cent of dissolved solids and the Dead Sea about 30 per cent. Over three-quarters of this dissolved material is sodium chloride, known as *common salt*. Figure 14.1 shows the average proportions of some of the other salts present in sea water.

	Grams of substance per tonne (10^6 g) of sea water
Sodium chloride	27 500 g
Magnesium chloride	4 000 g
Magnesium sulphate	2 500 g
Calcium sulphate	1 500 g
Potassium chloride	750 g
Calcium carbonate	100 g
Potassium bromide	100 g

Figure 14.1

Sodium chloride is an essential part of the diet of man and other animals. Since primitive times it has been obtained from sea water by evaporation. In hot countries this is still done with the aid of the heat of the sun. Sea water is trapped in a *salt pan* and when all the water has evaporated away, the salt is collected. In colder climates, sea water was evaporated in pans over a fire.

Figure 14.2   Salt obtained by evaporation of sea water

14.1   Salt from the earth

Rock salt

About 150 million years ago Britain was joined to the rest of Europe by a narrow neck of swampy sand dunes (Figure 14.3). A landlocked sea covered the central part of the British Isles. At that time the European climate was hot and dry, so that this inland sea gradually dried up. The salt concentration became higher and higher until all the water had evaporated, leaving deposits of salt up to 50 metres thick. (At the present time the Dead Sea, which is also an inland sea, is drying up in exactly the same way.)

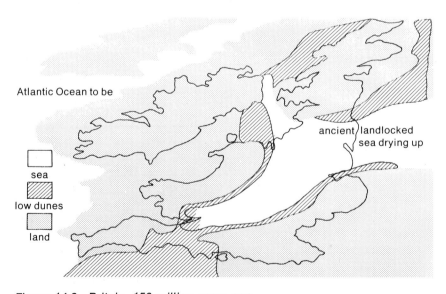

Figure 14.3   Britain, 150 million years ago

Hot, dusty winds from the East blew continuously and covered the salt beds with a red clay dust called *marl* to a depth of hundreds of metres. During the Ice Age, 20000 years ago, the layers of salt and marl were covered with thick ice. The weight of the ice compressed the marl into rock. In this way, by a process lasting 200 million years, beds of *rock salt* were deposited which now form the basis of the industry which provides all the salt used in this country. Rock salt deposits were formed in many other parts of the world in much the same way. 75 million tonnes of salt are used throughout the world each year. About two-thirds of this is obtained from rock salt deposits and the rest from the sea.

Mining rock salt

In Britain today there is only one working salt mine, at Winsford in Cheshire, where solid rock salt is obtained by ordinary mining methods. This one mine produces over one million tonnes a year, most of which is used, without refining, for clearing roads of ice and snow (page 172). It is also used as an ingredient in animal feed and as a fertiliser for sugar beet and mangolds.

During the eighteenth century salt mining in Britain was a very haphazard business. Mines were continually falling in, leaving huge holes underground. Rain water soaked down and filled up these holes, producing underground lakes of sodium chloride solution, known as *brine*. This made possible another method for obtaining salt in place of direct mining. The brine is pumped up to the surface, in effect mining salt in liquid, rather than in solid form. Nowadays most salt is obtained in Britain by this method. Water is pumped down into the rock salt deposits and brine is pumped back up to the surface.

Salt from brine

Up to the beginning of the twentieth century the method of obtaining salt from brine had remained the same for 3000 years. The boiling pans increased in size from less than a metre square to tanks the size of swimming baths, but the method was still to heat the pans, however large, by a fire underneath. This method required the heat from half a tonne of coal for each tonne of salt produced, which made salt rather expensive.

The first modern plant, using a different method, was built in 1905 and now almost all salt from brine is made in this way. The plant consists of three closed cylindrical vessels (Figure 14.4). Each vessel contains a steam chamber surrounded by brine. Steam from an outside source (often waste steam from electric power station turbines) is piped into the first steam chamber, where it boils the brine. The boiling brine in the first vessel is now generating its own steam, which is piped into the steam chamber of the second vessel. This steam gives up its heat to the brine and in doing so condenses, thus reducing the pressure (a similar effect to the 'collapsing can' experiment which you may have seen). At the lower pressure the brine boils

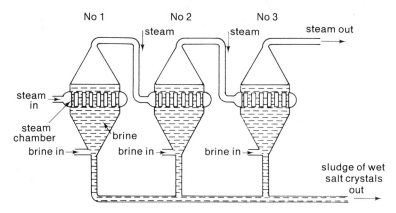

Figure 14.4 The manufacture of salt in a steam evaporating plant

at a lower temperature (page 50). Steam from the boiling brine in the second vessel is piped into the third steam chamber. Once again the condensation of the steam produces a pressure drop so that in the third vessel the brine boils at about 40°C. As the water evaporates, some salt crystallises out, and a wet sludge of salt crystals is tapped off from the bottom of each vessel. This is pumped to filters or centrifuges to separate the crystals. This method of evaporating brine is obviously much more economical in fuel consumption than the old open-pan method.

The salt produced by evaporation is much purer than rock salt, which contains about 5 per cent of marl. When brine is pumped up from underground the marl, which is insoluble in water, is left behind. Any remaining impurities are left in the brine as the salt crystallises out.

Salt is used in producing all kinds of foodstuffs. Bread, butter, margarine, and cheese all require salt for flavour. The first stage in curing bacon is sprinkling the sides of meat with salt. Herrings and kippers also require salt for curing. In canning and pickling vegetables salt is used as a preservative. Salt is also used to preserve hides in tanning. A quite different use for salt is in the manufacture of soap (page 306).

Salt is one of the most important raw materials in the heavy chemical industry. It is required for the manufacture of sodium, chlorine, hydrogen, sodium hydroxide, hydrochloric acid, sodium carbonate, sodium bicarbonate, bleaches, and weedkiller.

Experiment 14.1   Pure salt from rock salt

One-quarter fill a 100 cm³ beaker with water. Add 6 spatulas-full of crushed rock salt.

Stand the beaker on a wire gauze on a tripod and warm the mixture with a medium bunsen flame, stirring with a glass rod until all the salt has dissolved.

When the salt seems to have dissolved filter the mixture while it is still hot, and collect the salt solution in an evaporating dish.

Stand the dish on a tripod and heat it with a medium bunsen flame until most of the water has been boiled away.

Stop heating when nearly all the water has been boiled away and the mixture is beginning to go solid. Transfer the dish to a steam bath as shown in Figure 7.6, page 201.

Boil the water in the beaker until the heat from the steam has driven the last traces of water from the salt in the evaporating dish.

Why do you think a steam bath is used to evaporate the last traces of water from the salt?

Tip your salt on to a piece of paper and examine the crystals carefully with a hand lens. Make a drawing to show the shape of the salt crystals that you see.

Questions

1 Salt (sodium chloride) is one of the most important materials used in the chemical industry in this country. Where is this salt found and how did it get there? *(South East)*

2 For this question you will need to make use of the data in Figure 14.1, page 401.

(a) The formula of sodium chloride is NaCl. What is the weight of a mole of sodium chloride? What fraction of sodium chloride by weight is sodium? What weight of sodium could be obtained from a tonne of average sea water? (Atomic masses: Na = 23, Cl = 35)

(b) From your knowledge of the periodic classification what must be the formula of potassium bromide, since the formula of sodium chloride is NaCl? What weight of bromine could be obtained from a tonne of average sea water? (Atomic masses: K = 39, Br = 80)

Try to find out something about how bromine is extracted from sea water.

3 Find the Dead Sea on a map.

(a) Why is the Dead Sea drying up whereas the Red Sea is not?

(b) From the geographical position of the Dead Sea, what do you think the climate in the area is like?

(c) Imagine you are living near the Dead Sea. Explain how you could set out to manufacture salt as a commercial enterprise. (Remember that fuel consumption is usually the main item of expenditure in salt production, and this must be kept to a minimum.)

(d) To what industries might you be able to sell the salt you produce?

14.2 The electrolysis of molten sodium chloride

Sodium chloride is an electrolyte, composed of sodium and chlorine ions:

$$NaCl = Na^+ Cl^-.$$

In the crystal the sodium and chlorine ions are held together in a regular pattern (Figure 5.9, page 117) by the strong electrical forces between ions with opposite charges. When the solid is heated the ions vibrate more and more vigorously until the vibration becomes so violent that the regular structure breaks up. The breakdown occurs at 808°C: the sodium chloride melts. The ions are then moving around freely and at random.

When an electric current is passed through molten sodium chloride, sodium ions are attracted towards the negative electrode (cathode). Their charges are neutralised by electrons and they become sodium atoms:

$$Na^+ + e \rightarrow Na.$$

(Remember that e means *one mole of electrons*.) The chlorine ions are attracted towards the positive electrode (anode). They give up their extra electrons, and join in pairs to form molecules of chlorine gas:

$$2Cl^- \rightarrow Cl_2 + 2e.$$

In this way sodium and chlorine can be manufactured from salt.

Unfortunately there are two problems. First, the melting point of sodium chloride (808°C) is near the boiling point of sodium (890°C). To conduct the electrolysis at a temperature high enough to keep sodium chloride molten would be very hazardous, since sodium vapour explodes violently on contact with air. Instead of pure sodium chloride a mixture of sodium chloride and calcium chloride is used. This electrolyte can be kept molten at 600°C. More electrical energy is needed to discharge calcium ions than to discharge sodium ions, so only sodium is produced at the cathode. The calcium ions remain in the molten electrolyte.

The second problem is that the sodium and chlorine produced at the cathode and anode must be kept separate from one another to prevent them recombining to form sodium chloride. The anode and cathode are therefore separated by a cylinder of steel gauze with a fine mesh. This allows ions to pass through but not the molten sodium or bubbles of chlorine gas.

Figure 14.5 shows the electrolysis cell used commercially for the manufacture of sodium and chlorine from sodium chloride. A current of 30 000 amps at about 6 volts is used so that the electrolysis proceeds very rapidly. Each cell produces about 25 kilograms of sodium per day and it is usual to have 50 cells connected in series operating simultaneously.

Sodium is used to make sodium cyanide which is a useful pesticide and is required in the extraction of gold and in electroplating (page 126). Sodium is also used for the manufacture of the petrol anti-knock additive

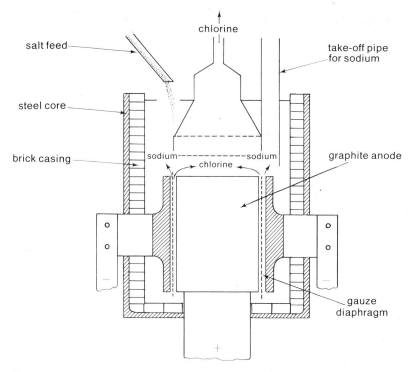

Figure 14.5 The electrolysis of molten sodium chloride

tetraethyl lead (page 280), and in the production of titanium (page 188), by reduction of titanium chloride:

$$TiCl_4(l) + 4Na(l) \rightarrow Ti(s) + 4NaCl(l).$$

A quite different use for sodium is in certain types of nuclear power stations to transfer heat from the reactor to the steam for the turbines (page 101).

3 The electrolysis of brine

Various kinds of electrolytic cell are used for the electrolysis of brine, but in Britain the commonest type is the *mercury cell* (Figure 14.6), which operates with a current of 30 000 amps at about 4 volts.

The chlorine ions are attracted towards the carbon anodes. They give up their extra electrons and join up in pairs to become molecules of chlorine gas, which is dried by passing it up a tower down which concentrated sulphuric acid is trickling. It is then liquefied by compressing it to about 3 atmospheres.

Salt   407

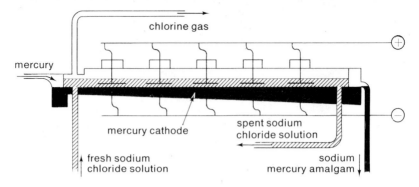

Figure 14.6 The electrolysis of brine in a mercury cell

The cathode is a layer of mercury which flows slowly across the bottom of the cell. Sodium ions are attracted towards the cathode, where they receive electrons to neutralise their positive charges, giving sodium atoms. Sodium is soluble in mercury, so a very dilute solution of sodium in mercury flows out of the bottom of the cell. This solution is reacted with water, giving hydrogen gas and a solution of sodium hydroxide:

$$2Na(\text{in mercury}) + 2H_2O(l) \rightarrow 2NaOH(aq) + H_2(g).$$

The mercury is returned to the electrolysis cell. For most industrial purposes solid sodium hydroxide is not required. The dilute solution is concentrated to 50 per cent by evaporation, using a multiple vacuum evaporator similar to that used in the production of salt from brine (Figure 14.4, page 404).

The electrolysis of sodium chloride solution is a very useful manufacturing process. The raw material is *brine* pumped up from rock salt beds. It needs only a little purification before use and is quite cheap. The expensive item is the electricity. 20 tonnes of coal are needed to produce enough electricity to make 10 tonnes of chlorine, $11\frac{1}{2}$ tonnes of sodium hydroxide, and $\frac{1}{4}$ tonne of hydrogen.

Making three useful products in a single process does, however, pose problems. Sodium hydroxide and chlorine have been manufactured since 1893 by the electrolysis of brine. When the process was first developed the product that was in great demand was the sodium hydroxide. At first there was little use for the chlorine. Gradually new uses were found for it and today the demand for chlorine is greater than the demand for sodium hydroxide. This is mainly the result of the tremendous expansion in chemicals manufactured from petroleum, many of which require chlorine at some stage in their manufacture (page 289). The electrolysis of brine becomes uneconomic if a lot of the sodium hydroxide produced cannot be sold. New processes for making chlorine are now being investigated, based on hydrogen chloride gas which is often a by-product of the petrochemicals industry.

Questions

1 Name *four* important chemicals obtained on an industrial scale from sea water or seaweed. (*Wales*)

2 Figure 14.7 is a simplified illustration of the cell used to manufacture sodium. The electrolyte is a molten mixture of sodium chloride and calcium chloride. A metal gauze separates the cathode and anode. Electrolysis is carried out with a current of 30000 amperes.

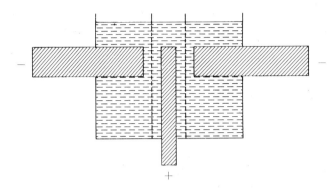

Figure 14.7

(a) The anode is made of graphite. Why?
(b) What is the purpose of the metal gauze?
(c) Why is a mixture of sodium chloride and calcium chloride, rather than pure sodium chloride, used as electrolyte?
(d) Sodium ions are discharged at the cathode. What other two types of ion are present in the electrolyte?
(e) One of the other types of ion is discharged at the anode. What happens to the third type of ion?
(f) The material in the vessel is kept molten at a temperature of about 600°C, yet no external heat is applied. Suggest a reason for this.
(g) There are approximately 90000 seconds in a day. How many coulombs would flow through the cell in a day?
(h) How many faradays is this? (1 faraday = 100000 coulombs.)
(i) A mole of sodium ions can be represented Na^+. Approximately what weight of sodium is produced per day in this cell? (Atomic mass of sodium = 23)
(j) The atomic mass of chlorine is 35. Approximately what weight of chlorine is produced per day?

3 Draw a labelled diagram of the apparatus you would use to collect the products of the electrolysis of strong brine, using carbon electrodes. (*Middlesex*)

4 The electrolysis of brine is a very important industrial process.
(i) What is brine?
(ii) Name the three main products from the industrial electrolysis of concentrated brine and give one important use of each product.
(iii) Special cells are used in the electrolysis of brine in industry. What is the essential purpose of such cells? (You are not asked to describe a cell.) (*West Midlands*)

5 Some submarines are propelled, under water, by electric motors supplied from batteries. Explain why just a little sea water, leaking into a submarine near the batteries, is dangerous. (*Wales*)

14.4 Chlorine

Experiment 14.2   Investigating salt gas

(a) Put a spatula-full of salt into a test tube. Add two or three drops (**not more**) of concentrated sulphuric acid. **This needs great care as the concentrated acid is very corrosive**. Describe what you see.
Hold a piece of wet universal indicator paper near the mouth of the tube. What happens? What does this tell you about the gas (*salt gas*) which is produced in the reaction?
Hold a lighted splint just inside the mouth of the tube. What happens? What does this tell you about salt gas?
(b) Add another two or three drops of concentrated sulphuric acid to the test tube. Wait for 10 seconds and then turn the tube upside-down and quickly dip the open end below some water in a beaker.
What happens? What does this tell you about salt gas?
From the result of the universal indicator test what class of chemical compounds do you think salt gas belongs to? What element do most compounds belonging to this class contain? Try to devise an experiment to prove the presence of this element in salt gas.

Experiment 14.3   Further investigation of salt gas

You have seen that one of the elements present in salt gas is hydrogen. The next step is to find out what else there is in this gas. The easiest way to do this will be to remove the hydrogen and see what is left. You may remember that hydrogen can easily be made to combine with oxygen to form water. Can you think of a substance which contains a lot of oxygen (an oxidising agent) which might remove the hydrogen from salt gas? (Try to remember which substances give off oxygen when heated.)
In experiment 14.2 you found that salt gas dissolves very easily in water. Since it is easier to do experiments with liquid substances than

with gases you can now use a solution of salt gas in water instead of salt gas itself.

Put a spatula-full of an oxidising agent into a test tube and add 10 drops of a solution of salt gas in water. If there is no immediate reaction warm the mixture very gently.

Describe what you see happening. Test any gas produced with a lighted splint and with wet universal indicator paper. Make a note of the results of these tests.

What gas do you know that has these properties?

What two elements can you now say are present in salt gas?

If these are the only two elements which go to make up salt gas, what will be the proper chemical name for salt gas?

If these are the only two elements in salt gas, can you suggest a way in which salt gas could be made (other than from salt and concentrated sulphuric acid)?

The properties of chlorine

Chlorine is a poisonous, greenish-yellow gas with a smell that is easily recognised (the smell of swimming baths and bleach). It is more than twice as dense as air.

Reaction with elements Chlorine is a very reactive gas. It combines directly with nearly all other elements, metals and non-metals, to form chlorides. The only exceptions are nitrogen, oxygen, carbon, and the noble gases. For example:

$$\text{iron} + \text{chlorine} \rightarrow \text{iron(III) chloride,}$$
$$2\text{Fe(s)} + 3\text{Cl}_2\text{(g)} \rightarrow 2\text{FeCl}_3\text{(s)};$$
$$\text{sulphur} + \text{chlorine} \rightarrow \text{disulphur dichloride,}$$
$$2\text{S(s)} + \text{Cl}_2\text{(g)} \rightarrow \text{S}_2\text{Cl}_2\text{(l)}.$$

The reaction between chlorine and metals is catalysed by water. Dry chlorine reacts much more slowly with metals than does wet chlorine. Liquid chlorine is transported in steel cylinders, drums, and road and rail tankers. Since the liquefied gas is dry there is no reaction with the steel. Moist chlorine quickly attacks steel at room temperature.

Reaction with hydrogen Chlorine combines directly with hydrogen to form HYDROGEN CHLORIDE gas:

$$\text{H}_2\text{(g)} + \text{Cl}_2\text{(g)} \rightarrow 2\text{HCl(g)}.$$

It is a colourless gas with a choking smell. It is very soluble in water, and the solution is known as HYDROCHLORIC ACID:

$$\text{HCl(g)} + \text{H}_2\text{O(l)} \rightarrow \text{H}_3\text{O}^+\text{(aq)} + \text{Cl}^-\text{(aq)}.$$

When hydrogen chloride gas meets water vapour in the air it produces a cloudy mist, consisting of fine droplets of hydrochloric acid. Concentrated hydrochloric acid is a saturated solution of the gas in water, containing about one-third of hydrogen chloride by weight.

Hydrochloric acid is the second cheapest commercial acid (after sulphuric acid). It is often used in place of sulphuric acid in the pickling of metals before galvanising, plating, or enamelling. The acid dissolves the oxide film off the surface, providing a perfectly clean metal surface for the coating to stick to.

Many drugs are organic bases which are insoluble in water. They are usually supplied in the form of their salts with hydrochloric acid, which are soluble in water and can therefore be injected. *Novocaine* is one of the commonest local anaesthetics used by dentists. It is injected as a solution of *novocaine chloride*.

Reactions with hydrocarbons Chlorine reacts easily with hydrocarbons, replacing their hydrogen atoms. With methane, CH_4, chlorine gives CH_3Cl, CH_2Cl_2, $CHCl_3$, and CCl_4 (page 290). Nearly three-quarters of all manufactured chlorine is used to make all kinds of organic compounds. These include: degreasing solvents (page 290); insecticides (page 290); plastics, particularly *PVC* (page 294); fire-extinguishing liquids such as bromochlorodifluoromethane, $CBrClF_2$, known as *BCF*, which is used in car and aircraft extinguishing systems; dyes and drugs.

Reaction with water Chlorine is quite soluble in water. $2\frac{1}{2}$ volumes of the gas dissolve in 1 volume of water at room temperature. The solution of chlorine in water first gives an acid reaction with indicator paper, then bleaches the dyes of the indicator. This is because chlorine reacts with water to form two compounds, hydrochloric acid and chloric(I) acid:

$$Cl_2(g) + H_2O(l) \rightarrow HCl(aq) + HClO(aq).$$

Both of these products are acids, but chloric(I) acid is also a bleach. Like all bleaches, it contains extra oxygen which combines with the dye to form a colourless compound:

$$HClO + \text{dye} \rightarrow HCl + \text{oxidised dye.}$$
$$\text{(coloured)} \qquad\qquad \text{(colourless)}$$

A solution of chlorine is used in many major industries, for instance to bleach paper, cotton, linen, and rayon.

The extra oxygen in chloric(I) acid also gives it disinfectant properties: the surplus oxygen is lethal to bacteria. It was in 1905 that drinking water was first sterilised with chlorine, to combat an epidemic of typhoid fever at Lincoln. Today chlorination of public water supplies is the commonest method of sterilisation all over the world. Chlorination is also used to purify the water in swimming baths and in sewage disposal.

Some important chlorides

POTASSIUM CHLORIDE, KCl, is used as a fertiliser to provide the potassium essential to the growth of plants.

AMMONIUM CHLORIDE, NH_4Cl, is formed by the reaction between hydrogen chloride and ammonia gases:

$$NH_3(g) + HCl(g) \rightarrow NH_4Cl(s).$$

When the gases react together a dense white smoke of ammonium chloride particles is produced. Ammonium chloride is used in dry cells (page 331).

ZINC CHLORIDE, $ZnCl_2$, is used as a flux in soldering. It reacts with metal oxides and so cleans the metal surfaces which are being soldered together.

Questions

1 Read the following report of a laboratory experiment and then answer the questions which follow.

A white crystalline sodium salt (A) was placed in a test tube with some concentrated sulphuric acid. A gas (B) was given off which fumed in moist air and turned damp blue litmus red. Some manganese(IV) oxide was added to the reaction mixture and the mixture was warmed. A greenish yellow gas (C) was given off which bleached damp litmus paper.

(a) Name the gases B and C and the sodium salt A. Write a chemical equation for the reaction between the sodium salt A and concentrated sulphuric acid.

(b) Why did gas B turn damp blue litmus paper red?

(c) What was the action of the manganese(IV) oxide in the experiment?

(d) Explain the bleaching action of gas C indicating why damp conditions are essential. *(West Midlands)*

2 Hydrogen from a cylinder is burnt at a jet which is lowered into a gas-jar of a green coloured gas. The resulting flame is pale grey and a few white fumes are seen while the green colour disappears. This gas-jar is then inverted over a gas-jar of ammonia. A swirling white smoke is immediately formed in both gas-jars.

(a) Name the green-coloured gas.

(b) State two important uses for this gas.

(c) Explain why the green colour disappears during the burning.

(d) Why are a few white fumes seen?

(e) Explain the immediate formation of the white smoke in the two gas-jars. Give an equation.

(f) State one important use of this white substance.

(g) Explain what would happen to a little of this white solid if it was gently heated in a test tube. *(Lancashire)*

3 Which of the following is formed when hydrogen chloride dissolves in water?

A   hydrogen peroxide   D   hydrochloric acid
B   chloric(I) acid   E   hydrogen gas.
C   an alkali

(North West)

4 Several domestic liquids are sold which contain chlorine. Give *two* properties of chlorine which account for its use in the home.

(*South East*)

5 A town's waterworks department buys bulk supplies of sodium carbonate and chlorine. For what purpose would each of the chemicals be used by the department? (*West Yorkshire*)

6 How would you use hydrochloric acid to make crystals of (i) copper(II) chloride, and (ii) zinc chloride? (*Wales*)

14.5 Bleaches

Chloric(I) acid, sometimes called *hypochlorous acid*, is an important bleach produced when chlorine dissolves in water. As a commercial bleach, however, it has two disadvantages. First, the concentration of the bleach is limited by the solubility of chlorine in water; the maximum concentration is about $\frac{1}{2}$ per cent of bleach by weight. Second, chloric(I) acid is not very stable. It slowly decomposes into hydrochloric acid and oxygen, and therefore loses its bleaching power:

$$2HClO(aq) \rightarrow 2HCl(aq) + O_2(g).$$

The salts of chloric(I) acid are much more useful as bleaches. They are more stable and can be dissolved in water to give a more concentrated bleaching solution. The two main salts used are sodium chlorate(I), NaClO, and calcium chlorate(I), $Ca(ClO)_2$.

Sodium chlorate(I)

A solution containing sodium chlorate(I) (also known as *sodium hypochlorite*) and sodium chloride is made by dissolving chlorine in cold dilute sodium hydroxide solution:

$$2NaOH(aq) + Cl_2(g) \rightarrow NaClO(aq) + NaCl(aq) + H_2O(l).$$

It is manufactured as a by-product of chlorine manufacture. Waste chlorine from the liquefaction plant is used, as well as from the concentrated sulphuric acid used in drying. It is dissolved in dilute sodium hydroxide solution direct from the mercury cells (page 408).

Solid sodium chlorate(I) cannot be made. If the solution is evaporated the sodium chlorate(I) decomposes into sodium chloride and oxygen:

$$2NaClO(aq) \rightarrow 2NaCl(aq) + O_2(g).$$

The solution of sodium chlorate(I) and sodium chloride is supplied for use as a domestic bleach under trade names such as *Domestos*, *Parazone*, and *Brobat*. A more dilute solution is sold as a disinfectant for babies' feeding bottles under the trade name *Milton*.

Calcium chlorate(I)

Calcium chlorate(I) is a solid bleach, commonly called *bleaching powder*. It is manufactured by passing chlorine through calcium hydroxide (*slaked lime*, page 385):

$$2Ca(OH)_2(s) + 2Cl_2(g) \rightarrow \underbrace{Ca(ClO)_2(s) + CaCl_2(s)}_{\text{bleaching powder}} + 2H_2O(l).$$

Bleaching powder is a white powder containing a mixture of calcium chlorate(I), calcium chloride, and unreacted calcium hydroxide. It was the first commercial bleach, and was of great importance in the nineteenth century in the manufacture of cotton and linen goods. Nowadays sodium chlorate(I) and calcium chlorate(I) are only used as bleaches for small-scale purposes. The major consumers of bleaches, the paper and textile industries, use chlorine itself, which is delivered liquefied in tankers.

Experiment 14.4 Looking at bleaches

(a) Investigate the effect of each of the following on indicator paper: (i) a solution of chlorine in water; (ii) a solution of sodium chlorate(I); and (iii) ½ a spatula-full of bleaching powder shaken with water. Make a note of what happens.

(b) Heat 1 cm depth of each of the above solutions in test tubes until they have boiled for about 10 seconds. Allow them to cool and again test their effect on indicator paper.
Make a note of what happens. Which of these bleaching solutions is the least stable and which is the most stable?

(c) Put a spatula-full of bleaching powder in a hard-glass test tube. Heat gently and see if you can detect oxygen gas being given off, using the usual test for oxygen.
Why will a bleach no longer work once it has lost this oxygen?

6 The manufacture of a weedkiller from salt

Chlorine reacts with sodium hydroxide solution in two ways, according to the conditions.

When a cold dilute solution of sodium hydroxide is used, the products are sodium chlorate(I) (a bleach) and sodium chloride:

$$2NaOH(aq) + Cl_2(g) \rightarrow NaClO(aq) + NaCl(aq) + H_2O(l).$$

When a hot concentrated solution of sodium hydroxide is used, the products are sodium chlorate(V), $NaClO_3$ (a weedkiller), and sodium chloride:

$$6NaOH(aq) + 3Cl_2(g) \rightarrow NaClO_3(aq) + 5NaCl(aq) + 3H_2O(l).$$

Sodium chlorate(V) is manufactured from sodium hydroxide and chlorine produced by electrolysis (page 408). It is one of the cheapest weedkillers. Unlike some of the more advanced weedkillers it is not selective: it kills everything. However, it is very useful for keeping paths, pavements, and roads free from weeds. It can also be used to prepare soil for planting. Treatment with a solution of sodium chlorate(V) will kill everything but after a month or two it is completely decomposed in the soil and the ground is ready for sowing.

Questions

1 (a) Name the substance manufactured by passing chlorine into (i) cold sodium hydroxide solution; (ii) hot sodium hydroxide solution.
  (b) Describe the chemical changes which take place when chlorine acts as a bleaching agent. *(East Anglia)*

2 Read the following passage and then answer the questions asked:

A concentrated solution of common salt was electrolysed in a suitable laboratory apparatus. Gases were evolved, one at each electrode, and the solution became alkaline around the cathode. This process is done on an industrial scale because the gases and the alkali are important commercial chemicals. Special cells are necessary, however, if these three products are required; otherwise the alkaline solution and one of the gases react together. In some cases the reaction is allowed to occur to produce two other important chemicals.

(a) Draw a diagram of a suitable piece of laboratory apparatus to illustrate how the electrolysis could be carried out. Label the anode and cathode, indicating the sign associated with each and the materials from which they are made. Name the gas evolved (i) at the cathode, (ii) at the anode. (You may find it helpful here to refer to part (e) of the question.)

(b) What chemical made the solution alkaline?

(c) Give one commercial use for each of the two gases and the alkali obtained from the electrolysis.

(d) Which of the gases would react with the alkaline solution? Name one product which could result, and give a use for it.

(e) Write out and complete the following statement concerning alternative methods of preparing the gases evolved at the electrodes:

The gas _____ could be made by adding pieces of zinc to _____ and the other gas _____ by pouring _____ on some crystals of potassium manganate(VII) (*potassium permanganate*).

(f) Name the two chemicals from which bleaching powder is made.

(g) Explain why chlorine requires the presence of water before it has bleaching properties. *(West Midlands)*

3 By means of a short essay or a pictorial chart show why rock salt is one of the most important raw materials of the chemical industry.

7 Limestone and salt get together

Sodium carbonate, Na_2CO_3, is required as a basic raw material in hundreds of industries. In some places, particularly Lake Magadi in Kenya, there are natural deposits of sodium carbonate, but these are not sufficient to meet the demands of industry. Sodium carbonate has to be manufactured, and the most important method, called the SOLVAY PROCESS, requires both *limestone* and *salt*.

The Solvay Process

The Solvay Process involves reaction between calcium carbonate (*limestone*) and sodium chloride (in the form of *brine*). The problem is to get the ions to change partners to produce sodium carbonate and calcium chloride:

$$CaCO_3 + 2NaCl \rightarrow Na_2CO_3 + CaCl_2.$$

This reaction will not take place directly. A roundabout method is necessary.

Stage 1 Limestone is heated in a lime kiln (page 384) to produce calcium oxide and carbon dioxide:

$$\boxed{CaCO_3(s)} \rightarrow CaO(s) + CO_2(g).$$

Stage 2 Brine (a saturated solution of sodium chloride, page 403) is saturated with ammonia by trickling it down a tower against an upward stream of ammonia gas.

Stage 3 In a second absorption tower the ammonia-saturated brine flows against a stream of the carbon dioxide produced in stage 1. The ammonia, sodium chloride, and carbon dioxide react together to produce sodium hydrogen carbonate and ammonium chloride:

$$\boxed{NaCl(aq)} + CO_2(g) + NH_3(g) + H_2O(l) \rightarrow NaHCO_3(s) + NH_4Cl(aq).$$

Sodium hydrogen carbonate is not very soluble in water and precipitates out in the form of a wet sludge which is tapped off from the bottom of the tower. Solid sodium hydrogen carbonate is obtained from this sludge by filtration.

Stage 4 Finally, the sodium hydrogen carbonate is heated to produce sodium carbonate and carbon dioxide:

$$2NaHCO_3(s) \rightarrow \boxed{Na_2CO_3(s)} + CO_2(g) + H_2O(l).$$

The carbon dioxide is returned for use in stage 3.

Stage 5 Ammonia is recovered from the ammonium chloride by reacting it with the calcium oxide obtained in stage 1:

$$2NH_4Cl(aq) + CaO(s) \rightarrow 2NH_3(g) + \boxed{CaCl_2(aq)} + H_2O(l).$$

The ammonia is returned for use in stage 2.

The end-products of the Solvay Process are therefore sodium carbonate (from stage 4) and calcium chloride (from stage 5). The ammonia is re-used over and over again, acting as a sort of catalyst. The whole process is illustrated in Figure 14.8.

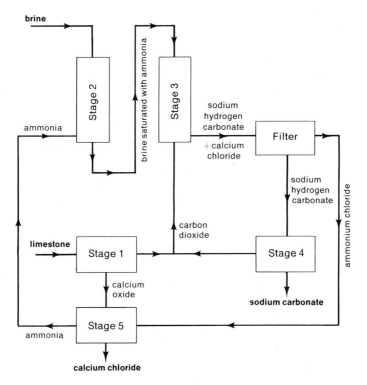

Figure 14.8 The Solvay Process

Some of the calcium chloride produced as a by-product is used in the manufacture of calcium (page 161). However, the demand for calcium chloride is not great and a good deal is wasted. This is the one disadvantage of the Solvay Process.

Experiment 14.5 Some properties of sodium carbonate and sodium hydrogen carbonate

(a) Quarter-fill two test tubes with water. To one add a spatula-full of sodium carbonate and to the other a similar amount of sodium hydrogen carbonate. Shake both tubes well.

What can you say about the solubilities of these two compounds?

(b) Drop a piece of universal indicator paper into each tube. Write down the pH of each solution.

(c) Put a spatula-full of sodium carbonate in one clean test tube and a spatula-full of sodium hydrogen carbonate in another. Add one-quarter of a tube of dilute hydrochloric acid to each.

Describe what you see. Use a suitable test to find out what gas is given off. What will the other product of the reaction be in each case?

(d) Put a spatula-full of sodium hydrogen carbonate in a small hard-glass test tube. Holding the tube in tongs, heat it gently at first and then gradually more strongly.

What do you see condensing in the cooler part of the tube? What do you think this might be? Carry out a suitable test to check if you are correct.

A colourless gas is also given off when the sodium hydrogen carbonate is heated. What gas do you think it is likely to be? Carry out a suitable test to check if you are correct.

Continue heating the residue in the tube strongly for about two minutes, by which time the decomposition should be complete. Allow the tube to cool and then add a few drops of dilute hydrochloric acid.

Describe what happens. Use a suitable test to check what gas is being given off. What do you think the residue from the decomposition of sodium hydrogen carbonate must be? Write a word equation for the action of heat on sodium hydrogen carbonate.

(e) Put a spatula-full of sodium carbonate in a small hard-glass test tube. Holding the tube in tongs heat gently at first and then more strongly.

What do you see condensing in the cooler part of the tube? Use a suitable test to check what this is.

Test to see if the same gas is given off when sodium carbonate is heated as when sodium hydrogen carbonate is heated. Is it?

Explain what has happened as a result of heating sodium carbonate.

Sodium carbonate

Sodium carbonate forms transparent, colourless crystals containing 10 moles of water of crystallisation per mole of sodium carbonate: $Na_2CO_3.10H_2O$. When exposed to the atmosphere they effloresce, losing 9 of these 10 moles of water of crystallisation to leave a white powder with the formula $Na_2CO_3.H_2O$. Heating speeds up this process.

Like all carbonates, sodium carbonate reacts with acids to form a salt, carbon dioxide, and water. For example, with sulphuric acid sodium sulphate solution is produced:

$$Na_2CO_3(s) + H_2SO_4(aq) \rightarrow Na_2SO_4(aq) + CO_2(g) + H_2O(l)$$

Unlike most carbonates, sodium carbonate is soluble in water and its solution is alkaline. Also it is not decomposed by heat.

Sodium carbonate is required in the manufacture of hundreds of different things. By far the biggest quantities are used in the manufacture of glass (page 343). Crystalline sodium carbonate, $Na_2CO_3.10H_2O$, is known as *washing soda*. In this form it dissolves easily in water and is used in softening hard water (page 396).

Sodium hydrogen carbonate

Sodium hydrogen carbonate is known by many names, including *sodium bicarbonate, bicarbonate of soda*, and *baking soda*.

It is powdery in appearance, quite different from sodium carbonate, because it does not have any water of crystallisation. It is also much less soluble in water.

Like sodium carbonate, it reacts with acids to form a salt, carbon dioxide, and water. For example, with hydrochloric acid it forms sodium chloride solution:

$$NaHCO_3(s) + HCl(aq) \rightarrow NaCl(aq) + CO_2(g) + H_2O(l).$$

Unlike sodium carbonate, it is decomposed by heating, giving off carbon dioxide and steam, and leaving a residue of sodium carbonate:

$$2NaHCO_3(s) \rightarrow Na_2CO_3(s) + CO_2(g) + H_2O(g).$$

(In this respect it behaves just like the calcium and magnesium bicarbonates responsible for the temporary hardness of water, page 391.)

Some of the sodium hydrogen carbonate produced in the Solvay Process is not converted into sodium carbonate. It is refined to a high degree of purity, over 99.99 per cent, mainly for use in the baking industry. Its mixture with a solid acid, tartaric acid, is known as *baking powder*. So long as the baking powder is dry there is no reaction between the carbonate and the acid, but when it is mixed into wet dough the reaction starts, producing carbon dioxide which causes the dough to rise. Self-raising flour is flour ready-mixed with baking powder.

Sodium hydrogen carbonate is also used as a cure for indigestion, which is often caused by excessive acidity in the stomach. Sodium hydrogen carbonate neutralises the excess acid. Health salts such as *Eno's* and *Fynnon's Salts* include a mixture of sodium hydrogen carbonate and a solid acid such as citric acid or tartaric acid. When they are dissolved in water the reaction of acid and carbonate generates carbon dioxide which causes the fizz.

Another use for sodium hydrogen carbonate is in certain types of fire extinguisher (Figure 14.9). When the plunger is knocked against the ground the bottle of sulphuric acid is broken and the acid mixes with the solution of sodium hydrogen

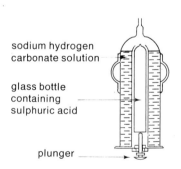

Figure 14.9

carbonate. The pressure of the carbon dioxide which is produced forces out the solution in a jet which can carry over 10 metres. Both the water and the carbon dioxide help to put out the fire. Water has a cooling effect and carbon dioxide, because of its high density, cuts off the supply of oxygen to the flames (page 389).

Some types of foam extinguishers also use sodium hydrogen carbonate. In this type aluminium sulphate, which is a weak acid, reacts with the sodium hydrogen carbonate to produce a jelly-like solid, aluminium hydroxide, and carbon dioxide:

$$Al_2(SO_4)_3(aq) + 6NaHCO_3(aq) \rightarrow 2Al(OH)_3(s) + 3Na_2SO_4(aq) + 6CO_2(g).$$

The carbon dioxide froths the aluminium hydroxide into a foam which blankets the fire. A foam stabiliser is usually added to prevent the frothy bubbles from collapsing once they have formed.

Questions

1 (a) (i) Name the *two* raw materials used in the manufacture of sodium carbonate by the Solvay process.
(ii) Name the only waste product of the process.
(iii) 'The Solvay process is based on the fact that sodium hydrogen carbonate is insoluble in water.'
To what extent is this statement true?
(b) State:
(i) one important use of sodium hydrogen carbonate (other than making sodium carbonate),
(ii) *two* important uses of sodium carbonate.   (*East Anglia*)

2 (a) What is the chemical name of (i) washing soda (ii) baking soda?
(b) Describe a chemical test which would serve to distinguish specimens of powdered washing soda and baking soda, and state what would be seen when the test is applied to each specimen.
(c) What are the starting materials used in the manufacture of baking soda?   (*East Anglia*)

3 A substance X is a well-known white powder which is moderately soluble in cold water. It dissolves with vigorous effervescence (fizzing) in dilute sulphuric acid, giving a colourless solution. When the powder X is heated in a dry test-tube it remains white even after prolonged heating. During the heating, moisture condenses inside the tube on the cooler parts, and a drop of clear lime water held at the mouth of the tube becomes cloudy. The cooled residue from the heating is also found to dissolve in dilute sulphuric acid with effervescence.

Name the substance X and justify your conclusion by equations *or* suitable statements to explain the above observations.   (*Wales*)

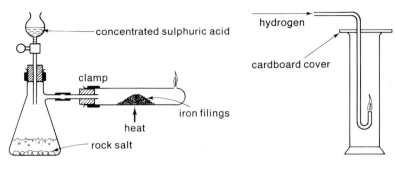

Figure 14.10   Figure 14.11

For the teacher

Experiment 14.1 A free sample of rock salt may be obtained from ICI. Alternatively, samples (*halite*) may be purchased from R. F. D. Parkinson or Lythe Minerals. These are worth having for class inspection, but for the experiment a mixture of 2 parts of cooking salt (not table salt, in which the crystals are too readily visible) and 1 part of sand can be introduced as 'crushed rock salt'.

Electrolysis of brine   It may be worth repeating experiment 5.5(c), page 112 at this point.

Experiment 14.2   For safety it is advisable to supply the concentrated sulphuric acid in 60 cm^3 dropping bottles or TK bottles.

Demonstration of the presence of hydrogen in salt gas   In experiment 14.2 pupils should have recognised that *salt gas* is an acid and therefore probably contains hydrogen. They know from earlier experience that acids generally react with metals high in the activity series to produce hydrogen. If salt gas contains hydrogen, therefore, it will react with a metal. The apparatus in Figure 14.10 can be used to confirm this. The iron filings are heated strongly and concentrated sulphuric acid is dropped slowly on to rock salt in the flask. It should be possible to ignite the hydrogen escaping from the small hole. Crystals of iron(II) chloride condense on the cooler parts of the tube.

Experiment 14.3   Concentrated hydrochloric acid labelled 'Solution of Salt Gas in Water' is required. Manganese(IV) oxide, potassium manganate(VII), and red lead oxide (dilead(II) lead(IV) oxide) are likely to be suggested as oxidising agents and pupils may be allowed to use the one they suggest. Potassium chlorate(V) is sometimes suggested by pupils who have seen this used (as demonstration on page 80) to generate oxygen, **but this must NOT be used.**

Synthesis of salt gas   Once pupils have recognised from experiments 14.2 and 14.3 that salt gas contains hydrogen and chlorine, it remains to show that these are its *only* constituents, using the apparatus in Figure 14.11. The glass tube is connected to a hydrogen cylinder and the gas supply adjusted to give a flame not more than 3 cm high. The burning hydrogen is then lowered into a gas jar of

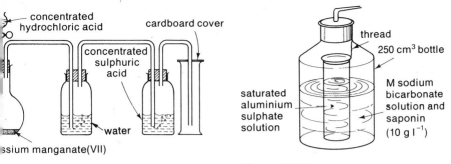

Figure 14.12   *Figure 14.13*

chlorine which is covered by a slotted cardboard lid. The hydrogen continues to burn. The greenish colour of the chlorine is replaced by steamy fumes of hydrogen chloride (tested with a splint, universal indicator paper, and a glass rod dipped in ammonia solution).

Laboratory preparation of chlorine If desired, the full laboratory preparation may be demonstrated using the apparatus in Figure 14.12. The preparation should be carried out in a fume cupboard. If this is not possible only one jar of chlorine should be prepared in the laboratory, the remainder being prepared in advance of the lesson. Jars can be tested with moist indicator paper, a lighted splint, and for solubility (the solution also being tested with indicator paper). Heated steel wool sparks brilliantly forming a smoke of brown iron(III) chloride. Gold leaf ignites spontaneously (extreme reactivity with elements, even those of low activity) as also does Dutch metal (a copper/zinc alloy). A wax taper continues to burn and warm cotton wool dipped in turpentine inflames spontaneously (reaction with hydrocarbons).

Laboratory preparation of hydrogen chloride If desired, the full laboratory preparation may be demonstrated using the generator shown in Figure 14.10. The gas may be dried through concentrated sulphuric acid and is collected by upward displacement of air. Jars may be tested with a lighted splint, universal indicator paper, for solubility in water, and with a glass rod dipped in ammonia solution. Because of the extreme solubility of hydrogen chloride, hydrochloric acid must be prepared using the apparatus shown in Figure 7.14, page 221.

Experiment 14.4 For the solution of chlorine in water use a saturated solution. For the sodium chlorate(I) solution use commercial sodium hypochlorite solution diluted $\times 2$.

Experiment 14.5 Carbon dioxide is best tested for by collecting a sample from the reaction tube in a teat pipette and expelling it through 1 cm depth of lime water in a 75×10 mm tube (see experiment 9.1, page 249).

Fire extinguisher The principle of the foam extinguisher may be demonstrated using the apparatus in Figure 14.13.

15 Chemical analysis

ANALYSIS is the process by which a chemist sets about finding out what a substance is made of. The analytical chemist is a detective whose skill is applied to many different problems. The geologist may want to know the chemical constitution of a sample of rock. The doctor often needs to know what substances are present in a patient's blood or urine in order to diagnose disease more accurately. The chemical industry has to keep a careful check on the purity of its products. The biologist wants to know what chemicals are present in living cells. The policeman may want to know whether the mud on someone's shoes came from a particular field or whether there are chemical traces on his clothes or hands which show he has recently fired a gun. A check must be kept on such things as the purity of our water supplies and atmospheric pollution.

Whatever the particular problem, analysis involves three stages.

1 Separation of the pure substance in which the analyst is interested from the other substances or impurities mixed with it.
2 Finding out *what* elements are present in the substance.
3 (Not always necessary) Finding out *how much* of each element is present in the substance.

15.1 Separation of pure substances

Figure 15.1 summarises the various methods available to separate substances from one another. In your experimental work you will have used some of these techniques many times. Others will be less familiar at present.

Experiment 15.1   Fractional crystallisation

You will be provided with a mixture of two salts: copper(II) sulphate and potassium dichromate(VI).

One-quarter-fill a 100 cm^3 beaker with water and add 3 spatulas-full of the mixture. Stand the beaker on a tripod and gauze and heat it

Substances to be separated	Separation technique	Example
A solid from a liquid in which it has *not* dissolved	FILTRATION or CENTRIFUGING	The preparation of copper(II) sulphate (page 200)
A solid from a liquid in which it is dissolved	EVAPORATION and/or CRYSTALLISATION	The preparation of copper(II) sulphate (page 200)
A liquid from dissolved solids	DISTILLATION	The purification of water (page 395)
Two or more liquids which mix completely with one another	FRACTIONATION (fractional distillation)	Separation of the components of crude oil (page 273) and of liquid air (page 360)
Liquids which do not mix with one another	SEPARATING FUNNEL	Separation of carbon tetrachloride from water
Two solids	SOLVENT EXTRACTION (if a solvent can be found in which one is soluble and one is not) FRACTIONAL CRYSTALLISATION (if both solids are soluble in the same solvent but one much less than the other) or SUBLIMATION (if one solid sublimes and the other does not)	Obtaining pure salt from rock salt (page 404) The manufacture of pure salt from brine (page 403) Separation of ammonium chloride from sodium chloride
Complex organic substances belonging to the same class of organic compounds and therefore having similar properties	CHROMATOGRAPHY	Dyes, sugars

Figure 15.1

slowly, stirring continuously until all the solid has dissolved. (If it is still not all dissolved when the water boils, add a little more water and continue heating.)

Allow the solution to cool to room temperature. Pour the solution into another beaker and examine the crystals which have been deposited. (If no crystals separate out on cooling heat the mixture to boiling, evaporate some of the water, and cool again.)

Which of the substances has crystallised out? Why has this substance crystallised whereas the other has remained in solution? How could the other substance be recovered?

Experiment 15.2   Separation of the components of ink

(a) Measure 10 cm³ of ink into a small conical flask and assemble the apparatus shown in Figure 15.2. Heat the flask very carefully with a *small* bunsen flame not more than 3 cm high. Take care to prevent the ink frothing over or splashing into the delivery tube.

What do you see collecting in the test tube? What do you think this is? Devise and carry out a test to prove the identity of this substance. (If you cannot think of one look back at experiments 7.8 and 7.9, page 207). What is this technique for separating a pure liquid called?

(b) Make two cuts in a filter paper as shown in Figure 15.3. Put enough water in a glass dish to just cover the bottom. Lay the filter paper over the dish with the tail dipping into the water. Using a teat pipette put *one drop* of ink in the very centre of the filter paper, at the end of the tail. Watch what happens to the ink during the next few minutes. When it has spread about half-way across remove the filter paper from the dish.

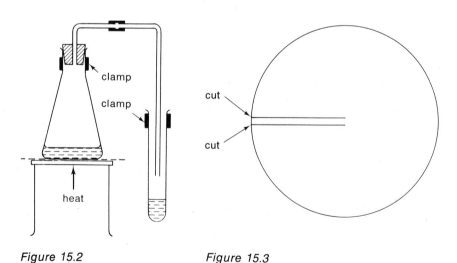

Figure 15.2   Figure 15.3

What does this experiment tell you about the coloured substance in the ink?

This technique for separating substances which would be very difficult, if not impossible, to separate by any other method is called CHROMATOGRAPHY, or chromatographic analysis. Your filter paper is an example of a CHROMATOGRAM.

(c) If you have time, repeat experiment (b) with some other colours of ink.

Experiment 15.3 Investigating the coloured substances in grass

Your teacher will prepare a solution of the coloured substances in grass in an organic solvent. Prepare a filter paper as in experiment 15.2(b). Put enough of the organic solvent in a glass dish to just cover the bottom, and rest the filter paper on the dish as before.

Using a teat pipette, put one drop of the grass extract on the exact centre of the filter paper. When the solvent has spread about half-way across, remove the filter paper from the dish and lay it on a clean piece of paper to dry.

Make a drawing to show the coloured rings you see on the filter paper. How many coloured substances can you identify in grass?

Questions

1 Describe carefully how you would separate a mixture of three solids, A, B and C and obtain a dry sample of each, given that:
A is soluble in water and alcohol
B is soluble in water but not in alcohol
C is insoluble in water and alcohol. (*North West*)

2 Draw a diagram of a suitable apparatus you would use to separate two liquids which do not mix. (*Wales*)

3 The different colours in a sample of school ink can best be separated by which of the following methods?
A   filtration
B   chromatography
C   fractional distillation
D   evaporation
E   crystallisation (*London*)

4 Crystallisation; sublimation; filtration; distillation; centrifuging. What of the above processes would be used to obtain
(a) ammonium chloride from a mixture of ammonium chloride and sodium chloride?

(b) methyl alcohol (methanol) from a mixture of methyl alcohol and ethyl alcohol (ethanol)?
(c) sulphur from a solution of sulphur in carbon disulphide?

(*East Anglia*)

5
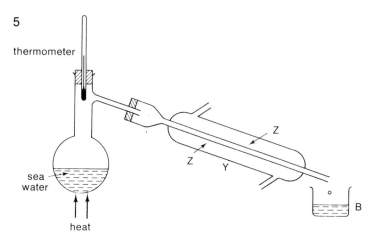

Figure 15.4

(i) Name the apparatus Y.
(ii) What liquid is normally used shown in part Z?
(iii) What purpose does this liquid serve?
(iv) Where does the liquid Z enter and where does it leave?
(v) What temperature should the thermometer read in the position shown in the diagram?
(vi) If the thermometer were replaced with its bulb in the boiling sea-water, about what temperature would be recorded on the thermometer?
(vii) What liquid is collecting at B?
(viii) If the sea water were replaced by blue ink, what colour would the liquid collected at B be?
(ix) The above apparatus would be a means of purifying the sea water, what other method could be used?

(*West Yorkshire*)

15.2 Tests for gases

Figure 15.6 (pages 430–31) summarises the more important properties of gases that you have studied. When you are trying to identify a gas it is best to examine as many of its properties as possible before coming to a final conclusion. However, the most characteristic property of each gas, which is sufficient to identify it in most cases, is shown in bolder type.

.3 Tests for metal ions

Flame tests

In section 5.4, page 127, we saw how the type of ion formed by an element could be explained by the idea of the electrons being grouped around the nucleus in SHELLS, some closer to the nucleus and some further away. When an atom is supplied with sufficient energy the electrons can pick up some of this energy and as a result move further from the nucleus into an outer shell. This energy is subsequently released as light, and the electrons drop back to their original position. Atoms of different elements are capable of absorbing different-sized *lumps* of energy. They therefore fire out different-sized lumps, which appear as light of different wavelengths. Sodium atoms, for instance, when *excited* by a high voltage (electrical energy) or a flame (heat energy) emit light at several different wavelengths, but of these only two are very intense, at 589.0 and 589.6 nm. Both of these wavelengths fall in the orange-yellow region of the spectrum (Figure 15.5).

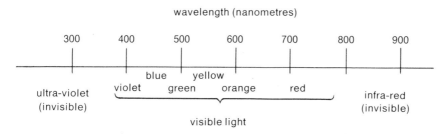

Figure 15.5 The spectrum

You have probably seen the yellow type of street lamp. These contain sodium vapour subjected to a high voltage. Excited sodium atoms emit this characteristic colour of light whatever chemical form they are in. Any sodium compound will therefore give a yellow colour in a FLAME TEST, details of which are given in experiment 13.1(b), page 380. Other elements which give characteristic flame test colours are as follows.

potassium	violet
copper	blue-green
barium	green
calcium	red

All elements emit a characteristic spectrum when their atoms are excited, but many unfortunately do not emit any one wavelength

Gas	Colour	Has it any smell?	Number of times denser than air	Solubility in water	Effect on indicator	Effect on lighted splint	Other properties
Ammonia	Colourless	Yes	0.6	Very soluble	**Strong base**	Splint out	Dense white fumes with hydrogen chloride
Carbon dioxide	Colourless	No	1.5	Slightly soluble	Weak acid	Splint out	**Turns lime water milky**
Carbon monoxide	Colourless	No	0.9	Almost insoluble	Neutral	Gas burns	**Burns to carbon dioxide** (test with lime water)
Chlorine	Yellow-green	Yes	2.4	Quite soluble	**Bleaches**	Splint out	

Hydrogen	Colourless	No	0.1	Almost insoluble	Neutral	Gas burns	
Hydrogen chloride	Colourless (steamy fumes in air)	Yes	1.2	Very soluble	Strong acid	Splint out	Dense white fumes with ammonia
Hydrogen sulphide	Colourless	Yes	1.1	Slightly soluble	Weak acid	Gas burns	Turns lead salts black
Nitrogen	Colourless	No	1.0	Almost insoluble	Neutral	Splint out	Totally unreactive (all tests negative)
Nitrogen dioxide	**Brown**	Yes	1.5	Very soluble	Strong acid	Splint goes on burning	
Oxygen	Colourless	No	1.0	Very slightly soluble	Neutral	**Glowing splint relit**	
Sulphur dioxide	Colourless	Yes	2.7	Slightly soluble	Weak acid	Splint out	Decolourises potassium manganate(VII)

Figure 15.6 Properties of gases

Chemical analysis   431

intensely enough to be seen in a flame test. For more sophisticated analysis a SPECTROMETER is used in place of the naked eye. This instrument splits light into its constituent wavelengths by means of a prism or diffraction grating and enables the wavelength of each of the spectral lines to be measured so that any element can be identified.

Precipitation of hydroxides

Nearly all metal hydroxides are insoluble in water, the exceptions being those of Group I of the periodic classification, together with some from Group II. Many of the transition metals have coloured hydroxides.

If, for example, sodium hydroxide solution is added to a solution of a copper(II) salt, the copper(II) ions from the copper salt and the hydroxide ions from the sodium hydroxide combine together to form a blue precipitate of insoluble copper(II) hydroxide:

$$Cu^{2+}(aq) + 2OH^-(aq) \rightarrow Cu(OH)_2(s).$$

The formation of this blue precipitate is a useful test for the copper(II) ion. Similarly iron(II) ions give a green precipitate of iron(II) hydroxide, iron(III) ions give a brown precipitate of iron(III) hydroxide, and calcium ions give a white precipitate of calcium hydroxide.

Aluminium and zinc ions also give white precipitates of aluminium hydroxide and zinc hydroxide, but in these cases addition of excess sodium hydroxide solution causes the precipitate to redissolve to give a colourless solution of sodium aluminate or sodium zincate:

$$Al(OH)_3(s) + NaOH(aq) \rightarrow \underset{\text{sodium aluminate}}{NaAlO_2(aq)} + 2H_2O(l),$$

$$Zn(OH)_2(s) + 2NaOH(aq) \rightarrow \underset{\text{sodium zincate}}{Na_2ZnO_2(aq)} + 2H_2O(l).$$

A solution of ammonia in water also contains hydroxide ions as a result of the reaction:

$$NH_3(g) + H_2O(l) \rightleftharpoons NH_4^+(aq) + OH^-(aq).$$

If ammonia solution is added to a solution of a metal salt the metal hydroxide may be precipitated. Iron(II) ions give a green precipitate of iron(II) hydroxide, iron(III) ions give a brown precipitate of iron(III) hydroxide, and aluminium ions give a white precipitate of aluminium hydroxide. Calcium hydroxide is not precipitated from solutions of calcium salts by ammonia solution because it is slightly soluble in water and the concentration of hydroxide ions in ammonia solution is not high enough.

Solutions of zinc salts and copper(II) salts behave rather differently with ammonia solution. In both cases the addition of ammonia solution at first produces a precipitate of the metal hydroxide:

$$Zn^{2+}(aq) + 2OH^-(aq) \rightarrow \underset{\text{white}}{Zn(OH)_2},$$

$$Cu^{2+}(aq) + 2OH^-(aq) \rightarrow \underset{\text{blue}}{Cu(OH)_2}.$$

But addition of excess ammonia causes the hydroxide precipitates to redissolve. Zinc hydroxide dissolves to give a colourless solution containing the complex ion $[Zn(NH_3)_4]^{2+}$ and copper(II) hydroxide dissolves to give a dark blue solution containing the complex ion $[Cu(NH_3)_4]^{2+}$ ion.

By testing solutions of an unknown metal salt both with sodium hydroxide solution and with ammonia solution it is generally possible to identify the metal ion present. Figure 15.7 summarises the results of these tests.

Metal ion	Sodium hydroxide solution added to solution of metal ion	Ammonia solution added to solution of metal ion	Flame test (on solid salt)
Aluminium	White precipitate which dissolves in excess to give colourless solution	White precipitate insoluble in excess	No characteristic colour
Barium	No precipitate	No precipitate	Green
Calcium	White precipitate insoluble in excess	No precipitate	Red
Copper(II)	Blue precipitate insoluble in excess	Blue precipitate which dissolves in excess to give dark blue solution	Blue-green
Iron(II)	Green precipitate insoluble in excess	Green precipitate insoluble in excess	No characteristic colour
Iron(III)	Brown precipitate insoluble in excess	Brown precipitate insoluble in excess	No characteristic colour
Potassium	No precipitate	No precipitate	Violet
Sodium	No precipitate	No precipitate	Yellow
Zinc	White precipitate which dissolves in excess to give colourless solution	White precipitate which dissolves in excess to give colourless solution	No characteristic colour

Figure 15.7

15.4 Tests for anions

Reactions with dilute hydrochloric acid

When dilute hydrochloric acid is added to salts containing certain anions a reaction takes place in which a gas is produced. Any gas evolved can be identified as in Figure 15.6, page 430, and in this way the anion can be identified.

The procedure is to add about 2 cm depth of dilute hydrochloric acid to one spatula-full of the solid salt in a test tube. If a gas is produced immediately this is tested for. If there is no immediate evolution of gas the mixture is warmed gently. This may initiate a reaction which would not occur cold, and again any gas is tested for.

CARBONATES react with dilute acid to produce carbon dioxide:

$$CO_3^{2-}(s) + 2H_3O^+(aq) \rightarrow CO_2(g) + 3H_2O(l).$$

SULPHITES react with dilute acid to produce sulphur dioxide:

$$SO_3^{2-}(s) + 2H_3O^+(aq) \rightarrow SO_2(g) + 3H_2O(l).$$

SULPHIDES react with dilute acid to produce hydrogen sulphide:

$$S^{2-}(s) + 2H_3O^+(aq) \rightarrow H_2S(g) + 2H_2O(l).$$

Other important anions

The most important anions which cannot be detected by the hydrochloric acid test are chloride, sulphate, and nitrate. To identify these, a solution of one spatula-full of the salt in half a test tube of distilled water is prepared and divided into three parts.

To test for a CHLORIDE a few drops of dilute nitric acid are added to one part of the solution, followed by an equal volume of silver nitrate solution. If the unknown substance is a chloride a white precipitate appears. This is silver chloride, which is extremely insoluble in water:

$$Cl^-(aq) + \underset{\substack{\text{(from} \\ \text{silver nitrate} \\ \text{solution)}}}{Ag^+(aq)} \rightarrow AgCl(s).$$

To test for a SULPHATE a few drops of dilute hydrochloric acid are added to one part of the solution, followed by an equal volume of barium chloride solution. If the unknown solution is a sulphate a white precipitate appears. This is the very insoluble salt barium sulphate:

$$SO_4^{2-}(aq) + \underset{\substack{\text{(from} \\ \text{barium chloride} \\ \text{solution)}}}{Ba^{2+}(aq)} \rightarrow BaSO_4(s).$$

To test for a NITRATE 2 cm depth of iron(II) sulphate solution is added to the final part of the solution of the unknown salt. The tube is held at an

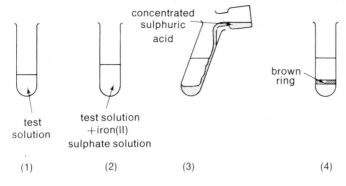

Figure 15.8

angle (Figure 15.8), and concentrated sulphuric acid is poured carefully down the side of the tube. The acid does not mix with the solution, and being more dense forms a separate layer at the bottom of the tube. Sufficient acid is added to give a layer about 1 cm in depth. If the unknown substance is a nitrate a brown ring appears within a few seconds at the junction between the two layers. The chemistry of this test is rather complicated and cannot easily be represented by an equation. A reaction takes place between the nitrate, the concentrated sulphuric acid, and the iron(II) sulphate to form nitrogen oxide gas, which then combines with more iron(II) sulphate to form the brown compound.

Experiment 15.4 Practising analysis

You will be provided with three compounds. Your job is to try to find out what they are. Carry out the following procedure with each one in turn.

(a) Can you make any helpful guesses from the appearance of the substance? For example, is its colour characteristic of a particular transition metal ion?

(b) Dissolve a spatula-full of the substance in half a test tube of distilled water. Divide the solution into two equal parts.

To one part add three drops of sodium hydroxide solution. Note any reaction, then add a quarter of a tube of sodium hydroxide solution.

To the other part add three drops of ammonia solution. Note any reaction, then add a quarter of a tube of ammonia solution.

Use Figure 15.7, page 433, to draw what conclusions you can about the substance.

(c) Carry out a flame test on your substance (for details see experiment 13.1(b), page 380). You should now be able to decide what cation is present in the unknown compound.

(d) To one spatula-full of the solid substance in a test tube add 2 cm depth of dilute hydrochloric acid. If there is no immediate reaction, warm the mixture gently. If there are any signs of a gas being given off find out what gas it is, using the information in Figure 15.6, page 430. If carbon dioxide is produced your substance is a carbonate, if sulphur dioxide is produced it is a sulphite, and if hydrogen sulphide is produced it is a sulphide (see page 434).

(e) If (d) is not sufficient to identify the anion in your substance, dissolve a spatula-full in half a test tube of distilled water, divide it into three equal parts and carry out the tests for a chloride, a sulphate, and a nitrate given on page 434.

Questions

1 When a certain gas was passed through lead acetate solution a black precipitate was formed. The gas was which of the following?
A   hydrogen sulphide   D   ethylene
B   steam   E   ammonia
C   oxygen   (North West)

2 The new British nuclear-powered submarine is equipped with an underwater distilling plant which will provide unlimited water supply on a long sea run. The main substance to be removed from the water is sodium chloride.
   (a) (i) How would the distilled water be tested to find if there is any chloride still present?
   (ii) What would be seen when this test is applied if there is still sodium chloride present?
   (b) Assuming that all the sodium chloride had been removed, how could it be found whether any other solid remained dissolved in the distilled water?   (East Anglia)

3 The yellow-orange flame seen when glass is heated in a bunsen flame is due to the presence of _____.   (East Anglia)

4 On heating, a blue compound gives off red-brown choking fumes, and when a glowing splint is plunged into the tube it is rekindled. The black residue left in the tube dissolves in warm, dilute nitric acid giving a blue-green solution. Ammonia solution added to this gives a pale blue precipitate which dissolves in excess ammonia to give a clear, deep-blue solution.
   (i) What was the compound?
   (ii) What was the black residue?
   (iii) What were the red-brown fumes?
   (iv) What caused the splint to relight?
   (v) Explain the reaction when excess ammonia solution was added.   (London)

5 Identify any *two* of the substances A, B and C, giving, in each case, full reasons for your conclusions. Explain the reactions which take place and name the substances underlined.

(a) Substance A is a silvery metal. When a little is placed in a test tube of water, an *inflammable gas* is given off and a *white precipitate* is formed. When heated in air, the metal burns with a red flame and forms a *white powder* which swells and gives off great heat when a little water is added, leaving a *white powder*.

(b) Substance B is a blue liquid. When some zinc is placed in a sample of the liquid, a *reddish-brown solid* is deposited and a *colourless liquid* remains. When a solution of barium chloride is added to B a *white precipitate* is formed.

When a solution of sodium hydroxide is added to another sample of B a *greenish-blue precipitate* results.

(c) Substance C is a white insoluble powder. When heated, it changes to a *yellow solid* and gives off a *gas* which turns lime water cloudy. On cooling this solid goes white.

C gives off the same gas when treated with dilute hydrochloric acid and leaves a clear *colourless liquid*. On adding a few drops of sodium hydroxide to this clear liquid, a *white precipitate* is formed.

(*South East*)

6 Identify the substances described below giving, in each case, reasons for your conclusions. Name the substances *underlined*.

(a) *Substance D* is a green powder.

When it is heated in a test tube a *black solid* remains and a *gas* is given off which turns lime water cloudy.

When D is treated with dilute sulphuric acid, the same gas is given off and a clear *blue solution* remains.

(b) *Substance E* is a silvery metal which tarnishes at once when exposed to air.

When heated, E melts and then burns with a yellow flame forming a *yellow-white solid*. When E is put into water it reacts violently, giving off an *inflammable gas* and leaving a *clear liquid* which turns red litmus blue.

E burns in chlorine giving a *white solid*.

(c) *Substance G* is a colourless gas with a sharp, choking smell.

G is very soluble in water forming a *solution* which turns blue litmus red.

G is rapidly absorbed by sodium hydroxide solution and, if the solution is evaporated, a *white solid* remains. This gives off G when treated with dilute acid and warmed.   (*South East*)

For the teacher

Experiment 15.2 For economy, ink made from ink powder is satisfactory for (a), though it does not give good results in (b). If pupils have not seen a full-scale distillation it is worth while distilling ink on a larger scale, using apparatus similar to that in Figure 15.4, page 428 (though Quickfit, if available, is much more convenient).

For (b) most brands of blue-black ink give a good separation. It is worthwhile trying other colours if time permits. Brown ink is often the most spectacular, and many brands of red ink provide an example of a single pure dye.

Experiment 15.3 An extract of grass for chromatography is prepared by grinding a handful of grass, cut up into pieces, with 5 g of fine sand and 10 cm^3 of acetone, then decanting or filtering. An outer yellow-orange ring of xanthophyll and an inner green ring of the chlorophylls should be seen. An alternative solvent is toluene which in addition to xanthophyll and chlorophyll gives an inner red ring of carotene.

The products can be separated by cutting the bands out of the filter paper and extracting with warm solvent.

Other separation techniques If pupils have not encountered them before, centrifuging, the use of a separating funnel, and sublimation are worth demonstrating.

Section 15.3 Pupils should have the opportunity of familiarising themselves with all the reactions in Figure 15.7 by carrying out all the tests for themselves using M solutions of salts of the metals, 2 M sodium hydroxide, and 2 M ammonia.

Section 15.4 Pupils should have the opportunity of carrying out the various anion tests on known substances before attempting experiment 15.4. Solutions required are 2 M hydrochloric acid, 0.1 M silver nitrate, 2 M nitric acid, and M barium chloride solution, all in distilled or deionised water. Iron(II) sulphate solution for the nitrate test must be freshly prepared: the concentration is of little consequence.

Appendices

Useful addresses

The following organisations are referred to in the text as sources of materials, etc.
Aluminium Federation, Broadway House, Five Ways, Birmingham 15
Association for Science Education, College Lane, Hatfield, Herts
British Drug Houses, Broom Road, Parkstone, Dorset
British Metal Corporation, 2 Metal Exchange Bldg., Leadenhall Ave., London EC3
British Oxygen Co. Ltd, Hammersmith House, London W6 9DX
Consortium of Local Education Authorities for the Provision of Science Equipment (CLEAPSE), Brunel University, Uxbridge, Middlesex
Distillers Company Ltd, Central Publicity Dept, 21 St James' Square, London SW1
Gallenkamp and Co. Ltd, Technico House, Christopher Street, London EC2
Griffin and George Ltd, Ealing Road, Alperton, Wembley, Middlesex
Griffiths (Alan) and Partners, 23 Harwood Road, Marlow, Bucks
Hopkin and Williams Ltd, Chadwell Heath, Essex
Industrial Diamond Co. Ltd, 88/90 Hatton Garden, London EC1N 8PN
Long Rake Spar Co., Youlgreave, Bakewell, Derbyshire DE4 1LW
Lythe Minerals, 36/38 Oxford Street, Leicester LE1 5XW
MacAdams and Co., 5 Lloyds Avenue, London EC3
Natural Rubber Producers Research Association, 19 Buckingham Street, London WC2
Parkinson (R. F. D.) and Co. Ltd, Doulting, Shepton Mallet, Somerset
Plastics Identification Service Ltd, 65 Western Road, Hove, Sussex
QVF (Quickfit), Duke Street, Fenton, Stoke-on-Trent, Staffs.
Radiospares, PO Box 2BH, 4/8 Maple Street, London W1
Spiring Enterprises Ltd, North Holmwood, Dorking, Surrey
Williams, Harvey and Co., Millancar Works, Charley Wood Road, Kirby Industrial Estate, Liverpool, L33 7SG

B Background reading and wall charts

This appendix gives addresses of industrial and other sources of booklets, charts and other background material, arranged for convenience by chapter. As many of the organisations named are constantly updating their material, detailed titles are not given, but rather an indication of the topics covered at the time of publication. In most cases the material is supplied free to teachers at a school address.

Chapter 1

Imperial Chemical Industries, Schools Liaison Officer, Imperial Chemical House, Millbank, London SW1P 3JF (Periodic table charts)
Longman Group, Longman House, Burnt Mill, Harlow, Essex (Small periodic tables)

Chapter 4

United Kingdom Atomic Energy Authority, Public Relations Branch, 11 Charles II Street, London SW1

Chapter 6

British Steel Corporation, 151 Gower Street, London WC1E 6BB
Copper Development Association, Orchard House, Mutton Lane, Potters Bar, Herts
Lead Development Association, 34 Berkeley Square, London W1X 6AJ
Royal Mint, Llantrisant, South Wales *or* Tower Hill, London EC3N 4DR (Coinage)
Schools Information Centre on the Chemical Industry, The Polytechnic of North London, Holloway Road, London N7 8DB (Steel)
Zinc Development Association, 34 Berkeley Square, London W1X 6AJ

Chapter 8

Chemical Construction Ltd, Regal House, London Road, Twickenham, Middlesex (Sulphuric acid manufacture)
Imperial Chemical Industries, Schools Liaison Officer, Imperial Chemical House, Millbank, London SW1P 3JF (Sulphuric acid manufacture)
Schools Information Centre on the Chemical Industry, The Polytechnic of North London, Holloway Road, London N7 8DB (Sulphuric acid)

Chapter 9

Allied Breweries Ltd, Box 205, Allied House, 158 St John Street, London EC1 (Beer)
British Man-Made Fibres Federation, Bridgewater House, 58 Whitworth Street, Manchester M1 6LF

British Petroleum Co. Ltd, Britannic House, Moor Lane, London EC2Y 9BU (Organic chemicals, plastics)
Courtaulds Educational Service, 18 Hanover Square, London W1A 2BB (Man-made fibres)
Dunlop Ltd, Dunlop House, Ryder Street, London SW1Y 6RA (Rubber)
Esso Petroleum Co. Ltd, Victoria Street, London SW1E 5JW (Petroleum refining)
Formica Ltd, De La Rue House, 84/86 Regent Street, London W1 (Plastic laminates)
Gas Corporation, 59 Bryanston Street, London W1A 2AZ (Natural gas)
Imperial Chemical Industries Fibres Ltd, Bowater House East, 68 Knightsbridge, London SW1X 7LN (Man-made fibres)
Monsanto Chemicals Ltd, 10 Victoria Street, London SW1 (Man-made fibres)
National Coal Board, Hobart House, Grosvenor Place, London SW1X 7AE
National Smokeless Fuels Ltd, Coal House, Lyon Road, Harrow, Middlesex
Natural Rubber Producers Research Association, 19 Buckingham Street, London WC2
Proctor and Gamble Ltd, GPO Box 1EE, Gosforth, Newcastle-on-Tyne 1 (Detergents)
RHM Foods Ltd, Daybrook, Nottingham NG5 6AG (Bread)
Sandeman (George G.) Sons and Co. Ltd, 36 Albert Embankment, London SE1 (Fortified wines)
Scotch Whisky Association, Information and Development Office, 17 Half Moon Street, London W1Y 7RB
Shell International Petroleum Co. Ltd, Shell Centre, London SE1 7NA (Petroleum, natural gas, petrol, detergents)
Tate and Lyle Ltd, 21 Mincing Lane, London EC4 (Sugar)
Unilever Education Section, Unilever House, Blackfriars, London EC4 (Detergents)

Chapter 11

Associated Portland Cement Manufacturers Ltd, Portland House, Stag Place, London SW1
Brick Development Association, 19 Grafton Street, London W1X 3LE
British Ceramic Manufacturers Federation, Federation House, Station Road, Stoke-on-Trent, Staffs. (Pottery)
Fibreglass Ltd, Ravenshead, St Helens, Lancs.
Glass Manufacturers Federation, 19 Portland Place, London W1N 4BH
Imperial Chemical Industries, Nobel Division, Stevenston Works, Stevenston, Ayrshire KA20 3LM (Silicones)
Shell International Petroleum Co. Ltd, Shell Centre, London SE1 7NA (Geology)
Vitreous Enamel Development Council, New House, High Street, Ticehurst, Wadhurst, Sussex

Wedgwood (Josiah) and Sons Ltd, 34 Wigmore Street, London W1 (Pottery)

Chapter 12

Longman Group, Longman House, Burnt Mill, Harlow, Essex (Charts on liquid air fractionation and manufacture of ammonia)

Chapter 13

Nalfloc Ltd, Ilford House, 133 Oxford Street, London W1 (Water treatment)

Chapter 14

Imperial Chemical Industries, Mond Division, The Heath, Runcorn, Cheshire WA7 4QF (Salt)
Imperial Chemical Industries, Schools Liaison Officer, Imperial Chemical House, Millbank, London SW1P 3JF (Chlorine)

C Films and film loops

Film libraries

The catalogues of the following libraries contain useful 16 mm films (optical sound), many of which are available on free loan.

British Petroleum Film Library, 15 Beaconsfield Road, London NW10 2LE
Central Film Library, Government Building, Bromyard Avenue, London W3 7JB
Education Foundation for Visual Aids, Paxton Place, Gipsy Road, London SE27
Golden Films, Stewart House, 23 Frances Road, Windsor, Berks
Guild Sound and Vision Ltd, Woodston House, Oundle Road, Peterborough, Hunts
Imperial Chemical Industries Film Library, Thames House North, Millbank, London SW1P 4QG
Shell Film Library, 25 The Burroughs, London NW4 4AT
Unilever Education Section, Unilever House, Blackfriars, London EC4

Films concerned with specific products are also available from the following organisations.

British Gypsum Ltd, 15 Marylebone Road, London NW1 (Plaster of Paris)
British Oxygen Film Library, 42 Upper Richmond Road West, London SW14 8DD
Cement Marketing Co. Ltd, Portland House, Stag Place, London SW1
De Beers, 40 Holborn Viaduct, London EC1 (Diamonds)
Electricity Council Film Library, Trafalgar Buildings, 1 Charing Cross, London SW1
Gas Corporation Film Library, 6 Great Chapel Street, London W1
High Commission for Kenya, 45 Portland Place, London W1 (Sodium carbonate)
National Coal Board Film Library, 68 Wardour Street, London W1
Sand and Gravel Association Ltd, 48 Park Street, London W1 (Cement and concrete)
Scotch Whisky Association, 17 Half Moon Street, London W1Y 7RB
United Kingdom Atomic Energy Authority, 11 Charles II Street, London SW1

Film loops

The following film loops may be found useful in relation to chapter 2.

Movement of Molecules, Liquid/Gas Equilibrium, Solid/Liquid Equilibrium (Longman Group, Longman House, Burnt Mill, Harlow, Essex)
Mass of an Atom (Ealing Scientific Ltd, 15 Greycaine Road, Watford WD2 4PW)
Aston's Mass Spectrograph (Macmillan Education Ltd, Houndmills, Basingstoke, Hants)

Index

Main references are shown in bold type

acetaldehyde 287–8
acetic acid 287–9
acetylene 285–6, 290, 319, 385
Acheson process 248
acids 195-203, 211-15, 224, 238, 295–7, 305, 308–9, 370, 411–12
actinides 20
activity series 137–41, 150–51, 163, 176, 178, 329
addition reactions 283–4, 287, 293–5, 303
age-hardening 184
air 4, 11, 176, 360–61
alchemists 12
alcohol 259-63, 287, 388
alkali metals 18, 129
alkalis 195, 213, 301, 305, 362–3, 385–6, 419
alkanes 277, 283
allotropes 233–5, 247–9
alloys 137, 168-73, 178, 182, 184, 186–9
alpha particles 83–4, 90–92
alpha rays 82, 105–6
aluminium 160-61, 176–7, 181, 183-4, 341, 374, 433
aluminium oxide 5, 160, 176
amine group 297
ammonia 256, 358-65, 369, 412, 417–18, 430–32
ampère 118
analysis 1, 424–35
anhydrous salts 208–9
anions 117, 119, 122, 127–8, 434–5
annealing 167, 174–5

anode 114, 117–20, 126, 161–3, 176, 332–3, 406–7
anodising 176–7
arc furnace 157
argon 11, 360–61
atom bomb 97–8, 102, 318
atomic mass 84–5, 88–91
atomic mass unit 38, 91
atomic number 84-5, 89–91, 99, 128
atomic theory 32, 60–61, 82
atoms 29-38, 169–71, 429

baking 261, 352
bases 195-6, 199, 201, 212-13, 224, 238, 362, 385, 412
batteries 186, 222, 330, 333, 336, 413
beer 261–3
Bequerel, Henri 82
Bessemer converter 155–8
beta particles 91-2, 99, 105–6
beta rays 82
blast furnace 153-4, 181, 267, 376
bleaches 239, 308, 386, 404, 411–12, 414-15
boiling 50, 395–6
boiling points 130, 133, 259–60, 273, 276, 404
Boyle, Robert 12
Bragg, Lawrence 169
brass 170, 184, 187
brewing 261–3, 388
bricks 340, 347-8, 351
brine 403, 407–8, 417
Brønsted, Johannes 213
bronze 12, 184

Brown, Robert 29
Brownian motion 29, 49

calcium 161, 376, 433
calcium carbonate 380-6, 392–3, 417
calcium chloride 161, 406, 415, 417–18
calcium hydroxide 161, 348, 363, 380, 383, 385-6, 396, 415, 432
calcium oxide 155, 380, 383-6, 417–18
calcium sulphate 225, 237, 255, 364, 391, 396
calorimeter 319–20, 322
cancer 106
carbohydrates 253, 259, 261, 335, 375–6, 388
carbon 150–52, 155, 161, 174, 246-54, 272, 274, 376
carbonates 203, 212, 214, 391, 419–20, 434
carbon cycle 252-3, 355, 388
carbon dioxide 153–4, 203, 252–3, 259–63, 360, 385-8, 419–21, 434
carbon fibre 248
carbon monoxide 153–5, 267, 280, 325-7, 429
Carothers, William 297
cast iron 181
catalysts 63-4, 229, 252, 279–81, 293–4, 360–61, 369, 375–6
cathode 114, 117–19, 126, 161–3, 177, 212, 332–3, 406–8
cations 117, 119, 122, 127, 363

444 Chemistry today

cells   178–9, **329–31**, 336–7
cellulose   252–3, **255–6**, **288–9**, 293, 375
cellulose acetate   288–90
cement   154, 237, **347–8**, 383, 385
centrifuging   255, 306, 404, **425**
Chadwick, James   84
chain reaction   97–8, 100
charcoal   249, 255
china   348
chlorides   411–13, 434
chlorine   10, 161, 294, 350, 404, 406–8, **410–12**, 414–16, 429
chromatography   425
clay   154, 303, 308, **343**, 347–8, 385, 397, 403
coal   240–41, **264–8**, 280–81, 317–19, 334–6
coke   153–5, 237, 241, 248, **266–7**, 319, 325–6, 349, 361, 385
Columbus, Christopher   301
compounds   5–6, 38–42
concentration of solutions   215–17
concrete   154, **348**, 351, 383
conductors   112–13, 162
contact process   228–30
copper   **155**, 162–3, 181, **183–5**, 433
copper(II) sulphate   **200**, 207, **225–6**, 321–2, 329–30
corrosion   **176–9**, 183–5, 189
coulomb   118
covalent compounds   133, 274
cracking   **278–9**, 283, 285, 304
critical mass   97
crystals   165–71, 175, 208–9, 233–4, 254–5, 404, 425
cyclotron   99–100

Dalton, John   32, 82
Daniell cell   330–31, 336
Davy, Humphry   12, 117
DDT   290
dehydrating agents   226, 229, 385
deliquescent substances   209, 371
Democritus   32
detergents   222, 246, **305–9**, 344, 396
diamond   247–9

Dickson, J. T.   297
die-casting   187
diffraction patterns   169–70
diffusion   31, 50
distillation   259–60, 263, 267, 395, **425**
double bond   **283**, 285, 294, 302–3
double decomposition   203, 242
dry cell   **331-2**, 364, 413
ductility   166–7, 184

earth   **9–10**, 144–5, 340–43
efflorescent compounds   209, 419
Einstein, Albert   91
electric furnaces   157–8
electricity   99–102, **110–26**, 178, 183–4, 212–13, 266, 336–7
electrode   **114**, 118, 157, 161, 333, 337, 406
electrolysis   **111–22**, 125, 151, 160–62, 176, 187, 406–8, 416
electrolytes   **113**, 117–18, 130, 161–2, 203, 330–32, 406
electrons   **82–91**, 113, 118–20, 125–33, 161–2, 176–7, 406–8, 429
electroplating   **125-7**, 188, 222, 406, 412
electrovalent compounds   130
elements   **1–29**, 39, 99, 355, 424, 429
enamel   222, **346**, 348, 412
endothermic reactions   321–7
energy   91–2, 97–106, 252–3, 259, 266, 277, **317–37**, 429
energy level diagrams   323–6
enzymes   64, 253, 259, 309, 355, **375–6**
equations   69–72, 323
equilibrium   77, 229
esters   **288**, 295, 305
etching   166
ethanal   287–8
ethanol   260, 263, **287-8**
ethyl acetate   288–9, 295
ethylene   **283–5**, 287, 293–5, 304
evaporation   162, 202, 208, 254–5, 401, 404, 408, 414, **425**
exothermic reactions   321–7

explosives   64, 97, 222, 226, 260, 271, 318, 358, 364, **370–75**, 382–3

faraday   **118**, 122–3, 336
Faraday, Michael   117
fast breeder reactor   100–101
fats   253, 285, 305, 308, 318, **376**
fermentation   **259**, 261–3, 287
ferro-manganese   156
fertilisers   222, 225, 270, 284, **356-8**, 364–5, 370–71, 376, 403, 412
fibreglass   345–6, 349
fibres, man-made   222, 289, **292-9**
filtration   417, 425
fire extinguishers   389, 412, 420–21
flame tests   429, 432
flash roasting   148
formulas   **41-2**, 122–3, 276
fossils   265, 270–71
fractionation   259–60, 263, 268, 272–4, 279–80, 350, 360, 388, **425**
Frasch, Hermann   231
Frasch process   231–2
froth flotation   148, 154–5
fuel cells   336–7
fuels   238–42, 249, 253, 260, 266–7, 270, 277–8, **318–22**, 334–7, 388

Galvani, Luigi   117
galvanised iron   165, **178–9**, 187–8, 222, 412
gamma rays   **82**, 92, 105–6
gases   **47**, 50, 52–3, 428, 430–31, 434
Geiger, Hans   83, 92
Geiger-Müller counter   **92**, 103, 105–6
giant structures   **113**, 117, 130, 298, 342, 352
glass   340, **343–6**, 383, 420
glucose   259, 293
glycerol   **297**, 305–6, 374
gold   137
graphite   99, **247–8**, 303, 332, 341

Haber, Fritz   358
Haber process   **359-61**, 369, 371
half-life   93–4, 99

Index   445

halogens **18-19**, 129, 284, 289-90, 294
hard water 225, 309, **390-97**, 420
heat treatment of metals **167**, 173-5, 181-2
helium 9, 11, 101-2, **360-61**
hydrates 208
hydrocarbons **272-80**, 283-6, 289-90, 297, 302-4, 308-9, 327, 375, 412
hydrochloric acid 397, 404, **411-12**, 414, 420, 434
hydrogen 9, 101-2, 150-51, 202, 276-7, 280-81, 326-7, **358-61**, 431
hydrogen bomb 102
hydrogen chloride 364, 408, **411-12**, 430-31
hydrogen sulphide **241-2**, 431, 434
hydronium ion **213-14**, 224, 238, 370, 388, 397, 411
hydroxides 432-3
hydroxyl group **288-9**, 295-7, 350, 352, 374-5
hygroscopic substances 209

ice 47-8
ignition temperature 61
indicators **196**, 212-13, 358, 362, 412
inhibitors 64
insecticides 290, 412
insoluble salts 203-4
iodine 376
ion-exchange 396-7
ionic compounds 130, 307-8
ionic equations 204
ionic theory 117-20
ions **117-33**, 161, 163, 203-4, 213, 329-33, 396-7, 406-8, 429, 432
iron 6, 71, **152-4**, 174, **181-2**, 186, 267, 346, 376, 383, 411, 433
iron(III) oxide **152-3**, 155, 157, 241, 382
iron(II) sulphate 201
iron(II) sulphide 6
isomers 276
isotopes **88-90**, 93, 96, 98, 102-7

joule 319

kaolin 342, 348
kilns **237**, 262-3, 347-8, 384-5, 417

Kipping, F. S. 349
knocking 21-2, 280
krypton 11, 360-61

Langmuir, Irving 128
lead **154**, 168, 181, **185-6**, 284, 333
lead accumulator 333
Liebig, Justus von 222
lime 155, 157-8, 195, 254, **380**, 417
limestone 153-5, 231, 344, 347-8, **380-83**, 385, 388, 391-2, 396, 417
lime water 386, 429
liquids 47-9, 52-3
litmus 196
lubricants 249

magnesium **161**, 358, 395
magnesium sulphate 199-200
malleability **166-7**, 172, 174-5, 181, 185
margarine 284-5, 308, 404
Marsden, Ernest 83
mass spectrometer **37-8**, 88, 91
Mège-Mouries 285
melting 48-9
melting points **47-9**, 130, 133, 172-3, 186, 276, 406
Mendeléev, Dmitri 19, 84
metals 14-15, 20-21, 113, **137-89**, 212, 376, 411-12, 429-33
methane 132, 266-7, **274-7**, 281, 285, 290, 318, 337, 350, 361, 412
Midgley, Thomas 21-2
mixtures 4-6
mole **40-42**, 51-2, 69-72, 118, 215-17, 276, 322-3, 406, 418
molecules 29, 52-3, 133, 272-85, 292-8, 302-9, 334-5, 351-2
monomers **292-5**, 297, 302-4, 352, 365, 375
mortar 348, 386
Müller, Hermann 92
Murdock, William 267

nanometre 33
native metals 144-5
natural gas 240-41, 246, 267, **277**, 327, 350, 361
neon 11, 360-61

neutralisation **195**, 199, 201-2, 212, 305, 321-3, 385, 420
neutrons **84-5**, 88-91, 96-8, 100, 102-3
Newton, Isaac 32
niobium 189
nitrates 370-71, 434-5
nitric acid 358, 365, **369-71**, 374-5, 434
nitrogen 11, 157, 267, 325-6 **355-60**, 365, 375-6, 388, 411, 431
nitrogen cycle 355-7
Nobel, Alfred 374
noble gases 11, 19, 52, 128, **360-61**, 411
non-conductors **112-13**, 303, 351-2, 394
non-electrolytes **113**, 117, 132-3
non-metals **14-15**, 20-21, 113, 411
nuclear fission 96-100
nuclear fusion 101-2
nuclear reactors **98-102**, 186, 189, 266, 407
nucleus **83-4**, 90, 92, 99, 101-3, 113, 132, 429
nylon 297, 365

octane number 280
oil 238, 240-41, 267, **270-90**, 318, 327, 335-6, 385, 388
open-hearth furnace 156-7
ores **144-5**, 147, 152-4, 160
oxidation 155, 287-8, 370
oxides **144-5**, 150-52, 160, 202-3, 214, 412-13
oxygen 9, 11, 158, 161, 252-3, 287-8, 371-4, 414, **431**

pacemaker 106-7
paper 226, 239, **255-6**, 412, 415
paraffins **277**, 283, 318
periodic classification **19-22**, 84-5, 90, 114, 122, 128-9, 358, 432
petrol 21-2, 260, 274, **279-80**, 284, 289, 318, 327, 334-5, 406
petroleum 222, 246, **270-90**, 304, 308, 319, 327, 337, 351, 408
pH 196, 365
phosphates 308

phosphorus 154–5, 375–6
photosynthesis 252–3, 319, 388
pickling 126, 222, 412
pig iron 154–6
Plaster of Paris 225
plastics 222, 246, 270, 284, **292–9**, 346, 365, 412
plutonium 99–101
pollution 238, 242, 267, 327, 424
polymers **292–8**, 302–4, 349–52, 355, 375
polystyrene 295, 298, 303
polythene **294–5**, 298, 304
pottery 340, 343, **348**
precipitation 203, 242, 309, 321, 391, 394, 396, 417, **432–4**
producer gas 325–7
proteins 293, **355**, 376
protons **84–5**, 88–91, 96, 102, 127, 213, 224, 362
PTFE 294–5
PVC **294–5**, 297, 412

quenching 167, 174–5
quicklime 380, **384–5**, 388

radioactivity **13**, **82**, 89–94, 99, 103–4, 186
radioisotopes 102–7
rare earths 20
rayon 232, **256–7**, 412
reaction rate 20–21, **58–66**, 306, 359–60, 372, 384, 419
reducing agents 151
reduction **150–52**, 154, 249
reforming 280
resins 297, 352, 397
respiration 253, 259, 375
reverberatory furnaces 148, 155
reversible reactions **75–6**, 333, 391, 393
roasting **148–9**, 154–5, 241
rubber 5, 64, 232, 270, **301–4**, 346, 352–3, 383
rusting 58, 178–9
Rutherford, Ernest 82–3

sacrificial protection 179
salt 10, **401–4**, **406**, 408, 415, 417
salts **199**, 202–9, 224–6, 305–9, 362–4, 370–71, 434–5

sea **9–10**, 102, 144–5, 161, 176, 184, 270, 382, 401–2
shells of electrons **128–9**, 132–3, 274, 429
Siemens, William 156
silica 341, 343
silicates 155, **341–4**, 346–7, 382, 396
silicon 154–5, 160, 182, **340–41**, 349
silicones 290, 340, **349–53**
slag **153–5**, 157–8, 348, 376, 383
slaked lime 161, 380, **385–6**, 396, 415
snow 48–9
soap **305–8**, 374, 395–6, 404
sodium 101, 161, 404, **406**, 408, 433
sodium carbonate 343–4, 346, 383, 396, 404, **417–20**
sodium chloride 10, 161, 200–201, 306, 397, **401**, 403, 414, 417, 420
sodium hydrogen carbonate 417, 420–21
sodium hydroxide 306, 308, 363, 397, 404, **408**, 414–15, 432–3
soft water 391
solder 172–3, 188
solids **47–8**, 51, 53
solid solutions **168**, 171–2, 174
Solvay process 161, **417–8**, 420
solvents 260, 288–90, 412
spray steelmaking 158
starch **252–3**, 293, 319
stars 102
states of matter 47–53, 71–2
steel **154–8**, 173–5, 178, 181–2, 187–8, 346, 348, 352, 383, 411
sublimation **51**, 364, 389, 425
sugar 226, 249, **252–5**, 259, 261–3, 293, 319, 343, 386
sulphates 224–6, 434
sulphides **144–9**, 154–5, 236, 241–2
sulphites 238–9, 434
sulphur 6, 155, 228, **231–5**, 241, 273, 302–3, 322, 365, 373, 376, 411
sulphur dioxide 148, 155, 228–9, **236–41**, 267, 273, 347, 373, 431, 434

sulphuric acid **222–31**, 236–7, 257, 308, 333, 364, 371, 374–6, 407
sulphurous acid 238
sulphur trioxide 228–9
sun 102, 252–3, 317, 319
surface tension 306–8

tantalum 189
tempering 174–5
Terylene 297
Thomson, Joseph 37
tin 154–5, 188, 345
titanium 188–9, 407
town gas 267, 277, **280–81**, 319, 327, 334–6, 358, 361, 385, 388
transistors 340
transition elements **20**, 122, 128–9, 432
tuyères 154

universal indicator 196, 201
universe 9–10
uranium 82, **96–102**, 106

vulcanisation 302–4

water 9, 213, 306–8, 347, 355, **390–97**, 412
water gas 325–7, 361
waterglass 343–4
water of crystallisation 207–9, 419–20
Watt, James 267
weedkillers 222, 404, **415–16**
Whinfield, J. R. 297
whisky 184, 261, **263**, 388
Wickham, Henry 301
wine 261, 263
wood 239, 249, **255–7**, 292, 318–9, 344
work-hardening 167
wrought iron 181

X-ray diffraction 169–70, 234
X-rays 82, **169–70**, 186, 225–6
xenon 11, 360–61

Yeast 259, 261–3

Zinc **154**, 165, **187–8**, 329–30, 433

Index 447

BOROUGH OF ENFIELD
EDUCATION COMMITTEE
—o—
EDMONTON
COUNTY UPPER SCHOOL